Energetics, Kinetics, and Life: An Ecological Approach

G. Tyler Miller, Jr.
St. Andrews Presbyterian College

D1502305

Wadsworth Publishing Company Inc., Belmont, California

ISBN-0-534-00136-X
L. C. Cat. Card No. 70-163983

Printed in the United States of America

1 2 3 4 5 6 7 8 9 10—75 74 73 72 71

Preface

How can anyone justify writing another book on chemical thermodynamics and kinetics when the field is apparently so overpopulated? My answer to this very relevant question lies in the groups to which this book is addressed and in the philosophy of approach to the subject.

The purpose of this book is not to teach the details of chemical thermodynamics and kinetics—but to give the reader an "intuitive" and "qualitative" feel for these important subjects and to illustrate their wide applicability. Particular emphasis is placed on the relationships of these subjects to the environmental problems of pollution and overpopulation. It is hoped that this book will prove useful to several groups: nonscience majors taking introductory courses in science, science majors taking introductory courses in chemistry and biology, high school students in advanced science courses, and interested laymen.

As a chemist, I have been concerned for some time about the teaching of thermodynamics and kinetics to undergraduate chemistry majors. Although we appear to be exposing freshmen students to heavy doses of these subjects, I have the feeling that the students don't know why thermodynamics and kinetics are important or relevant. They respond to our force-feeding but, too frequently, they do not see the woods for the trees. It is the purpose of this book to provide such an overview, so that students may profit more from the excellent and more quantitative approaches of authors such as Mahan and Nash.

I am also concerned about the biology major and premedical student. It is now essential that they have an understanding of thermodynamic and kinetic ideas and their application to life processes. Hopefully, this introductory approach can allow and encourage them to study more advanced treatments.

The science education of the nonscience major is of equal importance. My experience indicates that the most fruitful approach for nonscience majors (and for science majors) is to deal in depth with a small number of scientific concepts that are applicable to a number of disciplines and problems—as opposed to bewildering the student with an array of scientific facts and ideas. We are slowly beginning to realize that "less may be more." The concepts of energy and entropy provide two of the most interesting and widely applicable themes for illustrating the nature and scope of science.

My second argument has to do with the approach to teaching these subjects—in particular, the area of chemical thermodynamics. There is considerable debate as to whether chemical thermodynamics should be presented *classically* or *statistically*. The approach used in this book is neither classical nor statistical, but rather, a blend of both. Since both approaches lead essentially to the same ideas, a careful blend of the parts of each that lend themselves to an "intuitive" understanding seems to be a legitimate and useful pedagogical approach.

The approach to thermodynamics in this book is to go directly to Gibbs Free Energy (ΔG), rather than building up to it via heat work, Carnot cycles, and the like. In effect this approach is an attempt to "get to the promised land without going through the desert." A true appreciation of thermodynamics requires a disciplined trek "through the desert," but, before one undertakes such a journey, it helps to have not only a general sense of direction but an understanding that the trip is worthwhile.

Finally, there is a need for more books that discuss chemical dynamics (energetics and kinetics) as a unified way of studying chemical reactions. The student should understand that a reaction will occur only if it is both thermodynamically and kinetically feasible.

This book is a "popular" approach and necessarily I have used a minimum of mathematics. No calculus whatsoever is used. The only mathematics needed includes simple arithmetic, solving of very simple equations ($y = ax$), and some appreciation of exponential numbers.

I am deeply indebted to the many scientists who have provided information and inspiration upon which much of this book is based. It is impossible to acknowledge all of these, but those listed in the bibliography at the end of the book were particularly helpful.

I also wish to thank all those students, professors, and reviewers who have taken the time to point out errors and suggest improvements. The errors and faults remaining are mine, not theirs. I am particularly indebted to Professor Donald H. Andrews, Professor Emeritus, Johns Hopkins University, and now Distinguished Professor of Biophysics, Florida-Atlantic University, Edward Kormondy, Director of the Commission on Undergraduate Education in the Biological Sciences, Professors Frank L. Lambert, Occidental College, P. P. Feeny, Cornell University, Jay M. Anderson, Bryn-Mawr College, David and Joyce Roderick of Foothill College and San Jose City College, and two of my colleagues at St. Andrews Presbyterian College, Donald G. Barnes and David E. Wetmore. Finally, my sincere thanks go to Mrs. Ruth Y. Wetmore for her patience and skill in converting my hieroglyphics into the final manuscript and to Mr. Jack C. Carey of Wadsworth Publishing Company for his encouragement and help in preparing this book.

G. Tyler Miller, Jr.

Contents

Part Two Chemical Kinetics

Part Three Thermodynamics, Kinetics, and Life

Chapter 1

Introduction

1-1 Why Study Energetics and Kinetics?

Why is your room normally in a chaotic state?

Why is it absurd to speak of a pollution-free environment, car, or product?

Can you cool a hot kitchen by opening the refrigerator door?

Why is it essentially impossible for most people in the world to eat steak?

Hydrogen and oxygen gases are mixed in a bottle and nothing happens. A match is lit near the mouth of the bottle and an explosion occurs. Why?

Why should you thank a green plant and an alligator today and every day?

How did our present atmosphere of nitrogen and oxygen evolve?

Which language contains more information per word—English, German, Latin, or Greek?

Why will the production of more food not solve the overpopulation problem?

A lump of sugar held in your hand at body temperature will not decompose, but if swallowed it is "burned" readily. Why?

How can you improve your study habits?

Is there a technological solution to pollution?

Hydrogen and bromine gases mixed in a brown bottle do not react, but when mixed in a clear bottle they can explode. Why?

How can we tell the direction of flow of time?

Which fuel provides more energy per pound—gasoline or coal?

How might we reduce air pollution from the deadly chemicals, sulfur dioxide and the oxides of nitrogen?

How would you organize a company or a production line for greater efficiency?

How do poisons like cyanide kill and how do sulfa drugs heal?

Can the underdeveloped countries become developed?

How were the chemicals necessary for life on this planet originally formed?

Will carbon monoxide poisoning from automobiles on a busy downtown street be more severe on a hot day or a cold day?

Why can the United States be considered as the most overpopulated country in the world?

Can science provide us with any ethical rules?

Answers or at least insights to these and many other questions can be obtained by an understanding of chemical energetics and kinetics or chemical dynamics. This book is designed to provide you with an "intuitive" and "qualitative" feel for the kinetics of chemical processes and for two great laws of science—the first and second laws of thermodynamics—and to show their relationships to life processes.

Fortunately, the basic ideas of energetics and kinetics are surprisingly simple, and they require no specialized knowledge of chemistry, biology, physics, or mathematics to see many of their applications. Only the simplest mathematics will be used throughout this text, and if you can add, multiply, divide, subtract, and solve the problem below, you will have no problem with the mathematics in this book.

Study Question 1-1[1]   How many cubic feet of dirt are in a hole that is 6 feet deep, 4 feet wide, and 10 feet long?

If your answer to Study Question 1-1 was 240 cubic feet, any difficulties you may have will be verbal, not mathematical, because a more careful analysis of the problem will reveal that the correct answer is *not* 240 cubic feet.

1-2 How Can We Describe the Universe?

Many people have the vague feeling that science is just a meaningless assortment of data and information. Scientists do gather data, but their primary concern is with organizing these data into meaningful patterns. Facts are merely stepping-stones to theories and scientific laws. We can begin our study by asking how the data and concepts of energetics and kinetics fit into our overall scheme of the physical universe.

In beginning any study of the physical universe, we can get our bearings by asking, "Am I studying something that is very small, very large, or of ordinary size?" The physical universe is, then, divided into three domains: the *micro world* of the very small (atoms, molecules, and subatomic particles), the *macro world* of everyday objects and events (baseballs, grains of sand, people, and so forth) and the *super-macro*, or *cosmic, world* of the very large (planets and galaxies), as summarized in Figure 1-1. The "normal" macro world is experienced directly with our senses, while the micro and supermacro worlds are for the most part experienced indirectly.

Consider the concept of length, or distance. What does it mean to say that the length of a book is 10 inches, or 25.4 centimeters? In effect, it means that we take a ruler that has been marked off in internationally-agreed-upon units of macro distance

---

[1] Throughout each chapter is a series of study questions designed to help you think about the material—to use it just after you have read it. In a sense they form a sort of programmed learning sequence that can be used to test your understanding of the material. These study questions will be of greater value if you answer them while proceeding through each chapter. Answers to most of these questions can be found at the end of the book.

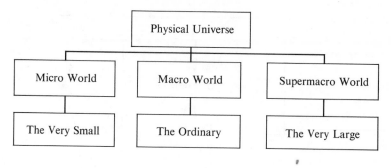

Figure 1-1   Classification of the Universe According to Relative Size.

(inches or centimeters) and determine that the book represents 10 or 25.4 of these units.

But what does it mean to say that the diameter of an atom is 10^{-8} cm (0.00000001 cm) or that the distance to the sun is 93 million miles? These micro and supermacro ideas of length, or distance, are quite different from our everyday distance concepts. To say that the diameter of an atom is 10^{-8} cm does not mean that someone took a very tiny ruler and measured the distance across the atom. Instead we measure this "distance" by indirect methods—for example, by analyzing what happens to matter when it is exposed to X-rays. Obviously no one has measured the distance to the sun using a yardstick. Again, indirect methods, such as using the velocity of light, enable us to measure very long distances.

We eat, live, and breathe in the macro world. What we call common sense[2] or logical usually means logical on a macro scale. One problem in studying the micro world of the very small or the supermacro world of the very large is that we tend to superimpose our macro world logic on them. Sometimes the answers we get thus seem "strange" or "illogical."

Study Question 1-2   Could you recall or "design" instruments and other devices used to enable us to obtain indirect evidence of phenomena in the micro and super-macro worlds?

It is more useful to break these three domains into sublevels. One approach is to classify matter in the universe according to *levels of organization*. Matter in the universe is found in various levels of organization, with each level an aggregate, or group, of the units of matter at the preceding level, as shown in Figure 1-2. Another classification describes various levels of organization of matter as either *living* or *nonliving*. Although some things can obviously be classified as either living or nonliving, scientists are still trying to define what is meant by life, and there is a borderline of uncertainty, as shown in Figure 1-2.

[2] Common sense is sometimes described as the good sense that horses have not to bet on people.

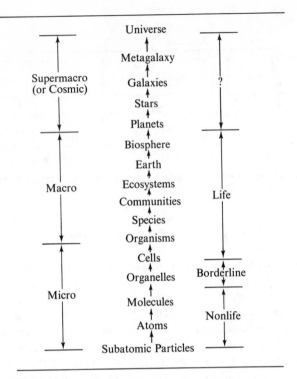

Figure 1-2 Levels of Organization of Matter.

Study Question 1-3 What is your definition of life? List carefully the qualities or criteria you would use to determine whether something is living or nonliving. Try to define some borderline cases. What is your definition of death?

The span of size and time from the extremely small world of subatomic particles and events to the very large domain of galaxies and cosmic time is summarized in Figure 1-3.

Using exponential numbers (powers of ten) to express quantities is convenient but deceptive. We tend to underestimate the vast differences between such numbers. Some implications of these differences are indicated in Figure 1-4.

Study Question 1-4 Exponential numbers sometimes represent people. It is estimated that from 1×10^7 to 2×10^7, or 10 to 20 million, people[3] now die each year

---

[3] P. R. Ehrlich and A. H. Ehrlich, *Population, Resources, Environment*, W. H. Freeman & Co., 1970. A lower but still horrifying estimate of the starvation rate is 4.5×10^6, or 4.5 million people per year. People speak of massive famines that may occur in the 1970's, 1980's, or 1990's. Yet a case can be made for the fact that we already have the greatest famine in the world's history, even using this lower estimate. Apparently we conveniently classify starvation as a famine only if it occurs in a particular country rather than on a global basis.

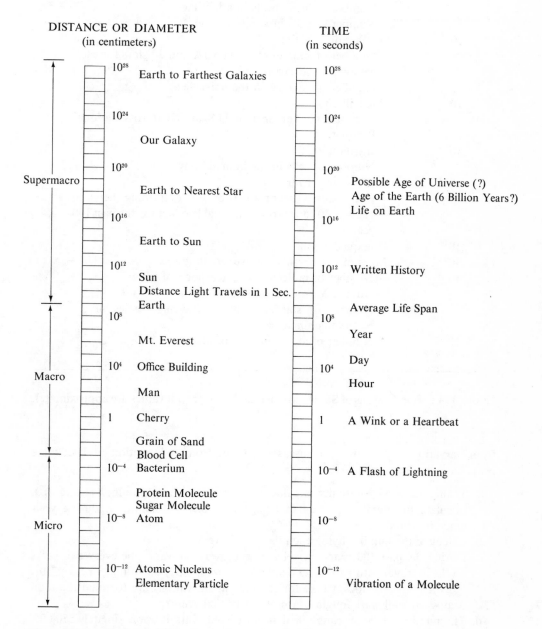

Figure 1-3   Scale of Distance and Time in the Universe.

10^6	One million
	—Number of distinct types of animals known to have evolved
10^9	One billion
3.5×10^9	Three and one-half billion
	—Number of people on earth
	—Number of heart beats in a lifetime
	—Number of cells in a man's brain
10^{11}	One-hundred billion
	—Number of stars in our galaxy (About 30 stars for each person on the earth)
	—Number of galaxies in the universe
10^{12}	One trillion
	—Number of dollars spent in U.S. in 1970 (Gross National Product)
10^{13}	Ten trillion
	—Number of cells in the human body
10^{14}	One-hundred trillion
	—Every dollar spent in U.S. for everything between 1829–1970 (our Gross National Product for the past 140 years)
10^{15}	One quadrillion
10^{21}	Mass of the Earth's atmosphere in grams
10^{23}	—Number of molecules in a teaspoon of water
	—Number of red corpuscles on earth
	—Number of stars and other heavenly bodies in the observable universe
10^{80}	—Number of atoms in the universe

Figure 1-4   Implications of Some Exponential Numbers (all values are approximate).

from starvation, malnutrition, or diseases resulting from malnutrition. Half of these people are children under five.

a. Taking the average casualty rate due to overpopulation as 15 million (1.5×10^7), calculate the number of human beings who died of starvation during your lunch hour today.
b. Calculate the number dying each day. Each week.
c. During the past 500 years there have been over 250 wars (one every two years on the average) with battlefield deaths of approximately 35 million, or 3.5×10^7. How many years does it take for the present starvation rate to equal the total deaths from all wars fought during the past 500 years?
d. The number of Americans killed in *all* of our wars is approximately 570,000, or 5.7×10^5. How many weeks does it take to equal this figure as a result of overpopulation?

Behind these appalling figures are unique individual human beings—not numbers —not things.

1-3 The Portion of the Universe to be Studied

How do energetics and kinetics fit into our overall classification of the physical universe according to relative size (Figure 1-1)? We will be concerned with chemical behavior in the *macro* world. Theoretical chemistry can be divided into three major categories: *chemical statics, chemical dynamics*, and *statistical mechanics*.

Look at a lump of sugar. The chemist is concerned with two questions: (1) what is its structure—that is, what are its basic units and how are they arranged? and (2) what holds these structural units together (chemical bonding)? These two questions of structure and bonding are the concern of chemical statics. It is a study of the micro world of individual atoms and molecules, and *quantum mechanics* is the branch of theoretical chemistry devoted to answering these questions.

The subject of this book is chemical dynamics, a study of chemical reactions at the macro level. It, in turn, is divided into two branches—chemical energetics and chemical kinetics.

Imagine mixing two colorless gases in a quart bottle. After several minutes a reddish color develops and the bottle has become warm. Why is heat given off? How much heat is given off? Why did it take several minutes for the reaction to occur? How did the two chemicals react to produce the red color?

Chemical energetics, or *chemical thermodynamics*,[4] is concerned with the energy (or heat) changes that occur. The branch dealing with the rate, or speed, of reaction and the stepwise mechanism, or path, by which a given reaction occurs is known as *chemical kinetics*. The general distinction between chemical thermodynamics and kinetics can be illustrated by the following analogy.

Imagine a person standing in the middle of the room with a stepladder beside him. You close your eyes and the person is now standing on the third rung of the ladder. Thermodynamics is concerned primarily with the difference in energy between the initial state (floor) and the final state (ladder), while chemical kinetics is concerned with the exact path, or mechanism, the person used to get from the initial state (floor) to the final state (ladder) and the speed, or rate, at which the change occurred.

A third branch of theoretical chemistry, called statistical mechanics, involves the calculation of the values of macroscopic properties by averaging the microscopic values. Statistical mechanics provides the bridge between the micro world of quantum mechanics and the macro world of chemical dynamics, as shown in Figure 1-5. These various branches of theoretical chemistry are summarized in Figure 1-6.

The thermodynamic properties of matter that are observed and measured are properties of aggregates, containing enormous numbers of atoms or molecules. These aggregates exist under dynamic conditions; atoms and molecules are moving at extremely high velocities, undergoing trillions of collisions each second, and they are subject to drastic "environmental" changes due to changes in temperature or pressure or to the addition or removal of vast "populations" of atoms and molecules.

---

[4] Historically, the term thermodynamics, as the prefix thermo implies, developed from a study of only one form of energy—heat. Since we are concerned with all forms of energy, the field is more aptly termed energetics, or chemical energetics. In this book these two names are used interchangeably.

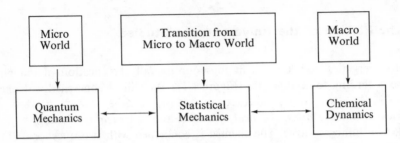

Figure 1-5   Classification According to Levels of Organization of Matter.

We would not expect to get an accurate view of American society by interviewing and thoroughly studying one American. By analogy the macro behavior and properties of aggregates of atoms or molecules cannot be adequately described and predicted from an understanding of the behavior of an individual atom or molecule. A quantum mechanical analysis of the microcosm will provide useful and necessary information but not sufficient information to describe macroscopic properties.

One approach to understanding American society would be to study and interview each American and then average the results statistically. Similarly, one could, in principle, determine the energy and position of all of the individual atoms and molecules in an aggregate of matter and then average the values to describe aggregate properties. This deterministic goal has been shown not to be feasible on experimental, theoretical, and practical grounds.

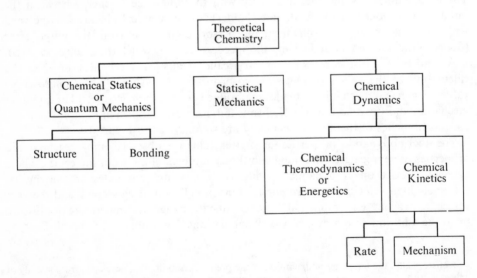

Figure 1-6   Major Branches of Theoretical Chemistry.

Experimentally and theoretically, according to interpretations of *Heisenberg's Uncertainty Principle*, we cannot obtain accurate and precise information simultaneously about both the position and velocity of very small particles such as electrons, atoms, and molecules. In other words, if we know the exact position of an electron, we have essentially no idea of its velocity; and if we determine its velocity very accurately, we do not know where it is located. One explanation for this is that in order to "look" at an atom or electron, we have to put so much energy into the system that we drastically change what we are trying to measure. In effect we are like an elephant trying to use his foot to pick up a particular marble from a pile of marbles. By putting his big foot into the system he scatters the marbles, so that the position of the particular marble has been changed radically.

Even if we could obtain complete and accurate information about atoms and molecules, the task of calculating macro properties from micro data could not be carried out on a practical basis. Suppose we try solving this problem for a tiny drop of water weighing approximately one thousandth of a gram. Such a drop would contain about 10^{20} molecules of water; by averaging the motions of this number of molecules, we could predict the properties of the water drop. A fast digital computer can carry out a simple mathematical operation (addition or subtraction) in one millionth of a second. Assuming (very optimistically) that the computer could calculate the motion of a single water molecule in this same time, then to calculate the motions of all 10^{20} molecules would take about 3 million years and would cost about 65 billion dollars in computer time—and this is for only one tiny drop of water.

The problem is really not so bad as this argument seems to suggest. At the macro level we do not make measurements on individual atoms or molecules, but on aggregates containing myriad atoms or molecules. For example, the pressure of the gas in an automobile tire depends on the average number of molecules hitting a particular area in a specified time and on the average energy of each collision. The temperature of a substance is a measure of the average energy of motion of its individual atoms, and the color of a substance depends on the average number of quanta or "energy lumps" of light of different wavelengths that it absorbs. Fortunately, chemical thermodynamics is concerned only with the difference in average energy values between the initial reactant chemicals and the final products. Because thermodynamic measurements of the initial and final stages can be made directly, it is probably the most successful branch of chemistry in providing the information we need to predict physical and chemical changes.

1-4 The Seven Fundamental Questions of Chemical Dynamics

Imagine that you have graduated from college and you are now a multimillionaire. You are actively concerned to use your wealth wisely to improve the lot of mankind rather than in buying gadgets and other trivia. In 1980, a young scientist approaches you with the idea of investing 20 million dollars in a plant for making a chemical substitute for food. If successful, this venture might enable you to solve the food problem in an overpopulated, starving world. What questions would you want to ask

the developer of this process before sinking your fortune into this "sure fire" investment that could make you a true world hero?

You recall dimly that the seven fundamental questions that a scientist should ask
about any chemical change were discussed in a portion of an exciting college science
course entitled *Energetics, Kinetics, and Life*. Looking at your notes, which you
carefully preserved, you find that you should ask the seven fundamental questions
of *chemical thermodynamics* (energy changes in reactions) and *chemical kinetics*
(speed, or rate, of reaction) as summarized below:

Questions of Chemical Thermodynamics (energetics)

1. Can the reaction occur spontaneously? (In other words, will it go
   naturally or will it require energy, and therefore money, to make it go?)
2. How much energy is released or absorbed when the reaction takes
   place? (That is, how much energy will be given off or how much energy
   will be required to force it to go?)
3. How far will the reaction go? (That is, what is the expected yield? Will
   the conversion of raw materials or reactants to the desired products be
   20% efficient? 90%?)
4. How can we make the reaction go further? (How can the % yield be
   increased?)

Assume that the answers to these thermodynamic questions are favorable: the
reaction does occur spontaneously, energy is released which can be recycled back into
the process to cut down on fuel bills, and the theoretical yield is high. Should you
invest your money?

Just because the reaction is *thermodynamically feasible* (that is, possible with a
reasonable input or output of energy) does not necessarily mean that it is *kinetically
feasible* (that is, it can occur in a measurable length of time). The reaction may
occur spontaneously with an evolution of energy and a high yield, but these favorable
thermodynamic answers do not provide any information about the rate of the reaction.
The reaction might occur spontaneously in one millionth of a second and, without
proper control, the entire plant could be blown apart one millionth of a second after
the grand opening. On the other hand, even though the theoretical yield might be
90%, the reaction could occur spontaneously at such a slow rate that it could take
200 years to make one ton of this amazing chemical. The following additional kinetic
questions must be answered:

Questions of Chemical Kinetics

1. What is the rate (speed) of reaction?
2. How can the rate of reaction be altered? (If it is too fast—explosive— how can it be slowed down? If it is too slow, how can its rate be increased? Either alteration will probably require considerable amounts of money.)
3. What is the reaction path, or mechanism, by which the reaction takes place? (Most reactions consist of a series of steps, or separate chemical reactions, that are collectively called the *reaction mechanism.* Knowledge of the mechanism can allow one to slow down or speed up particular steps or to alter the path to produce new products that might be better or less expensive to produce.)

Study Question 1-5   What other questions not related to chemistry would you want to answer before investing your money?

Study Question 1-6   Thermodynamic calculations indicate that when hydrogen gas and oxygen gas are mixed at room temperature, they should react spontaneously to form water with a large evolution of energy. When hydrogen and oxygen gases are actually mixed in a container at room temperature, no water is observed to form. Explain.

It is important to note that chemical thermodynamics and kinetics are concerned with distinctly different questions. Thermodynamics tells us what *can* happen; kinetics tells us whether it will take a millionth of a second or a million years. Yet, we must answer all of these questions in order to describe a chemical change adequately. The plan of this book is to take up each of these seven questions (Chapters 2 through 7) and then, in the remaining chapters, apply the answers to a study of life processes.

> Not being able to describe the Second Law of Thermodynamics is the equivalent of admitting that you have never read a work of Shakespeare's.
>
> *C. P. Snow*
> *Distinguished literary figure and scientist*

1-5 Review Questions

1-1. Distinguish between the micro, macro, and supermacro worlds and give examples of each domain.

1-2. Distinguish between
    a. quantum mechanics and statistical mechanics,
    b. quantum mechanics and chemical dynamics,
    c. statistical mechanics and chemical dynamics,
    d. chemical statics and chemical dynamics.

1-3. Classify each of the following terms according to whether it applies primarily to the micro world or macro world, or to the transition between the micro and macro worlds.
    a. Classical mechanics.
    b. Statistical mechanics.
    c. Thermodynamics.
    d. Chemical thermodynamics.
    e. Chemical dynamics.
    f. Quantum mechanics.
    g. Chemical energetics.
    h. Chemical kinetics.

1-4. a. What are the seven fundamental questions to be answered about a chemical reaction?
    b. Indicate the four questions answered by chemical thermodynamics.
    c. Indicate the three questions answered by chemical kinetics.

1-5. Can a given reaction be both
    a. thermodynamically and kinetically feasible?
    b. thermodynamically and kinetically unfeasible?
    c. thermodynamically feasible and kinetically unfeasible?
    d. thermodynamically unfeasible and kinetically feasible?

1-6. What is a reaction mechanism?

Part One

Chemical Energetics
(Thermodynamics)

Many people persuade themselves that they cannot understand
mechanical things or that they have no head for science or mathematics.
These convictions make them feel enclosed and safe, and of course
save them a great deal of trouble. But the reader who has a head for
anything at all is pretty sure to have a head for whatever he really
wants to put his mind to.

Jacob Bronowski

One must learn by doing the thing; for though you think you know it,
you have no certainty until you try.

Sophocles

Chapter 2

Question No. 1—Can a Given Reaction Occur Spontaneously?

2-1 Spontaneous, Nonspontaneous, and Forbidden Reactions

The first question of chemical thermodynamics is "Can a reaction occur spontaneously?" To discuss this question, it is necessary to define the term spontaneous reaction. Unfortunately, the common usage of the term to describe reactions which take place immediately after the reactants are mixed, is misleading. For example, a substance such as phosphorus that immediately bursts into flame when exposed to the air is commonly said to be "spontaneously flammable."

The meaning of spontaneous reaction as we shall use it in this section is "having the potential to proceed naturally without an input of energy from the outside." Note that this statement says nothing about the rate or speed of reaction. We shall be concerned here with *thermodynamic spontaneity*, not *kinetic spontaneity*. It is possible for a reaction to be thermodynamically spontaneous or feasible (having the potential to react), but it may react at such a slow rate that it is kinetically nonspontaneous. It can occur in principle though in practice it does not occur to any measurable extent. A familiar example is the reaction of hydrogen and oxygen gas to produce water. It is thermodynamically spontaneous at room temperature, in that hydrogen and oxygen tend to unite to form water; but the rate of reaction is so imperceptibly slow (probably a few molecules of water per century) that it is kinetically nonspontaneous and, for all practical purposes, does not occur.

Actually we need to distinguish among three questions.

1. Can the reaction occur under any circumstances?
   Some chemicals do not react even if energy is supplied. Reactions that are impossible and cannot occur under any circumstances are *forbidden reactions*.

2. Can the reaction occur spontaneously?
A reaction that potentially can occur naturally without the assistance of an external input of energy is a *spontaneous reaction.* A *nonspontaneous reaction* is one that does not occur naturally, but it can potentially be made to occur by supplying outside energy. It is important to note that we are really concerned with the direction of spontaniety. For example, consider the reaction of carbon (C) and oxygen gas (O_2) to produce carbon dioxide gas (CO_2).

$$C + O_2 \rightleftharpoons CO_2$$

Assume that C and O_2 react spontaneously to form CO_2, that is, the direction of spontaneity is from left to right to favor the formation of CO_2. What about the reverse reaction from right to left? Would CO_2 spontaneously decompose to carbon and oxygen? If the forward reaction is spontaneous, then the reverse reaction of CO_2 to form C and O_2 must necessarily be nonspontaneous.

3. Will the reaction actually occur?
A reaction that is thermodynamically spontaneous *can* occur and a nonspontaneous reaction can be made to occur. The reaction will actually occur, however, only if the reaction is also kinetically feasible and occurs at a measurable rate.

We can see the importance of studying both the thermodynamics and kinetics of a chemical reaction. Thermodynamics answers the question *Can it occur?* but only a combination of thermodynamics and kinetics answers the question *Will it occur in a reasonable time?*

Reaction Feasibility

Thermodynamically spontaneous	—Reaction can occur without outside energy
Thermodynamically nonspontaneous	—Reaction can occur only with outside energy
Forbidden	—Reaction cannot occur under any circumstances
Kinetically spontaneous	—Reaction can occur in a reasonable time
Thermodynamically and kinetically spontaneous	—Reaction *will* occur in reasonable time

---

Study Question 2-1

a. Can a reaction be both forbidden and thermodynamically nonspontaneous? Explain.
b. Can a reaction be thermodynamically spontaneous and still not occur? Explain.
c. Can a reaction be thermodynamically nonspontaneous and still occur? Explain.

2-2 Energy Changes in Spontaneous Mechanical Processes

Why are some reactions spontaneous while others either are forbidden or can occur only when energy is added? We are really asking the deeper question *What is the driving force for a chemical reaction?* Only by understanding this driving force can we determine whether a particular reaction or process can occur spontaneously.

Let's begin by looking at some spontaneous mechanical processes that we observe in everyday life. Water spontaneously flows downhill, and apples or other objects that are dropped fall spontaneously to the ground. These naturally-occurring processes are irreversible in that they never reverse themselves of their own accord. They can be reversed, but only by putting in energy from the outside to pump the water uphill or raise the object to its original position.

The feature common to all of these spontaneous processes is that the system loses energy and goes to a lower energy state. But what kind of energy is involved? Remember that there are two kinds of energy, *kinetic energy*, due to motion, and *potential energy*, due to position. When water flows downhill or objects fall, the system moves from one *position* to another. Thus, the driving force for these spontaneous changes appears to involve a net decrease in the *potential energy* of the system.

But what happens to the energy that is released? Part of it may be converted to kinetic energy of motion as the objects fall and part may be transferred as kinetic energy or heat to the surroundings.

The driving force for spontaneous processes in these mechanical systems appears very simple—*mechanical systems tend spontaneously to move to the state of minimum potential energy consistent with their surroundings.* To understand the necessity for adding the phrase "consistent with their surroundings," consider an apple falling to a table rather than to the floor. The apple reaches a minimum potential energy state consistent with its surroundings (the table); but if new surroundings are provided—if the table were tilted or the apple pushed—it would fall to the floor and achieve a new state of minimum potential energy consistent with the new surroundings, as shown in Figure 2-1.

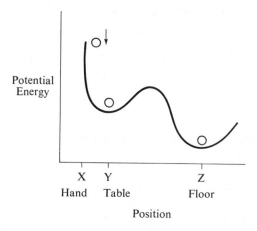

Figure 2-1   An Apple Falls Spontaneously to a State of Minimum Potential Energy Consistent with Its Surroundings.

Obviously, these and other examples on earth are based on the earth's gravitational force.

It is important to distinguish between the *system* (that is, the collection of matter under study) and its *environment or surroundings* (i.e., the rest of the universe). During any process energy may pass from the system to the surroundings or from the surroundings to the system.

According to the *Law of Conservation of Energy*, which is the simplest form of the *First Law of Thermodynamics*, in any process energy is neither created nor destroyed but is merely transformed from one form to another, such as heat, light, mechanical, electrical, or chemical energy. If the system loses energy, then an equal amount of energy must be transferred in some form to the surroundings. In any process, spontaneous or nonspontaneous, the total energy of the system plus surroundings remains constant, as shown in Figure 2-2.

First Law of Thermodynamics—Law of Conservation of Energy

In all macroscopic chemical and physical changes, energy is neither created nor destroyed but merely transformed from one form to another.

or

In any process the total energy of the system plus its surroundings remains constant.

Our primary focus, however, is on the system, and we appear to have found a fundamental driving force for spontaneous changes in mechanical systems. We shall call this *the minimum potential energy principle*:

The Minimum Potential Energy Principle

Mechanical systems tend spontaneously to a state of minimum potential energy consistent with their surroundings.

Study Question 2-2   Can you think of a number of examples which illustrate the minimum potential energy principle? In each case trace the changes in energy that occur in the system and its surroundings and show how the Law of Conservation of Energy is obeyed. Can you think of any examples that apparently don't obey the minimum potential energy principle?

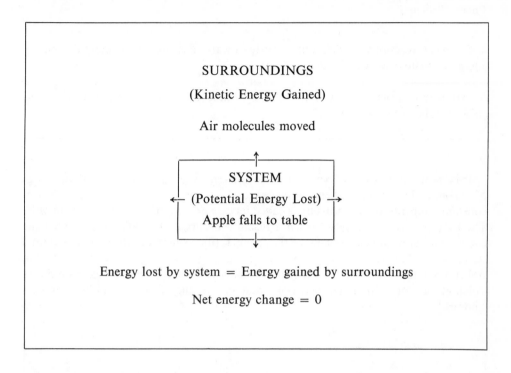

SURROUNDINGS

(Kinetic Energy Gained)

Air molecules moved

SYSTEM

(Potential Energy Lost)

Apple falls to table

Energy lost by system = Energy gained by surroundings

Net energy change = 0

Figure 2-2   Illustration of the Law of Conservation of Energy—The First Law of Thermodynamics.

2-3 Enthalpy—A Possible Criterion for Spontaneity

The minimum potential energy principle apparently applies to spontaneous processes in mechanical systems. Can we by analogy apply it to chemical reactions? Chemical energy may be considered as the potential energy inherent in the chemical bonds that hold a particular chemical structure together. By analogy, we might postulate that any chemical reaction will be spontaneous if the average chemical (potential) energy of the products is lower than the average chemical (potential) energy of the reactants— in other words, reactions that "go downhill" in terms of chemical energy should be spontaneous.

A given chemical reaction either gives off energy to or absorbs energy from its surroundings. A reaction like the burning of coal, gasoline, or other fuels that evolves heat is called an *exothermic reaction*, while one that absorbs or requires heat from its surroundings is called an *endothermic reaction*. In any spontaneous chemical reaction, energy should be evolved to the surroundings as the system goes to a state of lower chemical energy. Thus, all exothermic reactions "should" be spontaneous and all endothermic reactions "should" be nonspontaneous.

A Tentative Hypothesis—The Minimum Chemical Energy Principle

Chemical reactions tend spontaneously toward a state of minimum chemical energy consistent with their surroundings.

or

Exothermic reactions "should" be spontaneous and endothermic reactions "should" be nonspontaneous.

This information can be expressed in a different way by introducing the concept of *enthalpy*. The average chemical energy or heat content of a chemical system at constant temperature and pressure is called its *enthalpy*, represented by the symbol H. The heat evolved or absorbed when a reaction occurs is the difference between the average enthalpy, or heat content, of the products (final state) and the reactants (initial state). This difference in enthalpy is called the *heat of reaction*. It is given the symbol ΔH (pronounced "delta H"), where Δ, or delta, is the Greek letter used to indicate "change in" or "difference between" and is normally expressed in the units of calories.[1]

Enthalpy Change in Reactions

Heat of reaction at constant pressure and temperature	=	Average enthalpy of products − average enthalpy of reactants
Heat of reaction at constant pressure and temperature	=	$\Delta H = H_{products} - H_{reactants}$

---

[1] The calorie is a standard unit for heat. It is equivalent to the amount of heat needed to raise the temperature of one gram of water from 14.5 to 15.5°C. For those of you that diet, it may be interesting (but depressing) to note that calorie food values are actually in kilocalories, often represented by using a capital C.

1 kilocalorie or kcal = 1 Calorie = 1000 calories

If a piece of pie is listed as 2000 Calories, this means that it is actually 2000 kcal or 200,000 calories. Another common energy unit, now accepted as the internationally approved unit for energy, is the joule (pronounced jool). Calories can be easily converted to joules and vice versa.

1 joule = 4.184 calories
1 kilojoule = 1000 joules = 4184 calories

An exothermic reaction such as the reaction between hydrogen gas (H_2) and oxygen gas (O_2) to form water, is accompanied by a decrease in enthalpy, and heat is evolved to the surroundings. Thus, the heat of reaction, ΔH, is negative, as shown in Figure 2-3. This indicates that average bond strengths in and between the product molecules (H_2O) are stronger than the average bond strengths in and between the reactant molecules (H_2 and O_2).

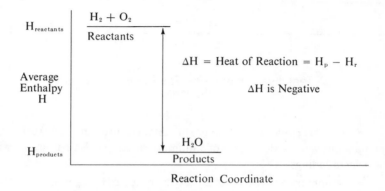

Figure 2-3   Enthalpy Change for an Exothermic Reaction. ΔH Is the Net Energy Evolved as Heat to the Surroundings as the Reaction Proceeds. (The term "reaction coordinate" is used to indicate the progress of the reaction from initial reactants to final products.)

In an *endothermic* reaction the average enthalpy of the final products is higher than that of the initial reactants and in terms of energy the reaction must "go uphill". Since outside energy from the surroundings must be put into the system to raise it to the higher energy state, ΔH for an endothermic reaction is positive.

Exothermic Reaction: Heat evolved to surroundings; ΔH_{system} is $-$
Endothermic Reaction: Heat absorbed from surroundings; ΔH_{system} is $+$

Recall, however, that according to the first law there is no net loss or gain of energy. The heat lost or gained by the system is equal but opposite to the heat gained or lost by the surroundings. $\Delta H_{system} = -\Delta H_{surroundings}$ at constant temperature and pressure.

Can a Given Reaction Occur Spontaneously?

Study Question 2-3   The decomposition of ammonia gas, NH_3, into gaseous nitrogen, N_2, and hydrogen, H_2, requires an outside energy source. Draw and label a diagram, similar to that in Figure 2-3, for this *endothermic* reaction.

The minimum chemical energy principle can now be restated in terms of enthalpy.

A Tentative Hypothesis—The Minimum Enthalpy Principle

Chemical reactions at constant temperature and pressure tend spontaneously towards a state of minimum enthalpy (H) consistent with their surroundings.

Spontaneous reactions are exothermic: ΔH_{system} is $-$

Nonspontaneous reactions are endothermic: ΔH_{system} is $+$

Study Question 2-4

a. Is the reaction of carbon dioxide (CO_2) and water (H_2O) plus heat to form glucose, a sugar, and oxygen, exothermic or endothermic?
b. Is ΔH + or $-$?
c. What is the sign of ΔH for the reverse reaction of glucose and O_2 to form CO_2 and H_2O?
d. According to the minimum enthalpy principle, which reaction should be thermodynamically spontaneous?

$$\text{Forward reaction: } CO_2 + H_2O + heat \longrightarrow glucose + O_2$$
$$\text{or}$$
$$\text{Reverse reaction: } glucose + O_2 \longrightarrow CO_2 + H_2O + heat$$

e. Which reaction should be nonspontaneous?
f. If the reaction is thermodynamically spontaneous in one direction, does this mean that this reaction *will* actually occur? Explain.

Have we found the driving force for spontaneous reactions so quickly and easily? One of the tests of any scientific hypothesis is whether or not it can account for known experimental data. When we go into the laboratory, do we find that all exothermic reactions are spontaneous and all endothermic reactions are nonspontaneous? Unfortunately, this simple hypothesis does not hold up under experimental testing. At room temperature, the majority of spontaneous reactions are indeed exothermic—but some endothermic reactions (for example, the reaction of carbon (C) with water to produce carbon monoxide (CO) and hydrogen gas (H_2) are spontaneous. At high temperatures there is an even larger number of spontaneous endothermic reactions.

What can we conclude? While the minimum potential energy or enthalpy principle may be a contributing factor to the driving force of a chemical reaction, it must not be the only factor involved.

2-4 Entropy Changes in Spontaneous Processes

Are there any other general trends or tendencies that seem to take place in the world around us? Consider the phenomena of order and disorder. In general, do you feel that things and events tend spontaneously to become more orderly or more disorderly? What is the most probable state of your room—order or increasing chaos? Doesn't it seem to tend spontaneously toward increasing randomness, disorder, or chaos? To keep it neat and orderly requires a continuous input of energy on your part.

Houses do not spontaneously acquire new coats of paint. They tend spontaneously to fall into disrepair. If you shuffle an ordered deck of cards, arranged by suits, you get a disordered deck. But if you shuffle a disordered deck, you do not expect it spontaneously to become arranged by suits. The probability of such an event is vanishingly small, because there is only one perfectly orderly state for the deck, but there are millions of possible disordered arrangements for the 52 cards. If you drop a vase to the floor and it shatters, you would under no circumstances expect the random array of fragments to return spontaneously to reform the vase. We cannot put Humpty Dumpty together again—at least not spontaneously.

The natural state of vegetation on the land is a higgledy-piggledy disorderly state. Lawns must be cut to keep them orderly. A garden or field, consisting of neat and orderly rows of plants, does not form spontaneously. Fly over the land and you can immediately see where man has been by the presence of straight lines—fields, roads, houses, and other marks of man. These lines represent an ordering process against the apparently natural tendency for the more random or disorderly growth of vegetation. These ordering activities of man are nonspontaneous processes requiring enormous and continuous inputs of energy for their maintenance.

Your body consists of a highly ordered arrangement of molecules. To preserve this order, you must supply it energy and raw materials. If you stop eating or supplying water to your system, death eventually occurs, and the ordered structure decays to smaller molecules and atoms that are spread out over the world to be used over and over. On the average, with every breath you inhale about 50 million molecules that were once inhaled by Plato—or any other person in history who lived for 65 years.

If a highly ordered crystal of water-soluble dye is added to a glass of water, the colored dye spontaneously spreads throughout the solution; in other words, the relative disorder of dye molecules increases. If a woman wearing perfume walks into the room, in a few minutes the molecules of the scent will diffuse throughout the room, even if the air is perfectly still.

Smoke from a smokestack and exhaust from an automobile spread out spontaneously to become more randomly dispersed in the atmosphere. Chemical or other wastes dumped into a river or other body of water are spontaneously dispersed and diluted into a more disordered or randomized state.

Indeed, our overall approach to pollution control to date has been based on the concept that "dilution is the solution to pollution." We dump our wastes into the air, water, or soil and assume that they will spontaneously be diluted to harmless amounts or at least dispersed far away from us. As we shall see later (Chapter 11), we are now beginning to realize that *dilution is not the solution to pollution.* It is a useful application of thermodynamics, but it contains some fatal flaws.

We have apparently uncovered a second principle, or driving force, for spontaneity—the tendency toward increasing disorder.

A Tentative Hypothesis—The Increasing Disorder
Principle

A *system* tends spontaneously toward increasing disorder, or randomness.

Again we must return to the laboratory to see if this hypothesis accounts for experimental data on spontaneous processes.

Can we find examples of spontaneous processes that do not involve an increase in disorder? Consider the evolution of life on this planet. It involves the creation of order, yet it appears to have occurred spontaneously from a thermodynamic standpoint.

Consider the relative disorder of ions, atoms, or molecules in the solid, liquid, and gaseous states, as shown in Figure 2-4. In solids, such as sugar or salt, the particles are highly ordered—they are close together and their motion is highly restricted. The particles in a liquid, such as water, are more disordered—they are still relatively close together but freer to move. But oxygen, water, or any substance in the gaseous state is a highly disordered, or random, array of particles relatively far apart and moving haphazardly at very high velocities.

According to the increasing disorder hypothesis, all substances should tend spontaneously toward the more disordered gaseous state. Yet we observe that many

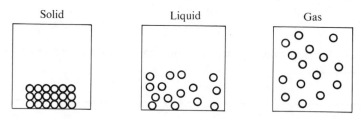

Figure 2-4   Relative Disorder of the Solid, Liquid, and Gaseous States of a Substance.

substances exist in the liquid or solid state. Certainly there is a lot of liquid water around. In some cases when relatively disordered solutions of two chemical substances such as sodium chloride (table salt) and silver nitrate are mixed, a highly ordered solid precipitate of silver chloride forms spontaneously and eventually settles to the bottom of the container—another apparent contradiction to the increasing disorder principle.

Although a large number of systems appear to tend spontaneously toward increasing disorder, we can observe many exceptions, and our tentative hypothesis must be either discarded or revised.

Suppose we modify the hypothesis by considering the net change in disorder not only in the system but also in its surroundings—just as we did in the mechanical example of an apple falling to a table and giving off energy to the surroundings. In the case of the evolution of life, the order created and maintained in the living system requires that disorder be created in the surroundings. For animals, living involves ingesting highly ordered complex molecules (starches, carbohydrates, proteins) and excreting usually smaller molecules, such as carbon dioxide and water, which are distributed more randomly throughout the universe. Measurements indicate that the increase in disorder in the surroundings is greater than the increase in order in the living system. Thus, life involves a net increase in disorder when we consider both the system and surroundings.

At normal atmospheric pressure and temperatures below 100°C, water vapor (steam) spontaneously changes into the more orderly liquid state so that there is a decrease in disorder in the system. As anyone who has been burned by steam can testify, the conversion of steam to water is an exothermic process, in which heat is evolved into the surroundings (including the skin). This addition of heat increases the relative disorder of the surroundings by increasing the average motion of the surrounding molecules. Again, measurements indicate that this spontaneous process produces a net increase in disorder—the increase in disorder of the surroundings is greater than the decrease in disorder in the system.

Experimental measurements of changes in disorder of the system plus its surroundings for any spontaneous process reveal that *there is always a net increase in disorder*. We must therefore revise our original hypothesis to include both the system and its surroundings.

Revised Hypothesis—The Increasing Disorder
Principle

Any system *plus its surroundings* tends spontaneously toward increasing disorder, or randomness.

Study Question 2-5   Can you think of other examples which illustrate the increasing disorder principle? Can you think of any exceptions?

The enthalpy function (H) was introduced to describe the average chemical energy of a system. Let us also define a function that is a measure of the average disorder in a system or its surroundings. Scientists have assigned this function the name *entropy* and given it the symbol S.[2] The more ordered a system, the *lower* its entropy and the more disordered, or random, the components of a system, the *greater* its entropy.

Entropy

Entropy ≡ S ≡ Measure of relative disorder

High entropy = Disorder

Low entropy = Order

Study Question 2-6   Which of the following *systems* has the higher entropy?

a. A brand new pack of cards arranged according to suit and number or the same deck after shuffling.

---

[2] It might seem more logical to assign it the name disorder and the symbol D or to give entropy the symbol E. At the time that entropy was first introduced by Clausius, the symbol E was already in use for energy and the relationship of entropy to disorder was not known.

b. A box full of sugar cubes or the sugar cubes thrown on the floor.

c. A solid sugar cube or the sugar cube dissolved in hot coffee.

d. A gas or a liquid at the same temperature.

e. A liquid or a solid at the same temperature.

f. A solid or a gas at the same temperature.

The overall change in entropy that takes place in the system when a reaction occurs is the difference between the average entropy of the final products and the initial reactants. It is called the *entropy of reaction*[3] and is given the symbol ΔS_{system}. Similarly, we can specify the entropy change in the surroundings as $\Delta S_{surroundings}$, and the net entropy change of the system plus surroundings as ΔS_{net} or ΔS_{total}.

Entropy Change

Entropy change $= \Delta S =$ Average entropy of final state
$\qquad\qquad\qquad\quad -$ average entropy of initial state

ΔS is $+ =$ Increase in disorder

ΔS is $- =$ Decrease in disorder (increase in order)

$$\Delta S_{system} = \Delta S_{products} - \Delta S_{reactants}$$

$$\Delta S_{net} = \Delta S_{system} + \Delta S_{surroundings}$$

The tendency toward increasing disorder principle can now be restated in terms of entropy.

The Increasing Entropy (Disorder) Principle

Any system plus its surroundings tends spontaneously toward a state of increasing entropy, or disorder.

---

[3] The entropy change, ΔS, has the dimensions of energy per degree and is usually expressed in calories/degree — or more accurately calories/degree Kelvin, where degrees Kelvin or $^\circ K = \,^\circ C + 273$.

Study Question 2-7   State whether ΔS, the entropy change of the *system*, for each of the following processes is + or − .

a. Initial state: students sitting in room listening to a lecture. Final state: 30 seconds after the end of the period.
b. Formation of a solid from a liquid at constant temperature.
c. Boiling of water at 100°C and constant temperature.
d. Melting of snow at a constant temperature.
e. Increasing the temperature of a gas.
f. Cooling a liquid.
g. A crowd leaving a football game.
h. Air escaping from a tire.
i. Decay of a dead organism.
j. Evolution of a living organism.

Study Question 2-8

a. ΔS_{system} for some of the processes in Question 2-7 is positive. Does that mean that these processes are necessarily spontaneous? Explain.
b. ΔS_{system} for some of the processes in Question 2-7 is negative. Does that mean that these processes are necessarily nonspontaneous? Explain.

   When applied to actual experimental processes, this principle does work, and we have found a criterion for determining whether a given reaction can occur spontaneously. ΔS_{net} will be positive for any spontaneous process and negative for any nonspontaneous process. This very fundamental statement about spontaneity in terms of entropy or disorder is known as the *Second Law of Thermodynamics*.

A Criterion for Thermodynamic Spontaneity—
The Second Law of Thermodynamics
(The Increasing Entropy Principle)

   Any spontaneous process results in an increase in the net entropy, or disorder, of the system plus its surroundings.

   Spontaneous   reactions = Net   increase   in   disorder: ΔS_{net} is +
   Nonspontaneous reactions = Net decrease in disorder: ΔS_{net} is −

Two examples of the second law are shown in Figure 2-5.

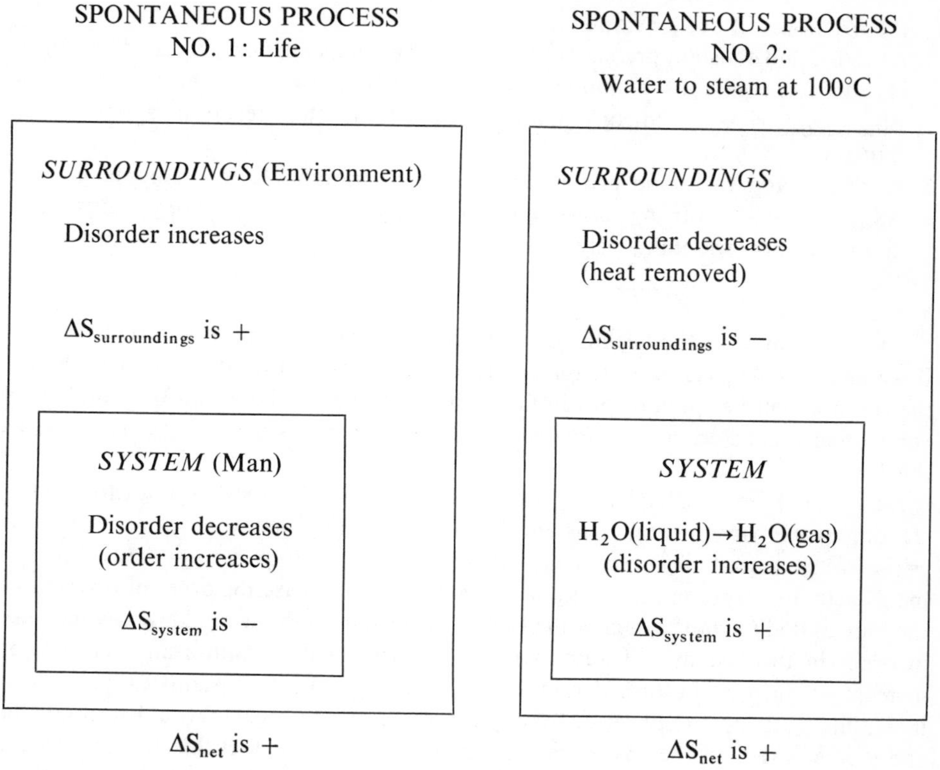

<table>
<tr><td style="text-align:center">SPONTANEOUS PROCESS
NO. 1: Life</td><td style="text-align:center">SPONTANEOUS PROCESS
NO. 2:
Water to steam at 100°C</td></tr>
<tr><td>

SURROUNDINGS (Environment)

Disorder increases

$\Delta S_{surroundings}$ is +

SYSTEM (Man)

Disorder decreases
(order increases)

ΔS_{system} is −

</td><td>

SURROUNDINGS

Disorder decreases
(heat removed)

$\Delta S_{surroundings}$ is −

SYSTEM

$H_2O(liquid) \rightarrow H_2O(gas)$
(disorder increases)

ΔS_{system} is +

</td></tr>
<tr><td style="text-align:center">ΔS_{net} is +</td><td style="text-align:center">ΔS_{net} is +</td></tr>
</table>

Figure 2-5   Two Examples of the Second Law of Thermodynamics. For any Spontaneous Process the Total Entropy of the System Plus Surroundings Increases.

Let us use the second law to determine the thermodynamic spontaneity of the reaction of gaseous hydrogen and oxygen to form steam at 100°C and constant pressure. Experimental measurements show that the entropy change for the system, ΔS_{system}, is approximately −11 units of entropy[4] while the entropy change for the surroundings, $\Delta S_{surroundings}$, is approximately +150 units of entropy. Thus,

$$\Delta S_{net} = \Delta S_{system} + \Delta S_{surroundings} = -11 + 150$$
$$= +139 \text{ units of entropy}$$

and according to the second law, the reaction should be thermodynamically spontaneous at 100°C.

---

[4] The units of entropy are normally expressed as calories per degree Kelvin.

Study Question 2-9

a. Does the fact that ΔS_{net} is positive for the formation of steam at 100°C mean that the reaction will occur? Explain.

b. At 100°C and constant pressure the reaction of gaseous nitrogen, N_2, and hydrogen, H_2, to form gaseous ammonia, NH_3, has $\Delta S_{system} = -24$ entropy units and $\Delta S_{surroundings} = +74$ entropy units. At 100°C is this reaction spontaneous? Explain.

c. At 900°C and the same pressure the same reaction to form ammonia gas has $\Delta S_{system} = -24$ entropy units and $\Delta S_{surroundings} = +15$ entropy units. Can ammonia form spontaneously at 900°C?

The significance of the Second Law of Thermodynamics cannot be overestimated. If we find that ΔS_{net} is positive for any reaction, we do not need to wonder whether the reaction will be spontaneous in Germany, in India, in the year 2010—we have a valid, unchanging criterion. How many practical conclusions are as universal as this?

As we shall see in Chapter 11, the second law provides the key both for understanding our environmental crisis and for understanding how we must deal with this crisis. We shall see that *in reality the environmental crisis is an entropy crisis*. Most of man's activities represent an attempt to maintain or increase the order of a system— the human body, a house, a city, the Earth. According to the second law any increase in order in the system will automatically and irrevocably require an even greater increase in entropy or disorder in the environment. Thus, paradoxically, as man increasingly attempts to order or "conquer" the Earth, he must inevitably put greater and greater stress on the environment. Failure to accept the fact that we can't repeal the Second Law of Thermodynamics and that we must learn to live with the restrictions it imposes on us can only lead to a steady degradation of the quality of life on this planet.

2-5 Enthalpy and Entropy Combined—The Gibbs Free Energy

Determining the sign of the net entropy change of the *system plus surroundings* does provide a criterion for answering the question: Can a reaction occur spontaneously? However, our primary focus is usually on the *system only*. Sometimes the surroundings are ill-defined, or it is difficult to measure entropy changes in the environment. It would be convenient to have an additional criterion for spontaneity that would work for the *system only* at constant temperature and pressure.

In our search for a spontaneity criterion we have found two tendencies. In a number of cases the system changes spontaneously to achieve a state of lower chemical energy, or enthalpy, and a state of higher disorder, or entropy. Sometimes these two tendencies work together, and sometimes they are opposed; the actual state, then, may be some compromise between these two opposing tendencies.

Perhaps the overall driving force for a chemical reaction is determined by some combination of enthalpy, or energy, and entropy. If the two factors were opposed, then the difference between them would determine overall spontaneity with the larger factor predominating. The difference between the two tendencies would be the energy or work that is free or available to drive the reaction. Earlier we found that at constant temperature and pressure, ΔH_{system} is a measure of the change in total energy of the system during a reaction. It is energy that is available to do work, while the energy change related to entropy, or disorder, is not available to perform useful work. If we burn ten pieces of coal, we get the same amount of heat regardless of whether they are in a disordered array or a neat, orderly pile. Thus, the energy that is free or available to drive a reaction might be the difference between the available energy or total energy change (enthalpy factor) and the "unavailable" energy change (entropy factor).

Free, or available, energy = Total energy — unavailable energy
(reaction tendency)   (enthalpy factor)   (entropy factor)

As a tentative hypothesis, let us assume that the reaction tendency, or free energy, might be obtained by subtracting ΔS_{system} from ΔH_{system}.

A Tentative Hypothesis—Reaction Tendency, or Free Energy Factor

Reaction tendency, or free energy = $\Delta H_{system} - \Delta S_{system}$

We can begin our evaluation of this factor by checking to see that it gives the dimensions of energy, such as calories. Recall that ΔH has the dimensions of energy in calories but ΔS has the dimensions of energy per degree Kelvin or calories/degree Kelvin. From this we see that our tentative hypothesis does *not* give the free energy, or reaction tendency, in energy units.

Free energy $\neq \Delta H_{system} - \Delta S_{system}$
calories $\neq$ calories — calories/degree Kelvin

Can a Given Reaction Occur Spontaneously?

Perhaps we can modify the expression to yield energy units. Since ΔS has the units of calories/temperature in °K,[5] if we multiply ΔS by the temperature in °K, the resulting factor should have the units of energy in calories.

$$\Delta S = \frac{\text{calories}}{°K}$$

$$T\Delta S = °K \times \frac{\text{calories}}{°K} = \text{calories}$$

Apparently, *the entropy factor should not be ΔS but $T\Delta S$.*

This seems reasonable because there are really two contributions to the overall disorder, or molecular chaos, in a system. One contribution involves the disorder due to the relative arrangement, or position, of the particles in the system and another results from the relative degree of motion, or thermal disorder, in the system. The temperature, T, in °K (degrees Kelvin) is a measure of the average kinetic energy of motion of the particles in a system. Thus, the relative degree of disorder in a system will depend on two things, the change in entropy of the system (ΔS_{system}) and its temperature (T) in °K.[6] The greater the temperature the greater the *intensity* of disorder, and the greater the entropy the greater the *capacity* of the system for disorder. Any energy factor is the product of an intensity factor and a capacity factor. For example, the kinetic energy of motion in a system is the product of the capacity factor of mass and the intensity factor of velocity (K.E. $= 1/2mv^2$). Thus, the molecular chaos or entropy factor in units of energy should be obtained by multiplying the temperature (T) by the entropy change (ΔS), as we have already predicted on dimensional grounds.

The Entropy Factor

$$\text{Entropy factor} = \text{Intensity factor} \times \text{capacity factor}$$

$$(\text{calories}) = \text{T in °K} \times \Delta S_{system} \text{ in calories/°K}$$

$$\text{Entropy factor} = T\Delta S_{system}$$

We can modify our hypothesis so that the free energy change[7] or reaction tendency is given by $\Delta H_{system} - T\Delta S_{system}$. This reaction tendency factor is given the

---

[5] In this text entropy has been treated statistically as a measure of disorder. Historically, entropy was introduced strictly as a mathematical definition, and its relationship to disorder was developed later. Mathematically the entropy change is defined as $\Delta S = q_r/T$, where q_r is the heat change for a reversible process (that is, one that is carried out in a series of infinitesimally small steps so that the process can be reversed at any time). For reversible processes at constant pressure and temperature, $\Delta H = q_r$ and $\Delta S = \Delta H/T$.

[6] The importance of temperature in changing reaction feasibility should already be apparent if you compare your answers to Study Question 2-9, parts b and c.

[7] Strictly speaking, ΔG is not energy because it is not conserved. Nevertheless, it has the dimensions of energy, and in common use it can be treated as energy.

symbol ΔG_{system} (pronounced "delta G") and is called the Gibbs free energy change, in honor of J. Willard Gibbs, an American physicist who developed much of the foundation of chemical thermodynamics.

A Revised Hypothesis—Reaction Tendency, or
Gibbs Free Energy Change

Free, or available, energy = Total energy − unavailable energy

Reaction tendency (free energy) = $\Delta H_{system} - T\Delta S_{system}$

$$\Delta G_{system} = \Delta H_{system} - T\Delta S_{system}$$

at constant temperature and pressure.

Since by definition ΔG is defined only for the system, we shall normally drop the subscript and use $\Delta G = \Delta H - T\Delta S$.

This very simple equation, involving no more than subtraction and multiplication, is one of the most important equations in science with respect to macroscopic chemical behavior. In the next section, we shall see that the sign of ΔG allows us to predict whether a reaction is spontaneous or nonspontaneous by considering only the system.

It should be noted that the temperature, T, in the ΔG equation is the Kelvin temperature, and it must be expressed in degrees Kelvin (°K), not degrees Centigrade (°C) or degrees Fahrenheit (°F). Kelvin temperature is obtained very easily by adding 273 degrees to the Centigrade temperature.

$$°K = °C + 273$$

Study Question 2-10

a. Room temperature is approximately 25 degrees Centigrade, or 25°C. What is this temperature in degrees Kelvin?
b. 100°C, the boiling point of water, = _____ °K.
c. T = 373°K; t°C = _____.

d. $300°K = $ _____ $°C$.

e. $127°C = $ _____ $°K$.

f. $0°C$, the freezing point of water, $= $ _____ $°K$.

Study Question 2-11

a. The change in enthalpy, ΔH, for a reaction at $27°C$ and normal atmospheric pressure is $-30,000$ calories. Does this enthalpy factor favor a spontaneous or a nonspontaneous reaction? Explain.

b. The change in entropy, ΔS, for the same reaction under the same conditions is -30 calories/$°K$. Does this factor favor a spontaneous or nonspontaneous reaction? Explain.

c. Calculate the change in Gibbs free energy, ΔG, for this reaction at $27°C$ and atmospheric pressure.

d. Compare the ΔH and $T\Delta S$ factors to determine which one predominates under these conditions. Would you expect the reaction to be spontaneous or nonspontaneous?

e. Based on your last answer, what would the sign of ΔG be for a spontaneous reaction? For a nonspontaneous reaction?

Study Question 2-12 ΔH for a reaction at $27°C$ and atmospheric pressure is -5000 calories and ΔG is -4400 calories. Calculate ΔS for the reaction at these conditions and indicate whether the reaction system undergoes an increase or a decrease in disorder.

2-6 Thermodynamic Spontaneity and Gibbs Free Energy

How can we use the Gibbs free energy change to determine whether a given reaction is spontaneous or nonspontaneous? Let us summarize the information already discussed about spontaneity.

Summary of Spontaneity Factors

	Sign	Spontaneous?	Change
ΔH	–	Yes	Lower energy
ΔH	+	No	Higher energy
$T\Delta S$	+	Yes	More disorder
$T\Delta S$	–	No	Less disorder (more order)

In other words, a spontaneous reaction is favored when the system changes to a state of lower energy (ΔH is $-$, exothermic) and higher disorder ($T\Delta S$ is $+$). A nonspontaneous reaction is favored when the system changes to a state of higher energy (ΔH is $+$, endothermic) and less disorder, or increased order ($T\Delta S$ is $-$). This information can be applied to determine the expected sign of ΔG for spontaneous and nonspontaneous reactions, where both factors are either favorable (yes) or they are both unfavorable (no).

Spontaneous reactions—Both factors favorable

$$\Delta G = \Delta H - T\Delta S$$
$$\Delta G = (-) - T(+)$$
$$\text{yes} \quad \text{yes}$$
$$\Delta G = (-) - (+) = -$$
$$\text{YES}$$

ΔG is $-$ for spontaneous reactions.

Nonspontaneous reactions—Both factors unfavorable

$$\Delta G = \Delta H - T\Delta S$$
$$\Delta G = (+) - T(-)$$
$$\text{no} \quad \text{no}$$
$$\Delta G = (+) - (-)$$
$$\Delta G = (+) + (+) = +$$
$$\text{NO}$$

ΔG is $+$ for nonspontaneous reactions.

Thus, for a spontaneous reaction the Gibbs free energy of the system decreases and for a nonspontaneous reaction it increases. A spontaneous reaction where ΔG is negative is called an *exergonic reaction*, and a nonspontaneous reaction with a positive ΔG is known as an *endergonic reaction*. (Figure 2-6).

Exergonic: ΔG is $-$, and the reaction can proceed spontaneously.
Endergonic: ΔG is $+$, and the reaction cannot occur unless outside energy is supplied.

Now we are prepared to formulate a second criterion for reaction spontaneity—one that works for the system only.

Criterion for Spontaneity—System Only:
The Minimum Gibbs Free Energy Principle

At constant temperature and pressure a system tends spontaneously toward a state of minimum Gibbs free energy;

ΔG_{system} is negative: Reaction can occur spontaneously.

ΔG_{system} is positive: Reaction cannot occur spontaneously.

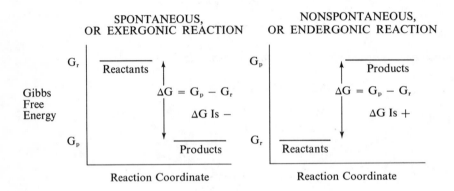

SPONTANEOUS,
OR EXERGONIC REACTION

NONSPONTANEOUS,
OR ENDERGONIC REACTION

Gibbs Free Energy

G_r Reactants $\Delta G = G_p - G_r$ ΔG Is − G_p Products

G_p Products G_r Reactants

G_p Products $\Delta G = G_p - G_r$ ΔG Is +

Reaction Coordinate   Reaction Coordinate

Figure 2-6   Gibbs Free Energy Changes for the System in Spontaneous and Non-spontaneous Reactions.

The power and utility of this simple principle—like that of the previous criterion for spontaneity using ΔS_{net}—cannot be overestimated. Since 1923, when it was first introduced, the Gibbs free energy function has been used extensively to answer the question: Can a reaction occur spontaneously?

Although ΔG is an extremely useful function, we should not lose sight of the fact that it is an artificial device introduced for convenience in dealing with the system only. Both ΔG_{system} and ΔS_{net} can be used to predict thermodynamic spontaneity, but the more fundamental approach involves ΔS_{net}.[8]

---

[8] Recently some chemists have advocated doing away with ΔG and returning to the more fundamental approach using ΔS_{net}. For an excellent discussion of this idea see L. E. Strong and H. F. Halliwell, Journal of Chemical Education, **47**, (1970), 347. This is a superb example of the utility of thermodynamics; a number of different and valid approaches can be used, all stemming from the same basic postulates—the laws of thermodynamics. Perhaps one reason for returning to or at least putting more emphasis on the ΔS_{net} approach is that it would force us to take into account the effect of any

Study Question 2-13 Does the fact that ΔG is $-$ for a reaction mean that the reaction *will* occur spontaneously? Explain.

Study Question 2-14 Suppose ΔG for a chemical reaction is zero. How would you describe this reaction? (This important question will be taken up in Chapter 4.)

In terms of ΔG, two cases have been analyzed for spontaneity: one in which both the enthalpy and entropy factors are favorable (yes) and lead to a spontaneous reaction and a second situation in which both factors are unfavorable (no) and lead to a non-spontaneous reaction. These two factors do not always work in harmony, and there are two additional possibilities when the enthalpy and entropy factors are opposed to one another. In one case the enthalpy factor is favorable (yes) and the entropy factor is unfavorable (no); in the other, the reverse situation occurs. These two yes-no and no-yes possibilities represent a struggle between two opposing tendencies, and the overall spontaneity depends on whether the enthalpy or the entropy factor predominates (Figure 2-7).

What determines overall spontaneity when the two factors oppose each other? Note that the Kelvin temperature, T, appears in the entropy factor, $T\Delta S$. The values of the enthalpy change, ΔH, and the entropy change, ΔS, do not change significantly if the temperature at which the reaction is carried out is changed.[9] However, the entropy factor, $T\Delta S$, does vary with temperature, and it becomes more and more

---

action on the environment. As we shall see in Chapter 11, our environmental crisis can be interpreted as our failure to consider man's impact on his surroundings. Indeed, we can obtain the equation, $\Delta G_{system} = \Delta H_{system} - T\Delta S_{system}$, rather easily from ΔS_{net}.

$$\Delta S_{net} = \Delta S_{surroundings} + \Delta S_{system}$$

Recall that at constant temperature and pressure ΔS is defined as $\Delta H/T$ and that according to the

first law, $\Delta H_{system} = -\Delta H_{surroundings}$. Thus, $\Delta S_{surroundings} = \dfrac{\Delta H_{surroundings}}{T} = \dfrac{-\Delta H_{system}}{T} + \Delta S_{system}$.

Substituting, we obtain

$$\Delta S_{net} = \Delta S_{surroundings} + \Delta S_{system} = \frac{-\Delta H_{system}}{T} + \Delta S_{system}.$$

Multiplying both sides of the equation by $-T$, we obtain

$$(-T)\Delta S_{net} = \Delta H_{system} - T\Delta S_{system},$$

which is equal to ΔG_{system}.

$$\Delta G_{system} = (-T)\Delta S_{net} = \Delta H_{system} - T\Delta S_{system}$$

at constant temperature and pressure.

If $\Delta G_{system} = (-T)\Delta S_{net}$, then $\Delta S_{net} = -\Delta G_{system}/T$ and we can see that $-\Delta G_{system}/T$ is a measure of the total entropy change when a process occurs, and we have again discovered that when ΔS_{net} is positive for a spontaneous change, then ΔG_{system} will be negative.

[9] This assumption is not always justified, depending on the substances involved and the overall range of temperature change. It does, however, provide useful approximations.

Reaction Type	Overall Spontaneity ΔG	$=$	Enthalpy Factor ΔH	$-$	Entropy Factor $T\Delta S$
1. Both factors favorable	$(-)$ YES	$=$	$(-)$ yes	$-$	$(+)$ yes
2. Both factors unfavorable	$(+)$ NO	$=$	$(+)$ no	$-$	$(-)$ no
3. ΔH favorable ΔS unfavorable	?	$=$	$(-)$ yes	$-$	$(-)$ no
4. ΔH unfavorable ΔS favorable	?	$=$	$(+)$ no	$-$	$(+)$ yes

Figure 2-7   Summary of Spontaneity Possibilities.

important in determining overall reaction spontaneity as the temperature is increased. At relatively low temperatures (for example, room temperature), the $T\Delta S$ factor will be relatively small and the enthalpy factor, ΔH, will normally predominate. *When the two factors are opposed, the relative temperature at which the reaction is carried out will determine overall reaction feasibility.*

Effect of Temperature on Spontaneity

1. At high temperatures, the entropy factor, $T\Delta S$, predominates.
2. At low temperatures, the enthalpy factor, ΔH, predominates.

Possibilities 3 and 4 in Figure 2-7 can now be resolved as shown in Figure 2-8, where the predominating factor in each case is shown in capital letters.

For reaction types 1 and 2 (Figure 2-7), a high temperature means that they become even more spontaneous (type 1) or more nonspontaneous (type 2).

Reaction Type	Temperature	Overall Spontaneity ΔG	=	Enthalpy Factor ΔH	−	Entropy Factor $T\Delta S$
3. ΔH favorable ΔS unfavorable	Low T	(−) YES	=	(−) YES	−	(−) no
	High T	(+) NO	=	(−) Yes	−	(−) NO
4. ΔH unfavorable ΔS favorable	Low T	(+) NO	=	(+) NO	−	(+) yes
	High T	(−) YES	=	(+) no	−	(+) YES

Figure 2-8 Effect of Temperature on Reaction Spontaneity.

Study Question 2-15   The following chemical changes occur at constant temperature and pressure. Predict whether each process is spontaneous or nonspontaneous, first at room temperature (25°C) and then at a high temperature.

a. Salt is dissolved in water, and it is found that the ΔH factor is negative and the $T\Delta S$ factor is positive.
b. The conversion of ozone (O_3) to oxygen (O_2) is exothermic, and the entropy of the system increases.
c. Reaction is endothermic, and the relative order in the system is increased.
d. Disorder increases, and heat is evolved.
e. Enthalpy increases, and the products have less disorder than the reactants.
f. When H_2O is decomposed into H_2 and O_2, the enthalpy of the products is higher than that of the reactants, and ΔS is positive.
g. The conversion of steam to water is exothermic, and entropy is decreased.
h. ΔH is negative, and disorder is decreased.

We began our search for a criterion for reaction spontaneity (Section 2-3) by proposing the minimum enthalpy principle—that all exothermic reactions should be spontaneous. We can now see that under the conditions of relatively low temperatures, this principle can be used as an approximate criterion for spontaneity, since at low temperatures the entropy factor, $T\Delta S$, normally makes relatively little contribution to ΔG and may be neglected. This is why at low temperatures beautiful, crystalline snowflakes form spontaneously, even though ΔS_{system} is unfavourable.

An Approximate Criterion for Spontaneity at Low
Temperatures—The Minimum Enthalpy Principle

At constant pressure and relatively low temperatures any exothermic reaction (ΔH is $-$) should be spontaneous and any endothermic reaction (ΔH is $+$) should be nonspontaneous.

Study Question 2-16   Under what conditions would you expect an endothermic reaction to be spontaneous? Explain.

2-7 The First and Second Laws of Thermodynamics

You should now have some understanding of two of the most comprehensive laws of science—the First and Second Laws of Thermodynamics. The *First Law of Thermodynamics* is the familiar Law of Conservation of Energy.

First Law of Thermodynamics—Law of Conservation of Energy

In all macroscopic chemical and physical changes, energy is neither created nor destroyed but merely transformed from one form to another.[10]

It requires that the energy change within a system be equal in magnitude but opposite in sign to the net energy change in the surroundings. It can be expressed in terms of the change in enthalpy, as $\Delta H_{system} = - \Delta H_{surroundings}$.

[10] Matter can be converted into energy (witness nuclear power), but this becomes measurable only in nuclear reactions, not ordinary chemical reactions. A more accurate and comprehensive statement would be that "matter-energy can be neither created nor destroyed."

Consider a balloon at a particular temperature. Suppose that heat energy is put into this system. This heat-energy input (or ΔH) will be used, or transformed, in two ways. Part of it will be used to increase the average internal energy (which will be given the symbol E) of the particles in the system, that is, to increase the average potential and kinetic energy of the components of the system. Another portion of the energy input will be used to do work on the surroundings, as the balloon exerts a force and expands against the surrounding atmosphere. According to the first law, the heat input (designated as ΔH) must be equal to the change in internal energy (designated as ΔE) plus the work done by the system on the surroundings (designated as w), as shown in Figure 2-9.

A Mathematical Form of the First Law of Thermodynamics

Heat energy input	=	Increased internal energy	+	Work done by the system on the surroundings
ΔH	=	ΔE	+	w

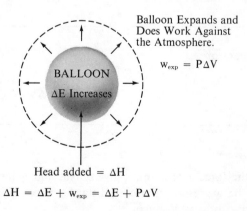

Balloon Expands and Does Work Against the Atmosphere.

$$w_{exp} = P\Delta V$$

BALLOON

ΔE Increases

Head added = ΔH

$$\Delta H = \Delta E + w_{exp} = \Delta E + P\Delta V$$

Figure 2-9   An Illustration of the First Law of Thermodynamics. When Heat Is Added to a Balloon, It is Used to Increase the Internal Energy of the System and to Perform Work on the Surroundings as the Balloon Expands. No Net Energy Is Lost or Gained.

Sometimes the symbol q is used in place of ΔH to represent the heat input. If this is put into the above equation, we obtain

$$q = \Delta E + w$$

or, rearranging,

$$\Delta E = q - w.$$

This is another mathematical form of the first law that is found in most textbooks on thermodynamics. It shows that the difference between the heat (q or ΔH) put into the system and the work done by the system on its surroundings is equal to the change in internal energy (ΔE).

Another Mathematical Form of the First Law

$$\Delta E = q - w$$

Work is done when one exerts a force to move an object over a distance. By definition, work is force multiplied by distance. In the case of the balloon in Figure 2-9, work is done when the gas in the balloon expands against the opposing atmospheric pressure, P_{opp}. The work of expansion, w_{exp}, is equal to the pressure, P_{opp}, multiplied by the change in volume of the balloon, ΔV.

Work of Expansion

$$w_{exp} = P_{opp} \, \Delta V$$

Pressure is the force exerted on a given area, and it has the units force/area, while volume can be expressed as area × distance. From this we see that $P\Delta V$ has the units of work—force × distance.

$$P\Delta V = \frac{force}{area} \times area \times distance$$

$$P\Delta V = force \times distance = work$$

Going back to our expression of the first law as $\Delta H = \Delta E + w$, we can substitute $P\Delta V$ for w and obtain

Enthalpy Change

$$\begin{array}{ccc}
\Delta H & = & \Delta E & + & P\Delta V \\
\text{Enthalpy} & = & \text{Energy} & + & \text{Work} \\
\text{change} & & \text{factor} & & \text{factor}
\end{array}$$

From this we see that the enthalpy factor, ΔH, is really a combination of two factors, an energy factor and a work factor. When there is no change in volume in a process (as is the case in most reactions occurring in living systems), then ΔV is zero, the term $P\Delta V$ becomes zero, and the enthalpy change (ΔH) is equal to total energy change (ΔE) of the system.

Constant Volume and Temperature Processes

$$\Delta H_{\text{system}} = \Delta E_{\text{system}}$$

$$\Delta G_{\text{system}} = \Delta E_{\text{system}} - T\Delta S_{\text{system}}$$

Study Question 2-17   A person who has not had the opportunity to study thermodynamics decides on a hot day that he will cool a hot kitchen (assume that all doors and windows are closed) by opening the refrigerator door. In terms of the laws of thermodynamics, explain why this scheme will not cool the kitchen and why it will actually result in the kitchen becoming warmer instead of cooler. (We shall see in Chapter 11 that an understanding of this problem is the key to understanding our environmental crisis.)

The *Second Law of Thermodynamics* states the already familiar tendency toward increasing entropy, or disorder.

Second Law of Thermodynamics

Any system plus its surroundings tends spontaneously toward a state of increasing entropy, or disorder; or

$$\Delta S_{net} = \Delta S_{system} + \Delta S_{surroundings}$$

is positive for any spontaneous process.

Again it should be emphasized that the second law makes no predictions about the system only. It must include an analysis of the entropy changes in both the system and its surroundings. Most "apparent" violations of the second law involve situations in which the observer fails to include the greater entropy increase in the surroundings where there is an increase in order in the system. The net result is still a spontaneous increase in total entropy.

We can summarize the first and second laws as follow:

The First and Second Laws of Thermodynamics

First law: The energy of the system plus surroundings remains constant.
Second law: The entropy of the system plus surroundings increases for any spontaneous process.

These are scientific laws—not theories. They are based solely on observation of what always happens in the world around us. You may doubt Einstein's theory explaining gravitation, but based on your experience you have no doubt whatsoever that if you drop a book or any object on the Earth, it will fall to the Earth in accordance with the law of gravity. Similarly, you may doubt the statistical mechanical theory used to explain the first and second laws, but no one has ever observed a violation of the laws themselves.

A more frivolous way of stating these two laws is

The Laws of Thermodynamics

First Law: You can't get something for nothing.
Second Law: If you think things are mixed up now, just wait.

Study Question 2-18

> ... you shall eat bread till you return to the ground, for out of it you were taken; you are dust and to dust you shall return.
>
> *Genesis 3:19*

Interpret this quote from Genesis in terms of the First and Second Laws of Thermodynamics.

This chapter has been devoted to seeking an answer to the first question of chemical thermodynamics: Can a given reaction occur spontaneously? We now have two ways of answering this question. One, involving the total change in entropy, ΔS_{net}, can be applied to the system plus its surroundings, while the second approach, involving the change in Gibbs free energy, ΔG_{system}, can be applied to the system only at constant temperature and pressure.

Two Criteria for Thermodynamic Spontaneity

System only at constant T and P: ΔG_{system} is negative
System plus surroundings: ΔS_{net} is positive

A theory is more impressive the greater is the simplicity of its premises,
the more different are the kinds of things it relates and the more
extended its range of applicability. Therefore, the deep impression
which classical thermodynamics made on me. It is the only physical
theory of universal content which I am convinced, that within the
framework of applicability of its basic concepts, will never be
overthrown.

Albert Einstein

2-8 Review Questions

2-1. What is the difference between a spontaneous process and a rapid process? In other words, distinguish between thermodynamic spontaneity and kinetic spontaneity. Give an example of each.

2-2. a. Distinguish between a spontaneous reaction and a nonspontaneous reaction. Can a nonspontaneous reaction occur? Explain. Give an example of each.
   b. Distinguish between a forbidden reaction and a nonspontaneous reaction. Give an example of each.

2-3. a. Explain why thermodynamics does not tell us whether or not a given reaction *will* occur.
   b. What does it tell us?
   c. How can we find out whether or not a reaction will actually occur?

2-4. a. What is the minimum potential energy principle? What is wrong with the statement: "In a spontaneous chemical process, the system goes to a state of lower energy?"
   b. Why must the surroundings be included in this principle?
   c. Would the principle apply on Mars? In outer space? Explain.

2-5. a. What is an exothermic reaction? An endothermic reaction?
   b. Draw and label diagrams representing the relative enthalpy levels for the reactants and the products for an exothermic reaction. Repeat for an endothermic reaction. In each case, what is the sign of ΔH?
   c. Are all exothermic reactions spontaneous? Under what conditions will an exothermic process normally be spontaneous?

2-6. What is enthalpy? Distinguish between H and the enthalpy change, ΔH.

2-7. For the hypothetical reaction of X + Y to form P + Q, ΔH is negative. True or false:
   _____ a. The average enthalpy (energy) of the reactants X and Y is lower than the average energy of the products P and Q.
   _____ b. The reaction can be represented as X + Y + heat = P + Q.
   _____ c. According to the minimum enthalpy principle, the reaction of X + Y to form P + Q should be spontaneous.
   _____ d. The reaction of P + Q to form X + Y should be spontaneous.

2-8. State the minimum enthalpy principle. Is it a valid criterion for spontaneity? Explain.

2-9. What is entropy? Distinguish between S and ΔS.

2-10. State the increasing disorder, or entropy, principle.

2-11. Criticize the following statement: "Any spontaneous process results in an increase in the entropy of the system."

2-12. State the First Law of Thermodynamics and give several examples.

2-13. State the Second Law of Thermodynamics and give several examples.

2-14. Criticize the following statement: "Life is an ordering process, and since it goes against the natural tendency for increasing disorder, it constitutes a violation of the Second Law of Thermodynamics."

2-15. Distinguish between entropy and enthalpy; between entropy change and enthalpy change.

2-16. What is the Gibbs free energy change? Why was it introduced?

2-17. a. Distinguish between degrees Kelvin and degrees centigrade.
    b. $10,000°K = \_\_\_\_°C$.
    c. $-10°C = \_\_\_\_°K$.

2-18. a. What are the two factors that determine whether a chemical reaction can occur spontaneously?
    b. How can they be combined to predict overall thermodynamic spontaneity?

2-19. Criticize the following statement: The change in entropy of the system is the factor determining reaction feasibility.

2-20. Explain why the entropy factor in the Gibbs free energy expression is T multiplied by ΔS.

2-21. What is the sign of ΔG for a spontaneous reaction? A nonspontaneous reaction?

2-22. Criticize: The entropy change for a process at 100°C is + 50 entropy units. The entropy factor for this reaction is (100) (50), or 5000, calories.

2-23. Is the entropy factor or the enthalpy factor more important at high temperatures in determining reaction spontaneity? Explain.

2-24. Comment on the following statement: "At low temperatures ΔG is approximately equal to ΔH, and the idea that at low temperatures all exothermic reactions are spontaneous is reasonable."

2-25. Under what conditions can an endothermic reaction be spontaneous?

2-26. State the minimum Gibbs free energy principle and draw and label diagrams of G vs. the reaction coordinate for spontaneous and nonspontaneous reactions.

2-27. State the First and Second Laws of Thermodynamics verbally and mathematically.

2-28. Both ΔS_{net} (second law) and ΔG can be used as criteria for predicting the spontaneity of a chemical reaction. Explain why many scientists prefer to use ΔG for this purpose.

2-29. Work is defined as force times distance. Show that the work of expansion of a gas, $P\Delta V$, has the units of work.

2-30. Explain how the expression, $\Delta E = q - w$, is an expression of the First Law of Thermodynamics.

2-31. What are the two factors that make up the enthalpy change, ΔH?

2-32. Which one of the following statements is true for a spontaneous chemical reaction?
    a. ΔS_{system} is always positive,
    b. ΔS_{system} is always negative,
    c. $\Delta G_{system} < 0$,
    d. $\Delta G_{system} > 0$,
    e. none of these.

2-33. According to the first law, energy can neither be created nor destroyed; yet when hydrogen and oxygen react to form water, energy is evolved. Is this a violation of the first law? Explain.

Chapter 3

Question No. 2—How Much Energy Can Be Released or Absorbed When a Reaction Occurs?—Free Energy Calculations

3-1   The Meaning of a Chemical Equation

In the last chapter we found that reactions that are spontaneous, or thermodynamically feasible, are those for which the sign of the Gibbs free energy change, ΔG_{system}, is negative. Now we want to determine how much energy could be given off in a spontaneous reaction or how much energy is required to force a nonspontaneous reaction to occur. First, it is necessary to review the meaning of a chemical equation.[1]

No one has directly observed individual atoms or small molecules.[2] Any piece of matter visible to the eye contains an unbelievably large number of atoms or molecules. In order to make the transition from the micro world to the macro world, chemists have chosen a "standard package" of particles (atoms, molecules, ions, electrons, quanta, or other entities) to use for comparison purposes. This "standard package" is called a *mole* and, by definition, it contains approximately 600,000,000,000,000,000,-000,000, or 6×10^{23}, units, or particles. This very large number, called *Avogadro's number*, is given the symbol N.[3]

---

[1] The reader who has a background in introductory chemistry may wish to proceed directly to Section 3-2.

[2] They have been observed indirectly with the field ion microscope and the electron microscope.

[3] Avogadro's number is more accurately expressed as 6.0225×10^{23}, but for our purposes 6×10^{23} is a satisfactory approximation.

$$1 \text{ mole} = 6 \times 10^{23} \text{ units, or particles}$$

By always using the same number of particles for comparison (whether they are atoms or molecules), we preserve the relative microscopic properties on the macroscopic level. For example, a single atom of oxygen, O, has a mass 16 times that of a single atom of hydrogen, H; if we take 6×10^{23} atoms of each element and compare the masses of this "standard package," the ratio of masses will still be 16 to 1.

Study Question 3-1

a. How many particles, or units, are there in
   1. 2 moles of marbles?
   2. 10 moles of dollars?
   3. 2 moles of people?
b. How many atoms are there in
   1. 5 moles of carbon, C?
   2. 5 moles of nitrogen, N?
   3. 3 moles of iron, Fe?
c. How many molecules are in
   1. 1 mole of water, H_2O?
   2. 10 moles of carbon monoxide, CO?
   3. 10 moles of nitrogen dioxide, NO_2?
d. How many moles do each of the following quantities represent?
   1. 12×10^{23} molecules of ammonia, NH_3?
   2. 3×10^{23} atoms of carbon, C?
   3. 120×10^{23} molecules of ozone, O_3?
e. Which of the following quantities represents the largest number of molecules?
   1. Avogadro's number of hydrogen gas, H_2, molecules.
   2. 10 moles of carbon dioxide gas, CO_2.
   3. 10 moles of hydrogen gas, H_2.
   4. 20 moles of water, H_2O.
   5. 30×10^{23} molecules of water, H_2O.

A chemical equation is a shorthand method for showing the initial reactants and final products of a reaction or sequence of reactions. It is a method of chemical bookkeeping to keep track of atoms. If we begin with a certain number of atoms of a element, then the products must contain this same number of atoms of the element—no more and no less. For example, solid carbon, C(s), when mixed with oxygen gas, $O_2(g)$, can yield gaseous carbon dioxide, $CO_2(g)$, which can be expressed by the following chemical equation:

$$C(s) + O_2(g) \longrightarrow CO_2(g).$$
Balanced Equation

This equation is *balanced* because all of the atoms are accounted for on each side of the equation. The symbols (s), (g), and (l) denote, respectively, the state of each substance in the equation as solid, gaseous, or liquid. In addition, (aq) is used to denote that a substance has been dissolved in water to form an aqueous solution; for example, a solution of table salt (sodium chloride) dissolved in water can be represented as NaCl(aq).

Hydrogen gas, $H_2(g)$, when mixed with oxygen gas, $O_2(g)$, can react to form liquid water, $H_2O(l)$, at room temperature. (If the reaction were carried out above 100°C, water vapor, $H_2O(g)$, would be formed.)

$$H_2(g) + O_2(g) \longrightarrow H_2O(l)$$
Unbalanced Equation

This equation, however, is not balanced. There are two hydrogen atoms on each side, but there are two atoms of oxygen on the left side and only one atom of oxygen on the right. This oxygen imbalance would be corrected if 2 molecules of $H_2O(l)$ were formed.

$$H_2(g) + O_2(g) \longrightarrow 2H_2O(l)$$
Unbalanced Equation

However, in correcting the oxygen imbalance, we have a hydrogen imbalance (4H on the right and 2H on the left). If 2 molecules of $H_2(g)$ are used, the equation is balanced.

$$2H_2(g) + O_2(g) \longrightarrow 2H_2O(l)$$
Balanced Equation

What information does this balanced equation provide? It can be interpreted on both the micro and macro levels. On the *micro* level, it indicates that when 2 molecules of hydrogen gas and 1 molecule of oxygen gas are brought together, they can somehow react to form 2 molecules of liquid water, if the reaction should go to completion. On the *macro* scale, we use our "standard package," or mole, of molecules; and the equation indicates that when 2 moles, or $2 \times (6 \times 10^{23})$ molecules, of hydrogen gas are brought together with 1 mole, or 6×10^{23} molecules, of oxygen gas, they can somehow react to form 2 moles, or $2 \times (6 \times 10^{23})$ molecules, of liquid water, if the reaction should go to completion.

Meaning of a Chemical Reaction

$$2H_2(g) \quad + \quad O_2(g) \quad \longrightarrow \quad 2H_2O(l)$$

Micro: 2 molecules of H_2 + 1 molecule of O_2 $\longrightarrow$ 2 molecules of H_2O

Macro: 2 moles of H_2 + 1 mole of O_2 $\longrightarrow$ 2 moles of H_2O

or

$2 \times (6 \times 10^{23})$ + $\quad 6 \times 10^{23}$ $\longrightarrow$ $\quad 2 \times (6 \times 10^{23})$

molecules of H_2 + molecules of O_2 $\longrightarrow$ molecules of H_2O

It is also important to note what a balanced equation *does not* tell us.

1. It does not indicate whether the reaction can or will occur. The fact that one can write an equation and balance it does not mean the equation represents a real or even a potential reaction. Whether the reaction *can* occur spontaneously—in time spans ranging from microseconds to eons—is determined by finding out whether ΔG for the reaction is negative. Whether the reaction *will* occur in reasonable time is determined by both thermodynamic and kinetic (rate) factors. *A balanced equation implies nothing whatsoever about either the free energy change or the rate of reaction.*

2. It does not indicate how the reaction occurs—that is, it does not give any information about the path, or *mechanism*, of the reaction. The above equation for water should *not* be interpreted to mean that 2 molecules of hydrogen and 1 molecule of oxygen collide to form 2 molecules of water. This is a very important point not emphasized by many textbooks. For example, the equation for the photosynthesis process is sometimes represented by

$$CO_2(g) + H_2O(l) \xrightarrow{\text{light}} (CH_2O)_x + O_2(g),$$

where CH_2O is a molecular unit that can be built up in multiple units to form a sugar such as glucose, $C_6H_{12}O_6$, which consists of 6 units of CH_2O, or $(CH_2O)_6$. The overall equation for photosynthesis can then be represented by

$$6CO_2(g) + 6H_2O(l) \xrightarrow{\text{light}} C_6H_{12}O_6(s) + 6O_2(g).$$

This equation does not represent the mechanism of photosynthesis, which is a very complex process consisting of a large number of small reaction steps. It is merely a bookkeeping device, showing the initial reactants and the possible final products, without any indication as to how products are converted to reactants. One advantage of thermodynamics is that it enables one to calculate energy differences between the initial and final states without worrying about the mechanism between the initial and the final states. The change in Gibbs free energy (ΔG), enthalpy (ΔH), and entropy ($T\Delta S$) are independent of path, or mechanism.

3. It does not indicate that the reaction goes all the way or any particular fraction of the way.

In summary, a balanced equation is a device for keeping track of atoms. It does *not* provide any answers to the seven fundamental thermodynamic and kinetic questions about a chemical reaction. We shall see, however, that an equation must be balanced if we are to obtain answers to these questions.

3-2 Some Important Chemical Equations

Fortunately, an extensive knowledge of chemicals or chemical reactions is not necessary to appreciate how one can answer the seven fundamental questions about reactions. For illustrative purposes, we shall deal with only eleven chemical equations, which we shall use over and over again in subsequent sections and chapters. The equations chosen represent important reactions involved in the origin of the earth's atmosphere, air pollution, and biological processes such as photosynthesis and respiration. An understanding of how thermodynamics can be applied to these examples can then be transferred to other reactions.

Study Question 3-2   The eleven equations to be studied are *not* balanced. As you go through this chapter, balance each equation. Balanced equations will be necessary for the discussion in a later section in this chapter.

Reaction 1—Ozone (O_3) Formation

A crucial reaction for the maintenance of life on this planet is the conversion of oxygen gas, $O_2(g)$, to ozone gas, $O_3(g)$, in the upper atmosphere. This process forms the *ozone layer*, which filters the sun's high-energy ultraviolet radiation that otherwise would destroy most of the life forms on this planet.[4]

No. 1

$$\text{oxygen(g)} \rightleftharpoons \text{ozone(g)}$$

$$\underline{\quad\quad} O_2(g) \rightleftharpoons \underline{\quad\quad} O_3(g)$$

Ozone is also being formed in the lower atmosphere by electrical discharge associated with some of our industrial processes. It is a dangerous pollutant in the air we breathe, because extremely small concentrations (a fraction of a part per million, ppm) irritate lung tissue and cause severe crop damage. Higher (but still minute) concentrations in the parts per million[5] (ppm) range can cause sterility and death—a severe way to

---

[4] A new type of "environmental warfare" has been proposed in which chemical rockets would be exploded in the upper atmosphere to create "holes" or "windows" in the ozone layer. This could then "burn out" life forms on selected parts of this planet. The threat to man comes from man himself. As Basil O'Connor said, "The world cannot continue to wage war like physical giants and to seek peace like intellectual pygmies."

[5] A part per million is equivalent to one ounce of vermouth in 7,530 gallons of gin.

solve the overpopulation problem. Some time ago U.S. health authorities set a recommended maximum limit for ozone at 0.1 ppm. This limit is now regularly exceeded in many major cities.

Reactions 2 and 3—Burning of Fossil Fuels and the "Greenhouse Effect"

Since the industrial revolution, there has been a dramatic increase in the burning of fossil fuels, such as coal, natural gas, oil, and gasoline. Fossil fuels are carbon (coal, graphite) or carbon-hydrogen compounds (such as methane, ethane, propane, butane, octane). When they undergo complete combustion with oxygen, the resulting carbon dioxide and water enter the atmosphere. The incomplete combustion of fossil fuels yields deadly carbon monoxide gas, $CO(g)$.

During the last one-hundred years the carbon dioxide (CO_2) content of our atmosphere has risen almost 10% because of the burning of fossil fuels. By the year 2000, it is predicted that the increase of CO_2 may reach 25%.

It is believed by some that this rise in CO_2 content could produce what is known as the " greenhouse effect." The CO_2 allows heat energy from the sun into the atmosphere, but it prevents lower energy (lower frequency) radiation from escaping back into space. As a result there could be a gradual warming of the earth's atmosphere. It is feared that over a period of years this could cause the Antarctic ice sheet to melt or slip into the ocean, thus raising the sea level by anywhere from 100 to 400 feet—just as a body raises the water level in a bathtub or melted ice raises the water level in a glass. Obviously, such a rise in the sea level would put most of the world's major cities under water (again, hardly a desirable "solution" to the overpopulation problem).

There is considerable debate in the scientific community on whether or not this process is occurring and on the number of years it might take.[6] Ironically, another form of pollution—the dust and soot we are pouring into the air—may be counteracting the warming trend. This debris in the atmosphere, it is argued, filters out the sun's energy and thus has a cooling effect on the atmosphere. If this form of pollution should predominate over the CO_2 pollution, then the atmosphere could cool enough to send us into a new ice age instead of a return to Noah's ark. Frankly, we don't know enough about these processes and others that are occurring in our atmosphere to make valid predictions.[7]

Two reactions will be used to represent the fossil burning reactions. One involves the complete combustion of coal or carbon (graphite) and the other the burning of n-octane, $C_8H_{18}(l)$, a major component of gasoline.

---

[6] Although this is apparently not an immediate environmental problem, it could become a very serious one in the future. For a summary of the problem see, "Report of the Study of Critical Environmental Problems," *Man's Impact on the Global Environment*, The MIT Press, 1970, and the discussion in Chapter 10, Section 10-3 of this book.

[7] The land investor who is cynical about whether man will really clean up his "environmental mess" in time might purchase land in the Midwest. It could then be used as beachfront property or as a cross-country ski resort, depending on which way the situation goes.

Reaction 2—Burning of coal

carbon (graphite) + oxygen(g) $\rightleftharpoons$ carbon dioxide(g)

_____C(graphite) + _____O_2(g) $\rightleftharpoons$ _____CO_2(g)

Reaction 3—Burning of gasoline (octane)

n-octane(l) + oxygen(g) $\rightleftharpoons$ carbon dioxide(g) + water(g)

_____C_8H_{18}(l) + _____O_2(g) $\rightleftharpoons$ _____CO_2(g) + _____H_2O(g)

Reaction 4—Nitrogen Dioxide—A Chemical Villain in Our Air

Large cities generally fall into one of two basic air pollution classes—the brown air cities and the gray air cities. Brown air cities, like Los Angeles, are relatively young cities, where the main source of air pollution is the internal (or infernal) combustion engine in the automobile. Every automobile spews several hundred different compounds into the atmosphere, including a number of substances capable of producing cancer.

One of these compounds is nitric oxide, NO(g). Although it is relatively harmless, it reacts with oxygen, O_2(g), in the air to form nitrogen dioxide, NO_2(g), a yellow-brown, poisonous gas that produces the brown haze over many cities. Nitrogen dioxide may be harmful to the lungs. Even more serious is the fact that in the presence of sunlight it combines with carbon-hydrogen compounds (hydrocarbons), also emitted in auto exhausts, to form a whole new brew of chemicals. This mixture, called photochemical smog, contains chemicals that irritate the eyes and damage vegetation.

To reduce this form of pollution[8] either the NO(g) in the exhaust must be sharply reduced (or eliminated) or a means must be found to prevent or slow down its

---

[8] Emission control devices now installed on cars do a relatively poor job of reducing oxides of nitrogen and may actually increase their concentration. Many experts argue that emission control devices provide only a temporary solution, which buys us a little time (perhaps 5–10 years) to find a replacement for the internal combustion engine. If, over a period of years, these devices reduce all dangerous air pollutants, on an average, by 50% and during the same time the number of cars doubles (as projected), then air pollution levels will return to present levels and even rise again as more cars are added. See Figure 11-8.

oxidation to nitrogen dioxide, $NO_2(g)$, in the atmosphere. A study of the thermo-dynamics and kinetics of this reaction could provide information on the feasibility of the latter approach.

No. 4

$$\text{nitric oxide(g)} + \text{oxygen(g)} \rightleftharpoons \text{nitrogen dioxide}$$

$$\underline{\quad}NO(g) + \underline{\quad}O_2(g) \rightleftharpoons \underline{\quad}NO_2(g)$$

Reaction 5—Sulfur Dioxide—Another Villain in the Air

The second type of cities, the gray air cities, are older cities, such as New York, Chicago, and London, that depend heavily on the burning of coal or oil for heating, manufacturing, and electric power generation. This burning produces two major air pollutants—*particulates* (dust and soot), which give the air over such cities its gray cast, and *sulfur dioxide*, one of the most dangerous pollutants in our air.

Coal and oil usually are contaminated with sulfur, $S(s)$,[9] and when they are burned, the sulfur combines with the oxygen of the air to produce sulfur dioxide, $SO_2(g)$. The level of sulfur dioxide in our air, already critical, is expected to double in the next 20 years. It can damage plants, dissolve marble and concrete, and eat away iron and steel—not to mention what it does to delicate lung tissue. It dissolves nylon, as women in a Florida city who wore blouses and stockings of nylon discovered to their dismay. Exposure for 24 hours to only 0.2 parts per million (the equivalent to finding two people with green hair in a city of ten million people—like New York) of sulfur dioxide, with particulate present, is considered to be a serious health hazard.

The $SO_2(g)$ can also react with oxygen in the air to produce sulfur trioxide, $SO_3(g)$, which when breathed can react with water to produce droplets of sulfuric acid on lung tissue. Unfortunately, traces of nitrogen dioxide, $NO_2(g)$, can serve as a catalyst to increase the rate of oxidation of $SO_2(g)$ to $SO_3(g)$, as we shall see in the chapter on kinetics.

The sulfur dioxide problem can be helped by using the more expensive fuels with low sulfur content, removing the sulfur from the fuels (also expensive), or by removing the SO_2 from the smoke stacks and possibly converting it to pure sulfur, which is in short supply. We shall study the thermodynamic feasibility of this reverse reclamation reaction.

No. 5

$$\text{sulfur dioxide(g)} \rightleftharpoons \text{sulfur(s)} + \text{oxygen(g)}$$

$$\underline{\quad}SO_2(g) \rightleftharpoons \underline{\quad}S(s) + \underline{\quad}O_2(g)$$

[9] Actually sulfur occurs as $S_8(s)$, but for simplicity we will use $S(s)$.

Reaction 6—Removal of Ammonia from the Primitive Earth Atmosphere

The question of the evolution of the earth's atmosphere is intriguing. One hypothesis is that the primitive earth atmosphere was believed to consist of such gases as methane (CH_4), ammonia (NH_3), hydrogen cyanide (HCN), and others. Our present atmosphere, consisting primarily of nitrogen (N_2) and oxygen (O_2), evolved at a later stage. After the evolution of the photosynthesis process, oxygen was added to the atmosphere where it possibly oxidized $NH_3(g)$, $CH_4(g)$, and $HCN(g)$ to produce our present atmosphere. Is this hypothesis reasonable? Part of the answer can be obtained by asking whether the oxidation of ammonia, methane, and other gases is thermodynamically and kinetically feasible. We shall look at the oxidation of ammonia as representative of this problem.

No. 6

$$\text{ammonia(g)} + \text{oxygen(g)} \rightleftharpoons \text{nitrogen(g)} + \text{water(g)}$$

$$\underline{\quad}NH_3(g) + \underline{\quad}O_2(g) \rightleftharpoons \underline{\quad}N_2(g) + \underline{\quad}H_2O(g)$$

Reaction 7—Formation of Water in the Primitive Earth Atmosphere

Obviously the formation of water, which is so essential to life, is an important reaction that deserves analysis. Could water have been formed in the primitive atmosphere on earth by the reaction of hydrogen and oxygen gas, or is it more likely that the reverse reaction, the decomposition of water into $H_2(g)$ and $O_2(g)$, served as a possible source of hydrogen and oxygen gases in our present atmosphere?

No. 7

$$\text{hydrogen(g)} + \text{oxygen(g)} \rightleftharpoons \text{water(l)}$$

$$\underline{\quad}H_2(g) + \underline{\quad}O_2(g) \rightleftharpoons \underline{\quad}H_2O(l)$$

Reaction 8—The Manufacture of Ammonia—An Important Chemical Industrial Process

Ammonia, $NH_3(g)$, one of our most important compounds, is widely used in the preparation of fertilizers essential for growing more food for our rapidly increasing population.[10] During World War I, Fritz Haber, a German scientist, developed an industrial process for combining nitrogen gas, $N_2(g)$, and hydrogen gas, $H_2(g)$, to form ammonia(g). This synthetic process is called the Haber process, and because of its importance, Haber received the Nobel prize for his work. This process is a particularly good example of how both thermodynamic and kinetic factors are combined to produce a chemical in an industrial process.

No. 8

$$nitrogen(g) + hydrogen(g) \rightleftharpoons ammonia(g)$$

$$\underline{\quad}N_2(g) + \underline{\quad}H_2(g) \rightleftharpoons \underline{\quad}NH_3(g)$$

Reaction 9—Photosynthesis in Plants

Life on earth is wholly dependent upon radiant energy (light) from the sun. Photosynthesis consists of the absorption of this radiant energy by chlorophyll (or other light-trapping molecules) in the cells of green plants. The absorbed light energy is then transformed into chemical energy, which is used to bring about the reaction of water and carbon dioxide from the atmosphere to produce a basic sugar unit, (CH_2O), plus O_2, which is released into the atmosphere. The basic sugar unit can be built up by a series of steps to form larger molecules, such as the glucose, $C_6H_{12}O_6(s)$. The entire process is extremely complex, involving a long sequence of reactions. Fortunately, thermodynamic feasibility can be predicted by looking at only the initial state (CO_2 and H_2O) and the final state (glucose and O_2). The *net* reaction for photosynthesis plus the succeeding steps to build up glucose can be represented as

No. 9

$$carbon\ dioxide(g) + water(l) \rightleftharpoons glucose(s) + oxygen(g)$$

$$\underline{\quad}CO_2(g) + \underline{\quad}H_2O(l) \rightleftharpoons \underline{\quad}C_6H_{12}O_6(s) + \underline{\quad}O_2(g)$$

Reaction 10—Aerobic Respiration in Cells

The chemical energy stored in glucose and other larger nutrient molecules produced in plants by photosynthesis is utilized by the plant cells for their life processes. Plants are also consumed by animals, where the same nutrient molecules are used to sustain

[10] Like many advances in technology, the use of fertilizer has both good and bad effects. On the one hand, it helps us feed more people, but on the other hand, it is the source of one of our most serious water pollution problems. Approximately half of the water pollution in the U.S. comes from farms in the form of fertilizer runoff and animal wastes. As a result, nitrate (which is poisonous, particularly to infants) levels in our water are rising and the excess nitrate threatens the water purification processes in some of our rivers and lakes. This is a good example of the collision course between overpopulation and pollution. We can grow more food by using more fertilizer on our declining soil—but the resulting pollution threatens our very survival.

life processes in the animal cells. In the plant or animal cells, the glucose can be burned aerobically (with oxygen) to produce carbon dioxide (CO_2) and water (H_2O) with a release of energy. This process is called respiration, and the CO_2 produced enters the atmosphere and oceans to enable the entire cycle to begin again. The *net* equation for the respiration process is the reverse of the *net* equation for the photosynthesis process, but this does not mean that the reaction mechanism for respiration is the reverse of that for photosynthesis. Remember that neither equation provides any information about the actual mechanism (series of steps) by which either photosynthesis or aerobic respiration takes place.

No. 10

$$\text{glucose(s)} + \text{oxygen(g)} \rightleftharpoons \text{carbon dioxide(g)} + \text{water(l)}$$

$$\underline{\quad}C_6H_{12}O_6(s) + \underline{\quad}O_2(g) \rightleftharpoons \underline{\quad}CO_2(g) + \underline{\quad}H_2O(l)$$

Reaction 11—Bioenergetics—ADP and ATP

Phosphorus occurs in plant and animal tissues as both inorganic and organic phosphates. In addition to their presence in the protoplasm of the cell, certain organic phosphates play vital roles in the chemical processes by which foods, in the form of carbohydrates, proteins, and fats, are metabolized in the body to give energy plus other products. Most organic phosphates react with water spontaneously with a maximum free energy change of about -2000 calories, or -2 kilocalories. However, several special organic phosphates, including adenosine triphosphate (ATP), adenosine diphosphate (ADP), and creatine phosphate, react with water (hydrolyze) to give much higher free energy changes, ranging from -7000 to $-12,000$ calories (-7 to -12 kilocalories). These phosphates, commonly called "high energy"[11] phosphates, provide the chemical energy for most processes in the cell. As representative of these important reactions, we shall study the thermodynamics of the hydrolysis of adenosine triphosphate, ATP, to release one of the three phosphate groups and yield adenosine diphosphate, ADP. As shown in Equation 11, ATP and ADP are rather complicated molecules consisting of an adenosine group and either 2 phosphate groups (ADP) or 3 phosphate groups (ATP).

3-3 Standard State Conditions

With some understanding of the meaning of a chemical equation and a description of eleven important reactions, we can begin to deal with the central question of this chapter: *How much energy can be released or absorbed when a reaction occurs?*

The thermodynamic spontaneity of a reaction is based on the sign of ΔG, while the energy that can potentially be released or absorbed is based on the actual value of ΔG. To answer the question, we calculate ΔG using the expression $\Delta G = \Delta H - T\Delta S$. If ΔG is negative, then its value represents the *maximum work* (energy is capacity to

---

[11] We shall see later that this is actually a misnomer.

No. 11

Adenosine triphosphate + water (l) $\rightleftharpoons$ Adenosine diphosphate + phosphoric acid

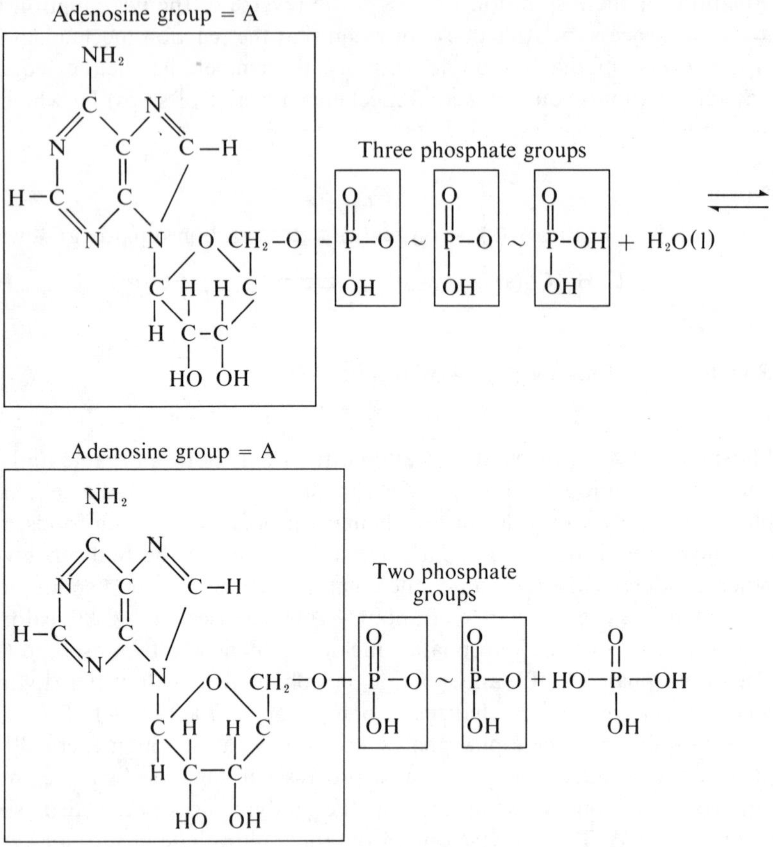

Adenosine group = A

Three phosphate groups

Adenosine group = A

Two phosphate groups

or

$$\underline{\quad} A \sim P \sim P \sim P + \underline{\quad} H_2O(l) \rightleftharpoons \underline{\quad} A \sim P \sim P + \underline{\quad} H_3PO_4$$

(ATP)   (ADP)

do work) available from the reaction. If ΔG is positive, its value represents the *minimum work* necessary to force the reaction to occur.

Note that ΔG represents the *maximum work* potentially available from a spontaneous reaction that goes to completion. In actual reactions the work available will always be less than this maximum. The maximum work can be attained only if the reaction is carried out reversibly. A thermodynamically reversible reaction would be carried out in a series of infinitesimal steps, so that the process could be reversed at any moment in time. This is an ideal that can only be approached—never reached— since a truly reversible process would require infinite time. The closer the reaction process approaches this ideal, the closer the actual work comes to the maximum represented by the value of ΔG. By a similar argument, we would expect that in nonspontaneous reactions the actual work needed to force the reaction would always be greater than the minimum value, represented by ΔG for a reversible process.

Study Question 3-3   A reaction is carried out at atmospheric pressure and 27°C. The Gibbs free energy change for the reaction is −10,000 calories, or −10 kilocalories. The reaction is exothermic and it evolves 19,000 calories, or 19 kilocalories.

a. Is the reaction spontaneous or nonspontaneous? Explain.
b. Draw and label a free energy vs. reaction coordinate diagram for this reaction.
c. Is the reaction exergonic or endergonic?
d. How much energy (work) could be available or required if this reaction occurs or is made to occur at 27°C? If you ran this reaction in the laboratory, would you expect to obtain this amount of energy (work) or have to put it in? Explain.
e. Calculate the entropy change, ΔS, in calories/°K for this reaction.
f. Which has the higher average degree of disorder, the reactants or products? Explain.
g. If the temperature at which the reaction is carried out is raised to 527°C, what is the new value of ΔG? Is the reaction spontaneous or nonspontaneous at this new temperature?
h. This reaction is spontaneous at 27°C and nonspontaneous at 527°C. Calculate in °C and °K the temperature at which this change from spontaneous to nonspontaneous would occur.

The free energy change, ΔG, must be calculated for a reaction at a fixed temperature and pressure. Pressure is the variable held constant for reactions involving gases, while concentration is the variable substituted for pressure if the reaction is carried out in an aqueous solution. The concentration of a substance is defined as the amount present (usually expressed as the number of moles) in a fixed volume (usually a volume of one liter).

$$\text{Concentration} = \frac{\text{Amount}}{\text{Fixed volume}} = \frac{\text{Moles}}{\text{Liter}}$$

A solution with a concentration of one mole of a substance dissolved in a liter of solution is called a 1-molar, or 1 M, solution, where molar, or M, stands for moles/ liter. A 2 M solution would then contain 2 moles of the substance dissolved in a liter of solution.

We can calculate the values of ΔG, ΔH, and ΔS at any specified conditions of constant temperature and pressure or concentration. However, we can compare the relative spontaneity and energy involved for different reactions only if they are carried out at the same temperature and pressure or concentration.

To allow this type of standard comparison, scientists throughout the world have agreed to report ΔG, ΔH, and ΔS values at a set of standard reaction conditions of temperature, pressure, and initial concentrations. These are known as the *standard state conditions*. The accepted values are as follows:

Standard State Conditions

Specified constant temperature—frequently 25°C
Gases:
$$Pressure = 1 \ atmosphere$$
Substances in aqueous solution:
   Initial concentration = 1 molar = 1 mole of each substance per liter of solution

---

To indicate that free energy, enthalpy, and entropy changes are being made under standard conditions, a small superscript letter " o " is used, and a subscript is used to indicate the temperature, T, in °K.

Change in Gibbs Free Energy at Standard State Conditions
$$\Delta G_T^\circ = \Delta H_T^\circ - T \Delta S_T^\circ$$

If the temperature is 25°C (298°K), then the Gibbs free energy change would be shown as

$$\Delta G_{298°K}^\circ = \Delta H_{298°K}^\circ - T \Delta S_{298°K}^\circ$$

ΔG_T°, ΔH_T°, and ΔS_T° are called, respectively, the standard free energy change, the standard enthalpy change, and the standard entropy change. Since the temperature, pressure, and initial concentrations are fixed at specific values for all substances, then ΔG_T is, in effect, a constant for a particular reaction, and it allows us to compare the relative thermodynamic feasibility of different reactions.

3-4 Standard States and Enthalpies of Formation

It is not possible to determine experimentally the absolute value for the enthalpy of a substance. We can, however, measure the change in enthalpy, ΔH, when a substance is formed from its constituent elements. *The change in enthalpy when one mole of a substance is formed from its elements in their standard states is called the standard enthalpy change of formation* (usually abbreviated to *standard enthalpy of formation* or *standard heat of formation*) and given the symbol $\Delta H^\circ_{f,T}$ where T is usually 298°K (25°C).

$\Delta H^\circ_{f,298°K} \equiv$ standard enthalpy of formation of a compound from its elements at standard conditions

---

The standard state is a substance in its most stable form at 1 atmosphere pressure and a temperature of 298°K (25°C). By convention, all elements in their standard states are arbitrarily assigned an enthalpy value of zero.

Over the years, scientists have made enthalpy measurements and calculations for thousands of compounds being formed from their elements at standard conditions. As a result, data for the standard heats of formation, $\Delta H^\circ_{f,298°K}$, for a number of compounds are available. Table 3-1 provides a small sample of these values, primarily for substances involved in the eleven reactions to be studied. For example, consider the formation of the compound carbon dioxide, $CO_2(g)$, from its elements, $O_2(g)$ and C (graphite), in their most stable states at 1 atmosphere and 298°K.

$$C(s) + O_2(g) \longrightarrow CO_2(g)$$

$$\Delta H^\circ_{f,298°K} = -94,050 \text{ cal/mole}$$

(from Table 3-1)

Study Question 3-4

a. How many calories of energy would be evolved if 5 moles of carbon dioxide were formed from its elements at standard conditions?
b. What is the standard enthalpy of formation for carbon monoxide, $CO(g)$? How many calories would be evolved if one-half of a mole of $CO(g)$ were formed at standard conditions? (Continued on page 65).

Table 3-1 A Selection of Standard Enthalpies of Formation ($\Delta H^\circ_{f,298^\circ K}$) and Standard Absolute Entropies ($S^\circ_{298^\circ K}$)

Name	Formula	$\Delta H^\circ_{f,298^\circ K}$ in cal/mole*	Approximate $\Delta H^\circ_{298^\circ K}$ in cal/mole**	$S^\circ_{298^\circ K}$ in cal/ deg-mole*	Approximate $S^\circ_{298^\circ K}$ Value in cal/deg-mole**
Carbon (graphite)	C(graphite)	0	0	1.36	1
Carbon monoxide(g)	CO(g)	−26,420	−26,000	47.30	47
Carbon dioxide(g)	CO$_2$(g)	−94,050	−94,000	51.06	51
Methane(g)	CH$_4$(g)	−17,890	−18,000	44.50	44
Ethane(g)	C$_2$H$_6$(g)	−20,240	−20,000	54.85	55
Propane(g)	C$_3$H$_8$(g)	−24,820	−25,000	64.5	64
n-Butane(g)	C$_4$H$_{10}$(g)	−30,150	−30,000	74.12	74
n-Octane(g)	C$_8$H$_{18}$(g)	−49,820	−50,000	111.55	112
n-Octane(l)	C$_8$H$_{18}$(l)	−59,740	−60,000	86.23	86
Hydrogen(g)	H$_2$(g)	0	0	31.21	31
Oxygen(g)	O$_2$(g)	0	0	49.00	49
Ozone(g)	O$_3$(g)	+34,000	+34,000	56.8	57
Water(g)	H$_2$O(g)	−57,800	−58,000	45.11	45
Water(l)	H$_2$O(l)	−68,320	−68,000	16.71	17
Nitrogen(g)	N$_2$(g)	0	0	45.77	46
Ammonia(g)	NH$_3$(g)	−11,040	−11,000	46.01	46
Nitric oxide(g)	NO(g)	+21,600	+22,000	50.34	50
Nitrogen dioxide(g)	NO$_2$(g)	+8,090	+8,000	57.47	57
Sulfur(s) (rhombic)	S(s) (rhombic)	0	0	7.62	8
Sulfur dioxide(g)	SO$_2$(g)	−70,960	−71,000	59.40	59
Sulfur trioxide(g)	SO$_3$(g)	−94,450	−94,000	61.24	61
Glucose(s)	C$_6$H$_{12}$O$_6$(s)	−304,600	−305,000	50.7	51

* Data from *Selected Values of Chemical Thermodynamic Properties* (Circular 500, National Bureau of Standards, 1952) and *Selected Values of Properties of Hydrocarbons* (Circular C461, National Bureau of Standards, 1947).
**The purpose of this book is to illustrate principles, not to have you make arithmetical calculations to several decimal places. In most cases (consult your instructor) you should be able to use the approximate values for S° and $\Delta H_f°$.

c. Approximately how many molecules of carbon monoxide would be present in one-half of a mole of the gas?

d. Standard enthalpy values can be taken as a crude measure of the relative stability of compounds under standard conditions. Which is the more stable compound at 25°C and 1 atmosphere, $CO(g)$ or $CO_2(g)$?

This table of standard enthalpy values can be used to calculate the standard enthalpy of reaction, $\Delta H^\circ_{298°K}$, for any reactions involving these substances. The standard enthalpy of reaction is equal to the difference between the total standard heats of formation of the products and the reactants.

Standard Enthalpy of Reaction

$$\Delta H^\circ_{298°K} = \text{sum of } \Delta H^\circ_{f,298°K} \text{ of products}$$

$$- \text{ sum of } \Delta H^\circ_{f,298°K} \text{ of reactants}$$

By using data from Table 3-1, we can calculate the approximate standard enthalpy of reaction for Reaction 4, the oxidation of nitric oxide to nitrogen dioxide:

$$\_\_\_NO(g) + \_\_\_O_2(g) \rightleftharpoons \_\_\_NO_2(g) \quad (4: \text{unbalanced})$$

First we must balance the equation. Standard enthalpies of formation are given in calories per mole, and the number of calories varies with the number of moles of each substance.

$$2\ NO(g) + O_2(g) \rightleftharpoons 2\ NO_2(g) \quad (4: \text{balanced})$$

$\Delta H^\circ_{298°K} = \Delta H_f^\circ(\text{products}) \quad - \Delta H_f^\circ(\text{reactants})$

$\Delta H^\circ_{298°K} = [2\Delta H^\circ_{f,NO_2(g)}] \quad - [2\Delta H^\circ_{f,NO(g)} + \Delta H^\circ_{f,O_2(g)}]$

$\Delta H^\circ_{298°K} = [(2)(8000)] \quad - [(2)(22,000) + (0)]$

$\Delta H^\circ_{298°K} = 16,000 \quad - 44,000$

$\Delta H^\circ_{298°K} = -28,000 \text{ calories } (-28 \text{ kcal})$

Enthalpy factor is favorable.

How Much Energy Can Be Released or Absorbed When a Reaction Occurs?

65

Study Question 3-5 The standard enthalpy change has been calculated for Reaction 4. Use Table 3-1 to determine approximate values of $\Delta H^\circ_{298^\circ K}$ for the other nine reactions (omit Reaction 11). Indicate whether the enthalpy factor for each reaction is favorable or unfavorable for a spontaneous reaction occurring at standard conditions.

3-5 Standard Absolute Entropies—The Third Law of Thermodynamics

As we decrease the temperature of a substance, the degree of disorder, or the entropy, decreases. Suppose we could cool a substance to the lowest possible temperature, $0^\circ K$ ($-273^\circ C$). At this point all molecular motion would cease. If the substance is also a perfect crystal, then perfect order exists and the entropy of the substance would be zero. This is a statement of the *Third Law of Thermodynamics*.[12]

Third Law of Thermodynamics

The entropy of a perfect crystal at $0^\circ K$ is zero. $S_{0^\circ K} = 0$ for a perfect crystal.

Since it is not possible in practice to get a *perfect crystal*, one can only approach this ideal state of zero entropy. All solids have at least a few impurities and defects. For example, there are always a few atoms missing (called vacancies, or holes) or a few atoms out of line wandering about the interstices between the rows of atoms in the crystal. Imagine a very large crowd watching a football game. There are always a few seats vacant (vacancies) and a number of people wandering about in the aisles. Conversely, since we can never get a perfect crystal, we can never absolutely reach $0^\circ K$—only approach it.

Since every substance has an absolute entropy value of zero at $0^\circ K$, we can calculate the absolute entropy for a substance at any other temperature above $0^\circ K$. The total entropy change the substance undergoes as its temperature is increased from $0^\circ K$ to some given temperature is its absolute entropy at that temperature, S_T.

---

[12] The first and second laws were formulated on the basis of our direct experience with the macroworld, but the third law is postulated—it is not experienced directly.

Absolute Entropy

$$S_T - S_{0°K} = S_T - 0 = S_T$$

The absolute entropy for a substance in its standard state at 298°K (25°C) and 1 atmosphere is called its standard absolute entropy (usually abbreviated to standard entropy), $S^o_{298°K}$. Standard entropy values have been determined for a number of substances, and representative examples are given in Table 3-1.

These standard entropy values, $S^o_{298°K}$, can be used to calculate the standard entropy change, $\Delta S^o_{298°K}$, for a particular reaction in a manner similar to that used to calculate standard enthalpies of reaction.

Standard Entropy of Reaction

$\Delta S^o_{298°K}$ = sum of $S^o_{298°K}$ of products − sum of $S^o_{298°K}$ of reactants

Data from Table 3-1 can be used to calculate the standard entropy of reaction for Reaction 4:

$$2\ NO(g) + O_2(g) \rightleftharpoons 2\ NO_2(g)$$

$$\Delta S^o_{298°K} = S^o_{298°K}(\text{products}) - S^o_{298°K}(\text{reactants})$$

$$= 2\ S^o_{NO_2(g)} - [2\ S^o_{NO(g)} + S^o_{O_2(g)}]$$

$$= (2)(57) - [(2)(50) + (49)]$$

$$= -35 \text{ calories/°K}$$

$$\Delta S \text{ is unfavorable.}$$

Study Question 3-6   Use Table 3-1 to calculate approximate values of $\Delta S^o_{298°K}$ for the other nine reactions (omit Reaction 11). Indicate whether the entropy factor, $T\Delta S^o$, for each question is favorable or unfavorable for a spontaneous reaction.

3-6 Calculation of Standard Free Energy Changes

The value of ΔG_T° can readily be determined by calculating the values of $\Delta S_{298^\circ K}^\circ$ and $\Delta H_{298^\circ K}^\circ$ from Table 3-1, and substituting these in the free energy change equation.

Standard Free Energy Change

$$\Delta G_T^\circ = \Delta H_{298^\circ K}^\circ - T\Delta S_{298^\circ K}^\circ$$

The value of ΔG_T° at any given temperature, T, in $^\circ K$ (not $^\circ C$ or $^\circ F$) can be determined. For all practical purposes, except for very exact calculations or extremely high temperatures, it can be assumed that the values of $\Delta H_{298^\circ K}^\circ$ and $\Delta S_{298^\circ K}^\circ$ do not vary significantly with temperature. They will have approximately the same values at temperatures other than $298^\circ K$.

The standard free energy change for Reaction 4 at any given temperature can be calculated by using the $\Delta H_{298^\circ K}^\circ$ and $\Delta S_{298^\circ K}^\circ$ values calculated in the two previous sections. We shall calculate $\Delta G_{300^\circ K}^\circ$ at $27^\circ C$ ($300^\circ K$).

$$2\ NO(g) + O_2(g) \rightleftharpoons 2\ NO_2(g)$$

$$\Delta G_{300^\circ K}^\circ = \Delta H_{298^\circ K}^\circ - T\Delta S_{298^\circ K}^\circ$$

$$\Delta G_{300^\circ K}^\circ = -28{,}000\ \text{calories} - (300^\circ K)\left(-35\ \frac{\text{calories}}{^\circ K}\right)$$

$$\Delta G_{300^\circ K}^\circ = -28{,}000\ \text{calories} + 10{,}500\ \text{calories}$$

$$\Delta G_{300^\circ K}^\circ = -17{,}500\ \text{calories}$$

Reaction can occur spontaneously at $300^\circ K$ ($27^\circ C$).

This means that if this reaction were to proceed from left to right *to completion* with the pressure held constant at one atmosphere and the temperature at $300^\circ K$, then a decrease in free energy of $-17{,}500$ calories would occur.

We can see that at the relatively low temperature of $300^\circ K$ ($27^\circ C$, around room temperature) the favorable enthalpy factor (ΔH is $-$) predominates over the unfavorable $T\Delta S$ factor (ΔS is $-$). As the temperature is increased, the unfavorable entropy factor (TΔS) will become more important and above some relatively high *crossover temperature*,[13] T_x (where $\Delta G = 0$, as it passes from − to + values), the reaction will become nonspontaneous.

The approximate crossover temperature can be calculated as follows:

$$\Delta G^\circ_{T_x} = \Delta H^\circ_{298^\circ K} - T_x \Delta S^\circ_{298^\circ K}$$

$$\Delta G_{T_x} = 0 \text{ at the crossover temperature } T_x$$

$$0 = \Delta H^\circ_{298^\circ K} - T_x \Delta S^\circ_{298^\circ K}$$

$$T_x \Delta S^\circ_{298^\circ K} = \Delta H^\circ_{298^\circ K}$$

$$T_x = \frac{\Delta H^\circ_{298^\circ K}}{\Delta S^\circ_{298^\circ K}} = \frac{-28,000 \text{ cal}}{-35 \text{ cal}/^\circ K}$$

$$T_x = \text{approximately } 800^\circ K$$

$$T_x \text{ in } ^\circ C = 800^\circ K - 273 = \text{approximately } 527^\circ C$$

Summary of Thermodynamic Factors

$$2\ NO(g) + O_2(g) \rightleftharpoons 2\ NO_2(g)$$

$$\Delta H^\circ_{298^\circ K} = -28,000 \text{ cal} \qquad \text{yes}$$

$$\Delta S^\circ_{298^\circ K} = -35 \text{ cal}/^\circ K \qquad \text{no}$$

$$\Delta G^\circ_{300^\circ K} = -17,500 \text{ cal} \qquad \text{yes}$$

Crossover temperature $T_x \cong 527^\circ C$.

$$\Delta G_T > 527^\circ C \text{ is } + \qquad \text{no}$$

Remember that NO(g) is a relatively harmless air pollutant, while NO_2(g) is very harmful. Assuming that the reaction has a reasonably rapid rate (a kinetic fact we

---

[13] It is important to note that calculations of values for the crossover temperature are very approximate. They are based on the assumption (which is not always valid) that ΔH° and ΔS° do not vary with temperature.

don't know at this point), we can see that unfortunately, from a thermodynamic standpoint, NO(g) should be spontaneously oxidized to the more deadly $NO_2(g)$ in the atmosphere at normal temperatures. If the atmospheric temperature increased to 527°C or higher (remember water boils at 100°C), the reaction could become non-spontaneous—a cure much worse than the illness. To eliminate or decrease pollution from $NO_2(g)$, we will have to drastically reduce the NO(g) content of the exhaust or find a substitute for the internal combustion engine.

Free energy changes can be calculated for physical changes as well as chemical changes. The freezing of water is an exothermic ($\Delta H^o_{298°K}$ is -1440 cal/mole) change, yet water does not freeze at temperatures above 0°C (273°K) and a pressure of 1 atmosphere. The entropy change for freezing involves a decrease in entropy because the solid state is more orderly than the liquid state. $\Delta S^o_{298°K}$ is therefore negative and has a value of approximately -5.3 cal/°K. The values of ΔH^o and ΔS^o can be considered to be *approximately* constant over a fairly wide temperature range.

At room temperature, 25°C (298°K), the standard free energy change for the freezing of water is

$$\Delta G^o_{298°K} = \Delta H^o_{298°K} - T\Delta S^o_{298°K}$$

$$\Delta G^o_{298°K} = -1440 - (298)(-5.3)$$

$$\Delta G^o_{298°K} = +140 \text{ calories.}$$

Since ΔG^o is positive at room temperature, the freezing of water should not occur spontaneously at 25°C. The reverse melting of ice, $H_2O(s) \rightleftharpoons H_2O(l)$, should occur spontaneously at 25°C, since $\Delta G^o_{298°K}$ would be -140 calories. At 0°C (273°K) and 1 atmosphere, $\Delta G^o_{273°K}$ is

$$\Delta G^o_{273°K} = \Delta H^o_{298°K} - T\Delta S^o_{298°K}$$

$$\Delta G^o_{273°K} = -1440 \text{ cal} - (273)(-5.3)$$

$$\Delta G^o_{273°K} = 0.00 \text{ cal.}$$

We see that 0°C (273°K) is the crossover temperature. At this point ice and water are in equilibrium.

If the temperature is decreased below 0°C, say to -10°C (263°K), then

$$\Delta G^o_{263°K} = \Delta H^o_{298°K} - T\Delta S^o_{298°K}$$

$$\Delta G^o_{263°K} = -1440 \text{ cal} - (263)(-5.3)$$

$$\Delta G^o_{263°K} = -50 \text{ cal.}$$

ΔG^o is negative and the liquid water should freeze spontaneously.

Thermodynamics of the Freezing—Melting of
Water

$$H_2O(l) \rightleftharpoons H_2O(s)$$

	$\Delta H^\circ_{298°K} = -1440$ cal/mole	Yes
	$\Delta S^\circ_{298°K} = -5.3$ cal/°K	No
25°C	$\Delta G^\circ_{298°K} = +140$ cal	No—ice melts spontaneously
0°C	$\Delta G^\circ_{273°K} = 0.00$ cal	Crossover point Equilibrium
-10°C	$\Delta G^\circ_{263°K} = -50$ cal	Yes—water freezes spontaneously

Study Question 3-7

a. Calculate $\Delta G^\circ_{300°K}$ values for the other nine reactions (omit Reaction 11) at 27°C (room temperature) and a pressure of 1 atmosphere (normal atmospheric pressure). Indicate whether each reaction is potentially spontaneous under these conditions.

b. Will each reaction become more or less spontaneous or nonspontaneous as the temperature is raised? Explain.

c. Will there be a temperature crossover point? Where this is a possibility, calculate the approximate temperature of the crossover point in °K and °C. What assumption is made concerning the validity of these calculations?

d. Summarize the $\Delta H_{298°K}$, $\Delta S^\circ_{298°K}$, ΔG_T°, and T_x data in a table for all ten reactions, including the $NO_2(g)$ reaction, as illustrated below.

Summary Table

Reaction	Balanced Equation	$\Delta H^\circ_{298°K}$ (cal)	$\Delta S^\circ_{298°K}$ (cal/°K)	$\Delta G^\circ_{300°K}$	approx. T_x (°C)	Spontane-ous above T_x
No. 4	$2NO(g) + O_2(g) = 2NO_2(g)$	$-28,000$ Yes	-35 No	$-17,500$ Yes	527°C	No
	etc.					

e. Study the summary table and answer the following questions about certain reactions. Assume that the reactions are kinetically feasible—an assumption which may or may not be valid.

1. *Reaction 1: Ozone formation.* Can ozone form spontaneously in the upper atmosphere, assuming the atmospheric temperature to be lower than 27°C? Assuming that the formation of ozone is nonspontaneous, explain how the ozone layer in our atmosphere can exist.

2. *Reactions 2 and 3: Burning of coal and gasoline.* On a per-mole basis, which fuel provides more energy, coal or gasoline? One mole of coal has a weight of 12 grams, while one mole of gasoline (octane) has a weight of 114 grams. On a per-weight basis, which fuel provides more energy?

3. *Reaction 5: Sulfur dioxide.* Is the oxidation of sulfur to SO_2 at 27°C spontaneous or nonspontaneous? As the temperature of a smoke stack increases, will the oxidation of sulfur to SO_2 become more or less spontaneous? From a thermodynamic standpoint is it feasible to get rid of SO_2 pollution by reversing the reaction and reclaiming the sulfur? Would this reclamation work better at high or low temperatures?

4. *Reaction 6: Removal of ammonia from the primitive earth atmosphere.* Is it thermodynamically feasible that ammonia could have been removed from a primitive earth atmosphere by this reaction? The temperature was probably much higher at that time. Would this help or hinder the reaction?

5. *Reaction 7: Formation of water in primitive earth atmosphere.* Is this a thermodynamically feasible way to get water into the atmosphere? Would a high temperature favor the reaction? When H_2 and O_2 are mixed at room temperature, no reaction is observed. Explain. Someone has suggested that the reverse decomposition of water reaction could have been the source of H_2 and O_2 in our present atmosphere. Comment on the feasibility of this suggestion, and indicate temperature conditions under which its feasibility would be enhanced.

6. *Reaction 8: Manufacture of ammonia.* Under what conditions of temperature would this reaction be thermodynamically feasible? It has been suggested that the reverse decomposition of ammonia could have been a source of N_2 and H_2 in our present atmosphere. Comment on the feasibility of this reaction and the atmospheric temperature conditions under which it might occur spontaneously.

7. *Reactions 9 and 10: Photosynthesis and respiration.* Which process is thermodynamically feasible? The process that is not feasible does occur and is essential for life. How can it occur?

3-7 Some Limitations on Our Predictions

In Chapter 2 we saw that the sign of the free energy change, ΔG_{system}, determines whether a given reaction can occur spontaneously when carried out at constant temperature and pressure (or concentration). In this chapter we have found that the value of ΔG_{system} is a measure of the maximum work or energy available if the reaction proceeds spontaneously and reversibly to completion.

Before we get carried away with our predictions based on ΔG and $\Delta G°$ calculations, we should be aware of several important limitations. First, values of ΔG or $\Delta G°$ can be determined only at a specified temperature and pressure (for reactions involving gases) or concentration (for reactions in solution). Thus, we can make comparisons between different reactions only when they are at the same conditions of temperature and pressure or concentration.

Predictions based on $\Delta G°$ (such as those you made in Study Question 3-7) are even more restricted since they are valid only when all of the substances involved (reactants and products) have *initial* concentrations of 1 molar or a pressure of 1 atmosphere at a given temperature. Very few reactions are actually carried out under such conditions, but the usefulness of $\Delta G°$ is that it allows us to compare the thermodynamic spontaneity for all reactions at a specified set of conditions.

Thus, if ΔG or $\Delta G°$ is positive or unfavorable, the reaction may still become spontaneous under other conditions of temperature, pressure, or concentration, as we shall see in the next two chapters.

Finally, we should remember that thermodynamics does not provide any information about the rate of reaction. A reaction will occur only if it is both thermodynamically and kinetically feasible.

> Science is a great many things but in the end they all return to this: science is the acceptance of what works and the rejection of what does not. That needs more courage than we might think.
>
> *Jacob Bronowski*

3-8 Review Questions

3-1. What is a mole?

3-2. What is Avogadro's number and what is its significance?

3-3. What information does a balanced chemical equation provide?

3-4. What information does a balanced chemical equation *not* provide?

3-5. What is wrong with this statement: The balanced equation for the photosynthesis process shows that the mechanism for the reaction involves the collision of 6 molecules of $CO_2(g)$ with 6 molecules of liquid water to produce one molecule solid glucose, $C_6H_{12}O_6$, plus 6 molecules of oxygen gas.

3-6. Why is ozone important to the maintenance of life on earth? How can it also become a threat to life?

3-7. a. What is the "greenhouse effect" and how is it produced?
b. What other pollution effect might be counteracting the greenhouse effect?

3-8. Distinguish between "brown air" and "gray air" cities.

3-9. Give at least two harmful effects of nitrogen dioxide.

3-10. What is photochemical smog?

3-11. How might NO_2 air pollution be decreased?

3-12. Explain why emission control devices provide only a temporary solution to pollution from the automobile.

3-13. What is the source of sulfur dioxide pollution? What are some of its harmful effects? How could its content in the air be reduced?

3-14. What gases might have been in the primitive earth's atmosphere? How could they have been removed and converted to our present atmosphere?

3-15. What is the Haber process? Why is it important?

3-16. Describe the overall (net) process of photosynthesis in green plants.

3-17. Describe the overall process of respiration in cells.

3-18. In what sense is respiration the reverse of photosynthesis? In what sense is this not true?

3-19. What is a "high energy" phosphate? How are they important to life processes?

3-20. Explain why the actual work available in a spontaneous process or reaction is always less than the maximum value.

3-21. What is a "reversible" process?

3-22. Why is it desirable to establish standard state conditions?

3-23. What is concentration? What is a 1.5 molar solution?

3-24. What is the standard enthalpy of formation? Distinguish between it and the standard enthalpy of reaction.

3-25. What is the Third Law of Thermodynamics? How does it allow one to obtain absolute entropy values?

3-26. Distinguish between standard absolute entropy and the standard entropy of reaction.

3-27. What is the standard free energy change? Why is it useful?

3-28. What is the crossover temperature?

Chapter 4

Question No. 3—How Far Can the Reaction Go?— Chemical Equilibrium

4-1 Do Reactions Go All the Way?

What happens if we mix N_2 and H_2 gases in a closed container at room temperature and pressure? Assuming that the reaction is thermodynamically and kinetically feasible under these conditions, would you expect the N_2 and H_2 to be converted completely to ammonia gas, NH_3?

$$N_2(g) + 3H_2(g) \rightleftharpoons 2NH_3(g) \quad \Delta G \text{ is } - \text{ at } 27°C$$

Under conditions of constant temperature, pressure, and volume (that is, in a closed system), most reactions, including this one, do not go "to completion." Instead they reach a state of dynamic equilibrium between reactants and products. Reactants form products, but, once formed, the products can also be converted back into the original reactants. When these two forward and reverse reactions occur at the same rate, the system will be in a state of *dynamic equilibrium*. In the equilibrium mixture there will be a fixed ratio of products (NH_3) and reactants (N_2 and H_2), and the ratio of the concentration of products to that of reactants is a measure of how far the reaction has gone to the right. In this case ammonia predominates and the equilibrium position lies to the right.

What would happen if we put ammonia only in the same container at the same temperature and pressure? Since ΔG is negative for the reaction to the right, we might expect the reaction to the left to be nonspontaneous, and we would predict that the NH_3 would not spontaneously decompose to form N_2 and H_2. Yet, if we measured the contents of the container after a reasonable time, we would find N_2, H_2, and NH_3

present in exactly the same ratio as before. This implies that both the forward and reverse reactions proceeded with a decrease in free energy—which seems impossible.

These results raise two questions: (1) What is the nature of chemical equilibrium? and (2) How can we determine how far a reaction can go (that is, what is the equilibrium position?)?

How can a reaction be spontaneous in both directions? Since we know that any spontaneous reaction proceeds spontaneously to a lower value of free energy, we conclude that the equilibrium mixture of N_2, H_2, and NH_3 must have a lower value of total free energy than either the product (NH_3) or the reactants (N_2 and H_2), as shown in Figure 4-1. Any system will spontaneously decrease in free energy until

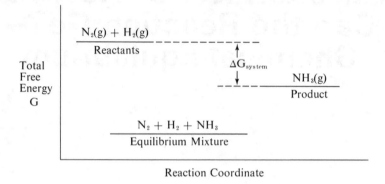

Figure 4-1 At Equilibrium the Total Free Energy Decreases to a Minimum.

it reaches its minimum possible value. If the products have the minimum value of G, then the reaction will go all the way to the right. However, when a mixture of reactants and products has a lower free energy value than either the reactants or products, then the system will change spontaneously until it reaches this equilibrium mixture—regardless of whether we start with pure reactants, products, or some non-equilibrium mixture of reactants and products. The equilibrium mixture is in a free energy trough, and once equilibrium is achieved the reaction will not spontaneously change to form more products or reactants, since this would involve an increase in free energy.

Thus, *when a reaction has reached equilibrium, the free energy has decreased to its minimum value.* When this state is reached, there is no more work available to drive the reaction spontaneously in either direction, and the free energy of the system is no longer changing, that is, $\Delta G_T = 0$. *Thus, one criterion for equilibrium is that the change in free energy of the system is zero, that is, $\Delta G_T = 0$.*

We could also use entropy as a criterion for equilibrium. According to the second law, the entropy of the system plus surroundings will increase for a spontaneous process. But under conditions of fixed pressure, volume, and temperature, the entropy of the system plus surroundings will eventually reach a maximum value. At this point of maximum entropy the system will be in a state of dynamic equilibrium—it has exhausted its capacity for doing work on its surroundings, and the entropy is no longer changing, that is, $\Delta S_{net} = 0$. It is this tendency to maximize total entropy that is the real driving force for any process. Note, however, that while the free energy of the system always decreases as a system approaches equilibrium, the entropy of the *system* may increase, decrease, or remain constant. But if the entropy of the system does decrease (that is, order increases, as in the growth of a living organism), then the entropy of the surroundings must increase so that ΔS_{net} is positive, as summarized in Figure 4-2.

SURROUNDINGS

$S_{surroundings}$: Can increase, decrease, or remain constant but $S_{system\ +\ surroundings}$ always increases to a maximum at equilibrium.

SYSTEM
(at constant P, T, V)

S_{system}: Can increase, decrease, or remain constant
G_{system}: Decreases to minimum at equilibrium,

Figure 4-2   Entropy and Free Energy Relationships in a Closed System (constant P, T, V) at Equilibrium.

In summary, we have found two criteria for determining when a system is in a state of dynamic equilibrium.

Criteria for Equilibrium in a Closed System
(system at constant P, T, V)

$$\Delta S_{system\ +\ surroundings} = 0$$
or
$$\Delta G_{system} = 0$$

For a system not at equilibrium, the value of ΔG is a measure of how far the reaction is from the equilibrium position, and the sign of ΔG determines whether the reaction will proceed spontaneously to the right (ΔG is $-$) or to the left (ΔG is $+$) to achieve the equilibrium position. The larger the numerical value of ΔG (whether it is $+$ or $-$), the further the reaction is from the equilibrium position.

Free Energy and Equilibrium

$\Delta G_T < 0$ (ΔG is $-$): Reaction can proceed spontaneously to the *right* to achieve the equilibrium position.

$\Delta G_T > 0$ (ΔG is $+$): Reaction can proceed spontaneously to the *left* to achieve the equilibrium position.

$\Delta G_T = 0$: Reaction is at equilibrium.

Remember that the fact that ΔG_T is negative does not mean that the reaction will necessarily reach an equilibrium state. This will happen *only* if the reaction is also kinetically feasible with a reasonable rate of approach to the equilibrium position.

Study Question 4-1   Which of the following statements are true for a process that is occurring spontaneously?

a. ΔS_{system} is always positive
b. $\Delta H_{system} = T\Delta S_{system}$
c. $\Delta G_T = 0$
d. $\Delta G_T < 0$
e. $\Delta G_T > 0$
f. $\Delta S_{net} > 0$

Study Question 4-2   Which of the following statements are true for a process at equilibrium?

a. $\Delta G_T < 0$
b. $\Delta G_T > 0$
c. $\Delta G_T = 0$
d. $\Delta G_T = \Delta H$
e. $\Delta H_{system} = T\Delta S_{system}$

4-2 Dynamic Chemical Equilibrium

What is happening at the *micro*scopic level of atoms and molecules when a reaction, such as that between N_2 and H_2, achieves a state of chemical equilibrium in a closed system? At equilibrium, billions and billions of reactant molecules are forming product molecules each second, and at the same time billions and billions of product molecules are breaking up and returning to the original reactant molecules. If we begin the reaction only with reactants (H_2 and N_2), the rate of the forward reaction is greater than that of the reverse, but eventually the rates of the two opposing reactions become equal, at which time the system is said to be in a state of *dynamic chemical equilibrium*.

What do we observe at the *macro*scopic level when a system is in dynamic equilibrium? When two opposite changes take place at the same rate, a fixed ratio of the concentrations of the products and reactants will be maintained and there will be a *constancy of macroscopic properties.* Properties such as temperature, concentration of reactants and products, pressure, and color remain constant when equilibrium is attained in a closed system.

Study Question 4-3   Which of the following choices are correct? A state of dynamic chemical equilibrium in a closed system implies:

a. that the system is at its maximum entropy
b. that the surroundings are at maximum entropy
c. that the system plus surroundings are at maximum entropy.
d. that the system plus surroundings are at minimum free energy
e. that $\Delta H_{system} = T\Delta S_{system}$
f. $\Delta G_T = 0$
g. $\Delta G_T < 0$
h. $\Delta G_T > 0$
i. that the rate of the forward reaction equals the rate of the reverse reaction
j. that macroscopic properties are changing continually.

Dynamic Chemical Equilibrium in a Closed System

Macro level: Constancy of macroscopic properties, $\Delta G_{system} = 0$, $\Delta S_{net} = 0$

Micro level: Rate of forward reaction = rate of reverse reaction

---

The world overpopulation problem can be analyzed in terms of dynamic equilibrium. We live in a finite, closed system—called spaceship Earth. At present there is

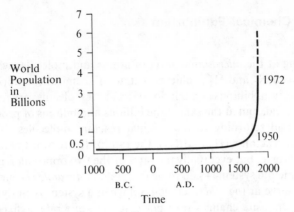

Figure 4-3 The Exponential, or "J," Curve of World Population Growth.

a sharp exponential ("J" curve) rise in the total population on our fragile spacecraft (Figure 4-3). This explosive population increase is occurring because our system is *not* in a state of dynamic equilibrium. The birth rate and the death rate are not equal.

Worldwide birth rates on an average have been declining slowly for the past fifty years. This being the case, how can we have a population explosion? Population change is not due to either birth rate (BR) or death rate (DR) but is the difference between birth rate and death rate (BR − DR). During the last fifty years, and particularly since World War II, the death rate has been falling sharply as a result of modern medicine and sanitation; hence the two factors are out of balance.

Population Dynamics

Population change (PC) = Birth rate (BR) − Death rate (DR)

Consider a metronome ticking away. The birth rate on this planet is now approximately 232 persons (ticks) per minute or approximately 334,000 per day, while the death rate is ticking at about 93 persons per minute or approximately 134,000 per day. The difference between BR (334,000) and DR (134,000) is approximately 200,000 additional passengers each day. In other words, we are adding approximately 1.4 million additional passengers each week—70 million each year. These are unique human beings—not numbers or things, but human beings who are to be fed, clothed, housed and given a share of our finite resources. This is occurring when almost two-thirds of the world's population is now either hungry or malnourished and more than three-quarters lack adequate housing and a safe or adequate water supply.

The Population Bomb

$$\text{Birth rate} \cong 334{,}000 \text{ per day}$$
$$\text{Death rate} \cong 134{,}000 \text{ per day}$$
$$\text{Population increase} \cong 200{,}000 \text{ per day}$$
$$(\text{BR} - \text{DR}) \cong 1.4 \text{ million per week}$$
$$\cong 70 \text{ million per year}$$

Actually there are three factors that determine whether population size will remain constant, grow or decline: (1) the birth rate, (2) the death rate and (3) the age structure of the population—in particular, the number of women of child-bearing age in the population. This third factor is often overlooked, and it is because of this factor that the population of the world will increase dramatically in the next 30 to 100 years, even if birth rates continue to decline. We now have a population with a young age structure. Forty-five percent of the people in the world today are under the age of 15. They are the highly combustible fuel, already on board, representing the real detonation of the population bomb that is yet to come. Even if every couple in the world magically decided to have only 2 children from now on—that is, to replace themselves—world population would not level off for at least 60–70 years, and we would add at least one and one-half billion additional passengers.

What will happen? The population story is told in two very simple but alarming curves—a "J" curve (Figure 4-3) and an "S" curve (Figure 4-4). In a closed system with finite resources, such as this planet, the explosive growth of any species (including man) cannot continue indefinitely, and the exponential, or "J," growth curve will be leveled off to an "S" curve. In other words, the system will eventually be brought into a state of dynamic equilibrium where birth rate equals death rate and we have zero population change.

Dynamic Population Equilibrium
or
Zero Population Change (ZPC)

$$\text{Birth rate (BR)} - \text{Death rate (DR)} = 0$$
$$\text{or}$$
$$\text{Birth rate (BR)} = \text{Death rate (DR)}$$

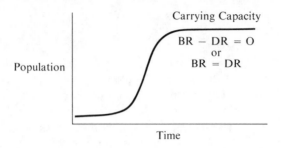

Figure 4-4   The "S" Curve of Dynamic Population Equilibrium, or Zero Population Change, Where Birth Rate (BR) = Death Rate (DR).

The equilibrium level is known as the *carrying capacity*, and it represents the population that can be supported adequately by the resources available. In actual practice, a species overshoots and undershoots this level and the population usually fluctuates around this equilibrium state, as shown in Figure 4-5.

We would do well to remember that on spaceship Earth *dynamic population equilibrium will be attained one way or the other*. This will happen either by a drastic rise in the death rate (due to war, famine, disease, local or global ecosystem collapse, or some combination of these) or by a drastic decrease in birth rate (not the slow decline we have now) or by some combination of the two. Remember that approximately 10–20 million human beings are already dying each year of starvation, malnutrition, or diseases resulting from malnutrition. The choice is ours. As rational human beings we can decide to lower the birth rate drastically (rather than slowly as we are doing now) or do nothing and let nature impose the automatic death rate solution. The task of bringing this situation into a state of dynamic equilibrium in a way that avoids the increased-death-rate alternative is the most serious problem now facing mankind. It hardly seems possible that an overpopulated world can ever be free or peaceful. As the theologian Harvey Cox said, *not to decide is to decide.*

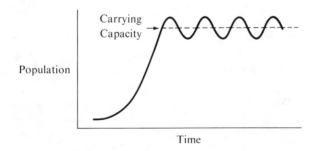

Figure 4-5   Modified "S" Curve to Show Population Fluctuations Around the Carrying Capacity.

4-3 The Distinction between ΔG_T and ΔG_T°

In order to better understand the concept of chemical equilibrium, it is very important to understand the difference between ΔG_T°, the standard free energy change, and ΔG_T, the *actual or measured free energy change*.

The standard free energy change, ΔG_T°, provides information about spontaneity and reaction driving force only when the reaction is carried out under a very specific set of conditions—a fixed temperature, a total pressure of 1 atmosphere, or when all substances are present initially at standard concentrations of one mole per liter. It is, in effect, a constant for any given reaction at a particular temperature. Thus, all of the rather glib predictions we have been making about reactions apply only under these rather stringent conditions. At the same time we have seen the necessity for having standard states, so that we may meaningfully compare the relative values of ΔG_T° for different reactions.

In most actual situations we do not have reactions occurring under these standard conditions of initial concentration or pressure. For example, with most reactions we do not start with initial concentrations of 1 mole per liter (1 M) for all reactants and products. The value of ΔG_T (not ΔG_T°) varies with concentration, and it provides information about the reaction at *nonstandard conditions* of concentration or pressure. Thus, under any actual reaction conditions where all initial concentrations of reactants and products are not one molar, it is ΔG_T, *not* ΔG_T°, that determines reaction spontaneity.

Distinction between ΔG_T° and ΔG_T

ΔG_T° = Free energy change under *standard conditions*: fixed temperature (T), 1 atmosphere, or initial concentrations of 1 mole/liter in aqueous solution.[1]

ΔG_T = Free energy change under *nonstandard conditions* of pressure or concentration.

The actual free energy change, ΔG_T, at nonstandard conditions can be obtained, therefore, by adding a concentration or pressure correction factor to the fixed value of ΔG_T°.

---

[1] Strictly speaking, activities (a) rather than concentrations (M) should be used, but at this introductory level the differences between activity and concentration will be ignored.

$$\boxed{\Delta G_T = \Delta G_T{}^\circ + \text{concentration or pressure correction factor}}$$

What will be the form of the concentration or pressure correction factor? We might expect it to involve some ratio of the initial concentrations or pressures of the reactants and products. Consider the hypothetical reaction with initial concentrations of A at 0.1 M (or 10^{-1} M) and B at 0.001 M (or 10^{-3} M).

$$A \rightleftharpoons B$$

$$10^{-1}\ \text{M} \quad 10^{-3}\ \text{M}$$

The ratio of the concentration of the product B to the reactant A could be used to describe these initial nonstandard conditions. We shall call this ratio the *reaction quotient* and give it the symbol Q. Q_c will be used to indicate that concentration in moles/liter is used, and square brackets, [ ], will also be used to indicate that the concentration of the substance shown inside the brackets is in moles per liter.

$$A \rightleftharpoons B$$

$$\text{Concentration reaction quotient} = \frac{\text{Concentration of product B}}{\text{Concentration of reactant A}}$$

$$Q_c = \frac{[B]}{[A]}$$

As the reaction proceeds, the concentrations of A and B will change and as a result, Q_c will change. Eventually, when the system reaches equilibrium, the concentrations of A and B will remain fixed at the equilibrium values and no further changes will occur ($\Delta G_T = 0$).

If A and B are gases, then the concentration can be expressed either in moles/liter or in terms of the partial pressure of each of the gases present. Each gas in a mixture of gases can ideally[2] be assumed to behave independently. The pressure which each gas would exert if it occupied the volume by itself is the *partial pressure*, p, of the gas. The total pressure, P_{total}, of a mixture of gases is the sum of the partial pressures of the gases in the mixture.

$$P_{total} = p_1 + p_2 + p_3 + \cdots + p_n$$

When concentration is expressed in partial pressures, a pressure reaction quotient, Q_p, can be used.

$$A \rightleftharpoons B$$

$$\text{Pressure reaction quotient} = \frac{\text{Partial pressure of B}}{\text{Partial pressure of A}}$$

$$Q_p = \frac{p_B}{p_A}$$

The probability of reaction will be a function of the reaction quotient, Q_p or Q_c. The greater the concentration or partial pressure of the reacting species present, the greater the probability of reaction. If it is a probability function, then it is a multiplicative function. In other words, the total probability of several events is obtained by multiplying the separate probabilities. For example, the probability of rolling a six with one dice is one in six, or 1/6. The total probability of rolling two sixes with a pair of dice is not 1/6 + 1/6 but 1/6 × 1/6, or 1/36 (1 in 36). Therefore

$$Q_{total} = Q_1 \times Q_2 \times \cdots \times Q_n.$$

On the other hand, $\Delta G_T{}^\circ$ and ΔG_T are additive functions. That is, the total value is the sum of the individual changes.

$$\Delta G^\circ_{total} = \Delta G_1{}^\circ + \Delta G_2{}^\circ + \cdots + \Delta G_n{}^\circ$$

We are looking for an equation in which the concentration or pressure correction factor is added to the $\Delta G_T{}^\circ$ factor. How can we change the multiplicative Q function to an additive function that can be added to the $\Delta G_T{}^\circ$ function? You may recall that the logarithm (log) of the product of two or more numbers or functions is the sum

---

[2] In practice gases are real and approach this ideal behavior only under certain conditions. For our purposes this involves only very minor corrections, and this deviation from ideal behavior will be neglected.

of the logarithms of each number or function. In other words, taking the logarithm of a multiplicative function such as Q converts it to an additive function.

$$\log Q_{total} = \log Q_1 + \log Q_2 + \cdots + \log Q_n$$

Thus, we might expect the concentration or partial pressure correction factor to be proportional to log Q, the logarithm of the reaction quotient. Any ΔG_T term also varies with temperature, so we might also expect temperature to be included in the correction factor. Since the new factor is an energy factor, we would expect it to be made up of the product of an intensity factor (T) and a capacity factor (log Q), as we found in Chapter 2 (Section 2-5) for the $T\Delta S$ entropy factor. Thus, a reasonable form for the correction factor might be

---

Correction factor $\propto$ T log Q

or

Correction factor $=$ kT log Q,

---

where the constant, k, is a proportionality constant. This constant has a value of approximately 4.6 when ΔG is measured in calories. Thus, the new factor is

---

Concentration or partial pressure
correction factor $=$ 4.6 T log Q,
where Q is either Q_c or Q_p.

We can now add this to $\Delta G_T°$ to obtain an expression for ΔG_T:

Free Energy Change at Nonstandard Conditions

$\Delta G_T = \Delta G_T° +$ concentration
correction factor

$\Delta G_T = \Delta G_T° + 4.6\ T \log Q$

The reaction A $\rightleftharpoons$ B is relatively simple, but what is the reaction quotient, Q, for a more complex reaction? For example, consider the hypothetical reaction

$$A + B \rightleftharpoons C + D.$$

To obtain the reaction quotient of the concentrations of two or more products or two or more reactants, we *multiply* their concentrations. Again, we might predict this on the basis of the probability of reaction. The probability that A and B will react is the product, not the sum, of their respective concentrations or partial pressures. Thus, the reaction quotient for this reaction can be expressed as

$$Q_c = \frac{[C][D]}{[A][B]} \quad \text{or} \quad Q_p = \frac{p_C p_D}{p_A p_B}.$$

Suppose we have the balanced equation

$$2A + D = E + 3F.$$

Again on the basis of probability, the concentration factor for A would be [A][A], or $[A]^2$, and for F it would be [F][F][F], or $[F]^3$. Q would then be

$$Q_c = \frac{[E] [F]^3}{[A]^2[D]} \quad \text{or} \quad Q_p = \frac{p_E p_F^3}{p_A^2 p_D}.$$

We can see that the concentration of each species is raised to the power corresponding to its coefficient in the balanced equation.

In the example equations used so far it has been assumed that each substance in the equation either was completely dissolved in aqueous solution (for Q_c) or was a gas (for Q_p). What happens if a substance is in the liquid state or solid state (for Q_p) or is an insoluble or slightly soluble precipitate, such as silver chloride (for Q_c)? In such cases, these species are not evenly distributed throughout the reaction medium. As a result, the probability of reaction is not directly dependent on their concentration or partial pressure. Most of their atoms or molecules are tied up internally in the liquid or solid structure, and only the molecules on their surfaces are actually available for reaction. For example, compare the burning of sawdust with the burning of a log where most of the wood is not available for immediate reaction with oxygen. *Thus, the only substances that will appear in the reaction quotient expression are those that are either in the gaseous state or completely dissolved in aqueous solution.* For example, consider

$$A(s) + 2B(g) = 3D(g) + E(l),$$

$$Q_p = \frac{p_D^3}{p_B^2}, \quad \text{or} \quad Q_c = \frac{[D]^3}{[B]^2}.$$

Note that species E(l) and A(s) do not appear in the reaction quotient expressions since they are in the liquid and solid states, respectively. Similarly, a reaction in aqueous solution (aq) involving a slightly soluble precipitate(s)

$$A(s) + B(aq) = C(aq) + 2D(aq)$$

has the concentration reaction quotient

$$Q_c = \frac{[C][D]^2}{[B]}.$$

For the generalized reaction

$$aA + bB \rightleftharpoons cC + dD,$$

where a, b, c, d represent the coefficients in the balanced equation, Q is as follows:

General Expression for Reaction Quotient
Reaction Not at Equilibrium

$$aA + bB \rightleftharpoons cC + dD$$

where A, B, C, D are either gases or substances completely dissolved in aqueous solution

$$Q_c = \frac{[C]^c[D]^d}{[A]^a[B]^b}$$

or

$$Q_p = \frac{p_C{}^c p_D{}^d}{p_A{}^a p_B{}^b}$$

Study Question 4-4   Write expressions for the reaction quotients, Q_c and Q_p, for reactions 1–10 discussed in Chapter 3 (Section 3-2).

4-4 Calculation of ΔG_T at Different Concentrations

In Chapter 3 (Section 3-6), we calculated the standard free energy change at 300°K, (27°C) for Reaction 4 involving pollution by the oxides of nitrogen. This gave us the thermodynamic feasibility for this reaction at 27°C when the initial concentration of all species is fixed at 1 mole/liter (1 M) or at partial pressures of 1 atmosphere.

$$2NO(g) + O_2(g) \rightleftharpoons 2NO_2(g) \text{ at } 300°K$$

Standard
Initial
Concentrations   1 M   1 M   1 M

$$\Delta G^{\circ}_{300^{\circ}K} = \Delta H^{\circ}_{298^{\circ}K} - T\Delta S^{\circ}_{298^{\circ}K}$$

$$\Delta G^{\circ}_{300^{\circ}K} = -17,500 \text{ calories (See Section 3-6)}$$

Thus the reaction is feasible under these standard initial concentration conditions.
    Suppose we vary the initial concentrations to the nonstandard conditions shown below but still maintain the temperature around room temperature, 27°C, or 300°K. What effect will this have on the thermodynamic feasibility?

$$2NO(g) + O_2(g) \rightleftharpoons 2NO_2(g) \text{ at } 300^{\circ}K$$

Nonstandard
Initial
Concentrations   0.01 M   0.1 M 10^5 M
                       or   or   or
                     10^{-2} M 10^{-1} M 10^5 M

The free energy change, ΔG_T, is calculated by using

$$\Delta G_T = \Delta G_T^{\circ} + 4.6 \text{ T log } Q_c .$$

From the calculation above, $\Delta G^{\circ}_{300^{\circ}K}$ is $-17,500$ calories and the reaction quotient, Q_c, is

$$Q_c = \frac{[NO_2(g)]^2}{[NO(g)]^2[O_2(g)]}$$

$$Q_c = \frac{(10^5)^2}{(10^{-2})^2(10^{-1})} = \frac{10^{10}}{(10^{-4})(10^{-1})} = \frac{10^{10}}{10^{-5}} = 10^{15}$$

and

$$\Delta G_{300^{\circ}K} = \Delta G^{\circ}_{300^{\circ}K} + 4.6 \text{ T log } Q_c$$

$$\Delta G_{300^{\circ}K} = -17,500 + (4.6)(300) \log 10^{15}.$$

By definition, the logarithm of any number raised to a power is equal to the power, $\log 10^x = x$. Thus,

$$\log Q_c = \log 10^{15} = 15$$

and

$$\Delta G_{300^{\circ}K} = -17,500 + (4.6)(300)(15)$$

$$\Delta G_{300^{\circ}K} = -17,500 + 20,700$$

$$\Delta G_{300^{\circ}K} = +3200 \text{ calories}$$

and the reaction is now *nonspontaneous* under these nonstandard initial concentrations. Under such conditions the oxidation of nitric oxide to the deadly nitrogen dioxide would not tend to occur unless outside energy is used. In Chapter 2 (Section 2-6), we saw that changing the temperature is one way of altering reaction feasibility, and now we see that altering the initial concentrations of reactants and products provides another means for altering thermodynamic feasibility. These effects will be discussed in more detail in the next chapter.

Study Question 4-5   The standard free energy change, ΔG_T°, for Reaction 8, the Haber process for the manufacture of ammonia, $NH_3(g)$, at 27°C (300°K) and with all substances at initial concentrations of 1 M was found to be approximately $-7,900$ calories (see Study Question 3-7). Calculate ΔG_T at 300°K and the two sets of initial concentrations listed below and indicate whether the reaction has become more or less spontaneous under these conditions.

a. $N_2(g) = 1$ M, $H_2(g) = 1$ M, $NH_3(g) = 0.01$ M, or 10^{-2} M
b. $N_2(g) = 0.01$ M, or 10^{-2} M; $H_2(g) = 0.01$ M, or 10^{-2} M; and $NH_3(g) = 100$ M, or 10^2 M.

4-5   The Equilibrium Constant (K_{eq}) and the Standard Free Energy Change ΔG_T°

As the reaction proceeds from the initial concentrations, and the system approaches the equilibrium state, the reaction quotient and the free energy change, ΔG_T, continually change. At equilibrium, ΔG_T is zero and the relative concentrations of the reactants and products remain constant at their equilibrium values. The reaction quotient at equilibrium is a constant number, and it can be designated as Q_{eq}. In most books it is called the equilibrium constant and given the symbol K_{eq}.

Equilibrium Reaction Quotient or Equilibrium Constant

$$aA + bB \rightleftharpoons cC + dD$$

At Equilibrium

$$\Delta G_T = 0$$

$$Q_{eq} = K_{eq} = \frac{[C]_{eq}^c [D]_{eq}^d}{[A]_{eq}^a [B]_{eq}^b}$$

where $[A]_{eq}$, $[B]_{eq}$, $[C]_{eq}$, and $[D]_{eq}$ are the fixed concentrations of the species at equilibrium.

How can we determine the value of the equilibrium reaction quotient (Q_{eq}) or equilibrium constant (K_{eq})? An expression that can be used to calculate K_{eq} can be obtained as follows:

At equilibrium,

$$\Delta G_T = 0 \text{ and } Q = Q_{eq}$$

and

$$0 = \Delta G_T^\circ + 4.6\ T \log Q_{eq} \quad \text{or} \quad 4.6\ T \log K_{eq}\,.$$

Thus,

$$\Delta G_T^\circ = -\,4.6\ T \log Q_{eq} = -\,4.6\ T \log K_{eq}$$

and we see that the equilibrium reaction quotient or constant (Q_{eq} or K_{eq}) can be calculated from the standard free energy change, ΔG_T°.

This very important relationship between ΔG_T° and K_{eq} can be used in two ways. It can be used to obtain values for ΔG_T° when the value of the equilibrium constant, K_{eq}, is known or to calculate the value of K_{eq} when the standard free energy, ΔG_T°, is known.

For example, consider Reaction 4, $2NO(g) + O_2(g) \rightleftharpoons 2NO_2(g)$, at equilibrium at a particular temperature. A chemical analysis of the system could provide the values for the equilibrium concentrations of each species and these could be used to calculate K_{eq}. This could then be used in the equation $\Delta G_T^\circ = -4.6\, T \log K_{eq}$ to determine the value for the standard free energy change, ΔG_T°. In Chapter 3 (Section 3-6) we presented a method for determining ΔG_T° from standard entropy and enthalpy values. Now we have a second method for determining ΔG_T° from the equilibrium constant, K_{eq}.

Two Ways of Determining ΔG_T°

1. From $\Delta H_{298^\circ K}^\circ$ and $\Delta S_{298^\circ K}^\circ$ values

$$\Delta G_T^\circ = \Delta H_{298^\circ K}^\circ - T\Delta S_{298^\circ K}^\circ$$

2. From K_{eq} or Q_{eq}

$$\Delta G_T^\circ = -4.6\, T \log K_{eq}$$

Again consider Reaction 4 involving the oxides of nitrogen. These substances are mixed and allowed to come to equilibrium at 27°C (300°K). The equilibrium concentrations of reactants and products are found to be

$$2NO(g) + O_2(g)\ \ 2\rightleftharpoons NO_2(g)\quad T = 27°C\ (300°K).$$

Equilibrium
concentrations $\quad 10^{-3}$ M $\quad 10^{-6}$ M $\quad 2.25$ M

$$Q_{eq}\ \text{or}\ K_{eq} = \frac{[NO_2(g)]_{eq}^2}{[NO(g)]_{eq}^2[O_2(g)]_{eq}}$$

$$K_{eq} = \frac{(2.25)^2}{(10^{-3})^2(10^{-6})} = \frac{5}{(10^{-6})(10^{-6})} = \frac{5}{10^{-12}}$$

$$Q_{eq}\ \text{or}\ K_{eq} = 5 \times 10^{12}$$

$$\Delta G_T{}^\circ = -4.6\ T \log K_{eq}$$

$$\Delta G^\circ_{300^\circ K} = -(4.6)(300)\log 5 \times 10^{12}$$

From tables[3] showing the values of logarithms of numbers, $\log 5 \times 10^{12}$ is found to be 12.7 and

$$\Delta G^\circ_{300^\circ K} = -(4.6)(300)(12.7)$$

$$\Delta G^\circ_{300^\circ K} = -17,500 \text{ calories,}$$

which is the same value of $\Delta G^\circ_{300^\circ K}$ attained earlier (Section 3–6) by using $\Delta H^\circ_{298^\circ K}$ and $\Delta S^\circ_{298^\circ K}$ values to determine $\Delta G^\circ_{300^\circ K}$.

Study Question 4-6   Among the products that come out of the exhaust system of an automobile are carbon dioxide (CO_2) and the extremely toxic gas, carbon monoxide (CO). In the presence of oxygen the following equilibrium is established

$$CO_2(g) \rightleftharpoons CO(g) + \tfrac{1}{2}O_2(g)$$

At $100°C$, K_{eq} for this reaction is 10^{-36}.

a. Calculate $\Delta G_T{}^\circ$ for this reaction at $100°C$ and comment on thermodynamic spontaneity under these conditions.
b. Use Table 3-1 of $\Delta H^\circ_{298^\circ K}$ and $\Delta S^\circ_{298^\circ K}$ values (see Section 3-4) to determine $\Delta G_T{}^\circ$ for this reaction at $27°C$.
c. Would you expect a greater proportion of deadly carbon monoxide gas in an equilibrium mixture at room temperature ($27°C$) or at the higher temperature of $100°C$? Explain.

What is the meaning of the relative values of the equilibrium constants for different reactions? We began the chapter by asking the thermodynamic question, "How far can the reaction go?" The answer to this question lies in the values of the equilibrium constants (K_{eq}) and the standard free energies ($\Delta G_T{}^\circ$) for various reactions. K_{eq} is related to the ratio of the concentrations of products to those of the reactants when the system attains equilibrium at a particular temperature. When K_{eq} has a relatively high value (greater than 1), the products predominate over reactants at equilibrium, $\Delta G_T{}^\circ$ is negative, and the equilibrium position favors the products, that is, it lies to the right ($\rightleftharpoons$). When K_{eq} is equal to 1, then the concentrations of all products and reactants must be equal to 1 M or some other combination of concentrations, so that the top half of the equilibrium expression equals the bottom half. In such cases $\Delta G_T{}^\circ = 0$ and the equilibrium contains a balance between reactants and products ($\rightleftharpoons$).

[3] Logarithm tables can be found in most introductory chemistry texts, but they are not necessary for the approximate calculations used in this text.

When K_{eq} has a relatively low value (less than 1), the equilibrium mixture favors the reactants and ΔG_T° is positive (Table 4-1).

Table 4-1   General Relationships between K_{eq}, Equilibrium Position, and ΔG_T°.

Relative Value of K_{eq}	Relative Equilibrium Position	Sign of ΔG_T°
High (>1)	⇌	−
1	⇌	0
Low (<1)	⇌	+

Study Question 4-7   Demonstrate the general data in Table 4-1 by making calculations to fill in the following table for reactions at equilibrium at a temperature of 27°C (300°K).

K_{eq}	Relative Equilibrium position (⇌ , ⇌ , ⇌)	$\Delta G_{300^\circ K}^\circ$ in cal/mole
10^{-100}		
10^{-10}		
10^{-1}		
0		
10^{1}		
10^{10}		
10^{100}		

Study Question 4-8   Look up the values for $\Delta G_{300^\circ K}^\circ$ that you calculated for the ten reactions (Section 3-6) and predict for each reaction

a. the relative equilibrium position at 300°K (⇌ , ⇌ , or ⇌).
b. the relative value of K_{eq} (>1, 1, or <1).

Earlier we saw that the significance of ΔG_T, the free energy change, is that it determines how far the reaction is from equilibrium and the direction in which the reaction goes to reach equilibrium. Now we see that ΔG_T°, the *standard free energy change*, determines how far the reaction has gone when it reaches equilibrium—it is a measure of the equilibrium position.

4-6   Calculation of Equilibrium Constant Values

The equation relating ΔG_T° and K_{eq} can also be used to calculate the value of the equilibrium constant, K_{eq}, when the value of the standard free energy change, ΔG_T°, is known.

$$\Delta G_T^\circ = -4.6\ T \log K_{eq}$$

or, by rearranging,

$$\log K_{eq} = \frac{-\Delta G_T^\circ}{4.6\ T}$$

As long as we deal with whole-number powers of 10 for equilibrium constant values, this is a very simple calculation because the logarithm of 10 to any power is equal to that power. For example, if

$$\log x = a\ \text{number}, \quad \text{then } x = 10^{number}$$

For example, if

$$\log x = 6, \quad \text{then } x = 10^6$$

or if

$$\log x = -3, \quad \text{then } x = 10^{-3}.$$

If the power is fractional, we will round it off to the nearest whole number. For the approximate calculations in this introductory treatment the error introduced will not be important. For example,

$$\log x = -3.6; \quad \text{then } x = 10^{-3.6} \cong 10^{-4}.$$

The relationship between ΔG_T° and K_{eq} can be used to calculate K_{eq} for Reaction 1, the formation of ozone.

$$3O_2(g) \rightleftharpoons 2O_3(g)$$

From Chapter 3 (Study Question 3-7) we found that $\Delta G^\circ_{300^\circ K}$ for this reaction is $+77,900$ calories. The value of K_{eq} at 27°C can be obtained as follows:

$$\log K_{eq} = \frac{-\Delta G^\circ_{300^\circ K}}{4.6T}$$

$$\log K_{eq} = \frac{-77,900}{(4.6)(300)} = \frac{-77,900}{1,380}$$

$$\log K_{eq} = -56.6 \cong -57$$

$$K_{eq} = 10^{-57}$$

Since K_{eq} is very small, the reactant, $O_2(g)$, is favored over the product, ozone, when the reaction reaches equilibrium at 27°C. This same information can also be obtained from the fact that ΔG_T° is positive and the net reaction to the right should be non-spontaneous.

Study Question 4-9   Calculate the approximate values of K_{eq} for the other 10 reactions at 27°C using the $\Delta G^\circ_{300^\circ K}$ values you calculated in Chapter 3 (Study Question 3-7). Use these to complete the following table.

Reaction No.	Balanced equation	$\Delta G^\circ_{300^\circ K}$ in cal	K_{eq}	Relative Equilibrium Position
1	$3O_2(g) \quad 2O_3(g)$	$+77,900$	10^{-57}	$\rightleftharpoons$
2				
etc.				

4-7 Summary

We have seen that the free energy change, ΔG_T, is a measure of the driving force for a system to change spontaneously to a state of *dynamic chemical equilibrium*, where the rates of the forward and reverse reactions are equal. As the reaction proceeds, ΔG_T becomes smaller and smaller and approaches zero as equilibrium is approached. ΔG_T is the free energy change under nonstandard conditions of concentration or partial pressure and is obtained by adding a concentration correction factor to the standard free energy change, ΔG_T°.

System Not at Equilibrium

$$aA + bB \rightleftharpoons cC + dD$$

$$\Delta G_T = \Delta G_T^\circ + 4.6 \, T \log Q_p$$

or

$$4.6 \, T \log Q_c$$

where

$$Q_c = \frac{[C]^c[D]^d}{[A]^a[B]^b}$$

$$Q_p = \frac{p_C^c p_D^d}{p_A^a p_B^b}$$

If initial $\Delta G_T < 0$ (that is, ΔG_T is $-$): Reaction can proceed spontaneously to the right to achieve equilibrium.

If initial $\Delta G_T > 0$ (that is, ΔG_T is $+$): Reaction can proceed spontaneously to the left to achieve equilibrium.

System at Equilibrium

$$aA + bB \rightleftharpoons cC + dD$$

$$\Delta G_T = 0$$

$$\Delta G_T^\circ = -4.6 \, T \log K_{eq}$$

where

$$K_{eq} = \frac{[C]_{eq}^c[D]_{eq}^d}{[A]_{eq}^a[B]_{eq}^b}$$

If $K_{eq} > 1$, then ΔG_T° is $-$ and equilibrium position lies to the right ($\rightleftharpoons$).

If $K_{eq} = 0$, then $\Delta G_T^\circ = 0$ and equilibrium position is balanced ($\rightleftharpoons$).

If $K_{eq} < 1$, then ΔG_T° is $+$ and equilibrium position lies to the left ($\rightleftharpoons$).

Once equilibrium is attained, the values of either $\Delta G_T{}^\circ$ or the equilibrium constant (K_{eq} or Q_{eq}) for a particular reaction can be used to answer the question, "How far can the reaction go?" If $\Delta G_T{}^\circ$ is negative and K_{eq} is > 1, the equilibrium position lies to the right ($\rightleftharpoons$) and products are favored over reactants in the equilibrium mixture.

We have also seen that the concept of dynamic equilibrium can be used to understand the world overpopulation problem that overburdens and threatens the life-support system on fragile spaceship Earth.

> The real problem is not how many people can live optimal lives on this
> planet. It is the quality of life that one needs to safeguard. The
> idea that we should see how many people one can possibly feed and
> keep alive on this planet would produce a degradation of humanity
> that we can't contemplate . . . . The question we should always ask is
> "Is it going to be good for children?"
>
> *George Wald*
> Nobel Laureate in biology

4-8 Review Questions

4-1. Why do most reactions in a closed system reach a state of equilibrium rather than going "to completion"?

4-2. What happens to ΔG_T as a reaction proceeds towards equilibrium?

4-3. If ΔG_T is $-$, does this mean that the reaction *will* necessarily proceed to equilibrium? Explain.

4-4. Define dynamic chemical equilibrium on the micro and macro levels.

4-5. How can one determine when a reaction mixture has attained equilibrium?

4-6. When chemical equilibrium is reached, the total concentrations of all species are static, in that they remain constant. Yet chemical equilibrium is said to be a dynamic equilibrium. Explain.

4-7. Use the concept of dynamic equilibrium to discuss the overpopulation problem. Is the world in a state of dynamic equilibrium with respect to population? Describe several ways in which a dynamic population equilibrium could be reached.

4-8. Distinguish between "J" and "S" curves in terms of population dynamics and indicate the meaning of the term carrying capacity.

4-9. Distinguish between ΔG_T and $\Delta G_T{}^\circ$.

4-10. What is the reaction quotient? Show two ways in which it can be expressed. Does it change or remain constant as a reaction proceeds?

4-11. Explain why the concentration correction factor should have the form, 4.6 T log Q.

4-12. What is the equilibrium reaction quotient or equilibrium constant? Distinguish between the reaction quotient and the equilibrium constant.

4-13. Which one of the following reactions has an equilibrium position lying to the right?
   a. Reaction 1: $\Delta G_T°$ is positive.
   b. Reaction 2: K_{eq} is 10^{-40}.
   c. Reaction 3: $\Delta H°$ is positive, $\Delta S°$ is negative.
   d. Reaction 4: $\Delta H°$ is $+100,000$ cal, $\Delta S°$ is $+$, and the temperature is very high.
   e. Reaction 5: $\Delta H°$ is -3000 cal, $\Delta S°$ is $-$, and the temperature is very high.

4-14. Can the equilibrium constant vary for a given reaction and set of reaction conditions? Explain.

4-15. Give two ways for determining the value of $\Delta G_T°$.

4-16. If the value of the equilibrium constant is high, does this mean
   a. that $\Delta G_T°$ for the reaction is positive?
   b. that the reaction *can* proceed spontaneously to the equilibrium position?
   c. that the reaction *will* attain the equilibrium position?
   d. that at equilibrium the concentration of products will predominate over reactants?

4-17. What is the significance of the relative value of the equilibrium constant? What is indicated by a relatively low value of K_{eq}? a large value? a value of 1?

Chapter 5

Question No. 4—How Can We Make a Reaction Go Further?

5-1 Free Energy Changes under Different Conditions

How can we make a reaction more thermodynamically feasible? How can we shift the reaction so that it goes further to produce a higher yield of products? There are two approaches to increasing the yield. One involves never letting the system attain thermodynamic equilibrium. If the products of a given reaction are continuously or periodically removed, more and more reactants will be used up and the percent yield increased. This approach to increasing yield cannot be studied by using equilibrium thermodynamics, since the system is not allowed to achieve or maintain equilibrium.

The second approach, which we shall discuss in this chapter, involves the equilibrium thermodynamics discussed in the previous chapters. We have already seen that ΔG_T changes with temperature (Section 2-6) and with concentration or partial pressure (Sections 4-3 and 4-4), so that a process that is nonspontaneous under one set of conditions may become spontaneous under other conditions. The answer to the question of increasing the yield has already been discussed—change the concentration or the temperature, or both. If they are varied in the proper manner, the new equilibrium mixture may contain a higher proportion of products. In this chapter, a few more examples illustrating these effects and a third approach, involving *free energy coupling*, is introduced.

5-2 The Effect of Concentration

In Chapter 4 (Section 4-4) we saw that ΔG_T can be varied by varying the initial concentrations or partial pressures of the reactants and products according to the equation

---

Effect of Concentration

$$\frac{\text{Free energy}}{\text{change}} = \frac{\text{Standard free}}{\text{energy change}} + \frac{\text{Concentration}}{\text{factor}}$$

$$\Delta G_T = \Delta G_T^\circ + 4.6\ T \log Q$$

Study Question 5-1 In Section 4-4, the free energy change was calculated for Reaction 4 involving the oxides of nitrogen at two sets of initial concentrations at 300°K, as summarized below. Note that with the first two sets of concentrations the reaction is not at equilibrium, while the third set represents a set of equilibrium concentrations, where ΔG_T is zero and $Q_c = K_{eq}$. Complete this table by calculating ΔG_T at the two new sets of initial concentrations:

$$2NO(g) + O_2(g) \rightleftharpoons 2NO_2(g).$$

$$\Delta G_{300°K}^\circ = -17{,}500 \text{ calories}$$

	Initial Concentrations				4.6 T log Q_c (cal)	$\Delta G_{300°K}^\circ$ (cal)	Sponta-neous to Right	System at Equilib-rium
Set	[NO(g)]	[O$_2$(g)]	[NO$_2$(g)]	Q_c				
1	1	1	1	1	0	−17,500	Yes	No
2	10^{-2}	10^{-1}	10^5	10^{15}	+19,200	+3200	No	No
3	1	1	4.3	18.6	−28,000	0	No	Yes
4	10^{-5}	10^{-1}	10^5	—	—	—	—	—
5	10^2	1	10^{-3}	—	—	—	—	—

The effect of changing the starting concentrations for Reaction 8, the Haber process for the manufacture of ammonia, $NH_3(g)$, at 27°C (300°K) is shown in Table 5-1. This shows the change in ΔG_T for 12 different experiments, each with a different set of starting concentrations.

$$N_2(g) + 3H_2(g) \rightleftharpoons 2NH_3(g)$$

$$Q_c = \frac{[NH_3(g)]^2}{[N_2(g)][H_2(g)]^3}$$

$$\Delta G^{\circ}_{300^{\circ}K} = -7900 \text{ calories}$$

$$\Delta G_{300^{\circ}K} = \Delta G^{\circ}_{300^{\circ}K} + (4.6)(300) \log Q_c$$

$$\Delta G_{300^{\circ}K} = -7900 + 1380 \log Q_c$$

Table 5-1   Variation of ΔG_T with Starting Concentrations for the Haber Synthesis of Ammonia at 300°K.

Set	Initial Concentrations $[N_2(g)]$	$[H_2(g)]$	$[NH_3(g)]$	Q_c	1380 log Q_c (cal)	$\Delta G_{300^{\circ}K}$ (cal)	Spon-tane-ous?	At equi-libri-um?
1	1	1	0	0	$-\infty$	$-\infty$	Yes	No
2	1	1	10^{-2}	10^{-4}	$-5,520$	$-13,520$	Yes	No
3	1	1	10^{-1}	10^{-2}	$-2,760$	$-10,760$	Yes	No
4	1	1	1	1	0	$-7,900$	Yes	No
5	1	1	10	10^2	$+2,760$	$-5,140$	Yes	No
6	1	1	10^2	10^4	$+5,520$	$-2,380$	Yes	No
7	1	1	735	$5.4 \times 10^5 = K_{eq}$	$+7,900$	0	No	Yes
8	10^{-1}	10^{-1}	7.35	$5.4 \times 10^5 = K_{eq}$	$+7,900$	0	No	Yes
9	10^{-1}	10^{-2}	0.232	$5.4 \times 10^5 = K_{eq}$	$+7,900$	0	No	Yes
10	10^{-2}	10^{-2}	10^2	10^{12}	$+16,560$	$+8,660$	No	No
11	10^{-2}	10^{-2}	10^4	10^{24}	$+33,120$	$+25,220$	No	No
12	0	0	1	∞	$+\infty$	$+\infty$	No	No

In Experiment 1 only N_2 and H_2 gases are mixed, and the reaction proceeds spontaneously to form ammonia with an infinite driving force. In Reactions 2 through 6 the starting concentration of ammonia is gradually increased. The reverse reaction can then occur more readily and the driving force, as reflected in the negative ΔG_T values, decreases as the system approaches equilibrium. In experiments 7, 8 and 9,

ΔG_T, the driving force, has become zero and the system is at equilibrium. At this point Q_c equals the equilibrium constant, K_{eq}.

$$\Delta G_T = 0$$

and

$$\Delta G_T{}^\circ = -4.6 \, T \log K_{eq}$$

In Experiments 10, 11, and 12 the initial concentration of ammonia predominates over those of N_2 and H_2, and the driving force for the reaction is to the left instead of the right.

Fortunately, a very simple principle, known as *Le Chatelier's principle*, can be used to predict the general shift in the composition of the equilibrium mixture when the initial concentrations of the reactants or products are altered. This allows us to predict whether the new equilibrium mixture will contain more or less of the products.

Le Chatelier's Principle

If a system in dynamic equilibrium is subjected to a stress, then the system will change, if possible, to relieve the stress.

Consider the reaction

$$A + B \rightleftharpoons C + D$$

at equilibrium. The concentrations of all species are fixed to preserve the value of the equilibrium constant.

$$K_{eq} = \frac{[C][D]}{[A][B]}$$

Now suppose that more of reactant A is added to the equilibrium mixture. This changes the concentration ratios so that they no longer equal K_{eq} and the system is no longer in equilibrium. Since the denominator has increased by the addition of A, the numerator must increase in order for the ratio to again equal the K_{eq} value. This means that when the system is stressed by adding A, the reaction mixture shifts to the right to use up the excess A and produce more C and D until equilibrium is established again. Addition of B would have the same effect. Likewise the addition of C and D to the equilibrium mixture would form more A and B to relieve the stress, and the new equilibrium position would be shifted to the left ($\rightleftharpoons$). Removing or decreasing the concentration of one of the products would shift the equilibrium position to the right as A and B would react to replace the C or D removed.

Study Question 5-2   Given Reaction 6 for the reaction of ammonia gas with $O_2(g)$ to form nitrogen and water vapor,

$$4NH_3(g) + 3O_2(g) \rightleftharpoons 2N_2(g) + 6H_2O(g),$$

predict whether the following concentration changes (stresses) will shift the equilibrium position to the right ($\rightleftharpoons$) or left ($\rightleftharpoons$).

a. Removal of oxygen.
b. Removal of water.
c. Increasing the concentration of ammonia.
d. Decreasing the concentration of ammonia and oxygen.
e. Increasing the concentration of nitrogen.

5-3 The Effect of Temperature

We saw in Section 2-6 that changing the temperature at which a reaction takes place will alter the value of ΔG_T and ΔG_T° and make the reaction more or less spontaneous. Recall that the enthalpy change, $\Delta H_{298^\circ K}$, and the entropy change, $\Delta S_{298^\circ K}$, do not vary significantly with temperature. However, the entropy factor ($T\Delta S$) is a function of temperature (T), and its change with temperature causes ΔG_T to vary with T. At high enough temperatures the entropy factor will predominate over the enthalpy (ΔH) factor.

$$\Delta G_T = \Delta H_{298^\circ K} \qquad - \qquad S\Delta S_{298^\circ K}$$

Predominates at   Predominates at
low temperatures   high temperatures

We can summarize the effect of temperature on free energy change as follows:

Temperature and ΔG_T

All chemical reactions that involve an increase in entropy, or disorder, (ΔS is $+$) occur spontaneously (ΔG_T is $-$) at high enough temperatures.[1]

---

[1] Assuming that some other reaction (for example, decomposition) does not occur at the high temperature.

Conversely, any reaction with a decrease in entropy (ΔS is $-$) will become nonspontaneous at a high enough temperature. These familiar yes-no and no-yes situations are summarized below.

Table 5-2   Effect of Temperature on ΔG_T and Thermodynamic Spontaneity.

Reaction Type	$\Delta H_{298°K}$	$T\Delta S_{298°K}$	ΔG_T		Reaction Proceeds Spontaneously
A	Negative (yes)	Positive (yes)	Negative	(yes)	At *all* temperatures
B	Positive (no)	Positive (yes)	Positive low T	(no)	At sufficiently *high* temperatures
			Negative high T	(yes)	
C	Negative (yes)	Negative (no)	Negative low T	(yes)	At sufficiently *low* temperatures
			Positive high T	(no)	
D	Positive (no)	Negative (no)	Positive	(no)	Nonspontaneous at *all* temperatures

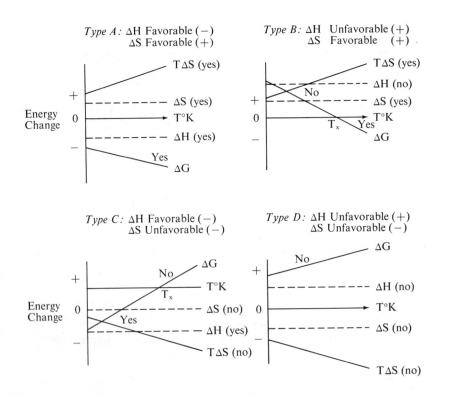

Figure 5-1   Variation of ΔH, ΔS, $T\Delta S$, and ΔG_T for Various Types of Reactions.

Figure 5-1 summarizes these effects graphically for various types of reactions. In Types B and C the temperature, T_x, is the crossover temperature at which there is a change from a spontaneous to a nonspontaneous situation, or vice versa.

Since ΔG_T and $\Delta G_T{}^\circ$ vary with temperature ($\Delta G_T{}^\circ = -4.6\ T \log K_{eq}$), the equilibrium constant, K_{eq}, also varies with T. Note that at a fixed temperature the equilibrium remains constant, but as T changes, the value of K_{eq} is altered. The effect of temperature on K_{eq} for the Haber process for the synthesis of ammonia is shown in Table 5-3.

Haber Synthesis of Ammonia

$$N_2(g) + 3H_2(g) \rightleftharpoons 2NH_3(g)$$

$$K_{eq} = \frac{[NH_3(g)]^2}{[N_2(g)][H_2(g)]^3}$$

$$\Delta G_T{}^\circ = -4.6\ T \log K_{eq}$$

or

$$K_{eq} = 10^{-\Delta G_T{}^\circ/4.6T}$$

$$\Delta H^\circ_{298\,^\circ K} = -22,000 \text{ cal (yes)}$$

$$\Delta S^\circ_{298\,^\circ K} = -47 \text{ cal/}^\circ K \text{ (no)}$$

Table 5-3 Variation of Equilibrium Constant, K_{eq}, with Temperature for the Haber Synthesis of Ammonia: $N_2(g) + 3H_2(g) \rightleftharpoons 2NH_3(g)$.

Temperature (°K)	Approximate ΔG_T (cal)	K_{eq}	Relative Equilibrium Position
100	−17,300	4×10^{37}	
300	− 8,000	6.3×10^5	
500	+ 1,500	2.2×10^{-1}	
700	+10,900	3.3×10^{-4}	
900	+20,300	1.3×10^{-5}	
1100	+29,700	1.0×10^{-6}	

Observe that as the temperature increases, the reaction becomes less feasible. From a thermodynamic standpoint, the commercial synthesis of ammonia should be carried out at a relatively low temperature in order to maximize the yield of NH_3.

Study Question 5-3

a. Using the data in Table 5-3 for the Haber process, carefully plot ΔG_T versus T on graph paper, and determine the approximate value of the crossover temperature, T_x, in °K from your graph.

b. Use $\Delta G_T = \Delta H_{298°K} - T\Delta S_{298°K}$ to calculate the approximate value for the crossover temperature, T_x, for the same reaction and compare the value with that obtained by the graphical method.

c. What is the significance of this crossover temperature?

In Study Question 5-3 you plotted ΔG_T versus T and used this curved line to interpolate and obtain a value for T_x. It is difficult to get an accurate value for T_x using a curved line—a straight line would provide a better value. Indeed, the analysis of scientific data is frequently an attempt to plot the data in a manner that will yield a straight line.

What should we plot to obtain a straight line? Recall the general equation for a straight line:

Equation for a Straight Line

$$y = mx + b$$

where m = slope of the line

and b = y intercept

Note that $\Delta G_T° = -4.6\ T \log K_{eq}$ can be rearranged to give

$$\log K_{eq} = \frac{-\Delta G_T°}{4.6\ T}, \quad \text{or} \quad \frac{-\Delta G_T°}{4.6}\left(\frac{1}{T}\right),$$

which is in the form

$$y = mx + b,$$

where

$$y = \log K_{eq}, \ m = \frac{-\Delta G_T{}^\circ}{4.6}, \ x = \frac{1}{T}, \text{ and } b = 0.$$

Thus we should obtain a straight line by plotting $\log K_{eq}$ versus $1/T$.

Study Question 5-4 Replot the data in Table 5-3 by plotting $\log K_{eq}$ versus $1/T$. Use this graph to obtain a better value of the crossover temperature, T_x. Compare this with the answers you obtained in Study Question 5-3, parts a and b.

Actually a plot of $\log K_{eq}$ versus $1/T$ can be used for even more important purposes. It provides one means for determining values of $\Delta H_{298^\circ K}^\circ$ and $\Delta S_{298^\circ K}^\circ$, such as those found in Table 3-1 in Chapter 3.

$$\Delta G_T{}^\circ = \Delta H_{298^\circ K}^\circ - T \Delta S_{298^\circ K}$$

$$\Delta G_T{}^\circ = -4.6 \ T \log K_{eq}$$

Equating these two equations, we have

$$-4.6 \ T \log K_{eq} = \Delta H_{298^\circ K}^\circ - T \Delta S_{298^\circ K}^\circ$$

Dividing both sides by $-4.6 \ T$, we obtain

$$\log K_{eq} = \frac{-\Delta H_{298^\circ K}^\circ}{4.6 \ T} + \frac{\Delta S_{298^\circ K}^\circ}{4.6}$$

$$y = mx + b,$$

which is in the form of a straight line, where $y = \log K_{eq}$, the slope $m = -\Delta H_{298^\circ K}^\circ/4.6$, $x = 1/T$, and the y intercept is $b = \Delta S_{298^\circ K}^\circ/4.6$. Thus, a plot of $\log K_{eq}$ versus $1/T$ can be used to calculate $\Delta H_{298^\circ K}^\circ$ and $\Delta S_{298^\circ K}^\circ$. $\Delta H_{298^\circ K}^\circ$ can be obtained from the slope of the line, and $\Delta S_{298^\circ K}^\circ$ can be obtained from the value of the y intercept.

Study Question 5-5 From the y intercept of your graph in Study Question 5-4, determine the value for $\Delta S_{298^\circ K}^\circ$ for the Haber process. Compare your answer with the value of $\Delta S_{298^\circ K}^\circ = -47$ cal/$^\circ$K calculated in Chapter 3 (study Question 3-6). Why would you expect the value calculated in this problem to be approximate? (*Hint*: What assumption have we made about ΔH° and ΔS°?)

By the definition of a logarithm, if

$$\log K = a + b,$$

then

$$K = (10^a)(10^b).$$

Thus, the equation

$$\log K_{eq} = \frac{-\Delta H^{\circ}_{298^{\circ}K}}{4.6\ T} + \frac{\Delta S^{\circ}_{298^{\circ}K}}{4.6}$$

can also be written as

$$K_{eq} = (10^{-\Delta H_{298^{\circ}K}/4.6T})\ (10^{\Delta S_{298^{\circ}K}/4.6})$$

or

$$K_{eq} = (\text{enthalpy factor}) \times (\text{entropy factor})$$

In our discussion of kinetics in Chapter 7, we shall use this form of the equation relating K_{eq} and $\Delta H^{\circ}_{298^{\circ}K}$ and $\Delta S^{\circ}_{298^{\circ}K}$.

$$\log K_{eq} = \text{enthalpy factor} - \text{entropy factor}$$

and

$$\Delta G_T^{\circ} = \Delta H^{\circ}_{298^{\circ}K} - T\Delta S^{\circ}_{298^{\circ}K}$$

K_{eq} expresses the reaction tendency in multiplicative form, since K_{eq} is a multiplicative function. ΔG_T° is another way of representing the same information using an additive function.

5-4 Free Energy Coupling

Suppose one has a nonspontaneous reaction A $\rightleftharpoons$ B with an unfavorable $\Delta G_{300^{\circ}K}$ of $+5000$ calories. A third way of making this reaction potentially spontaneous is to couple it with one or more additional spontaneous reactions that have a favorable free energy change. For example, assume that a spontaneous reaction B $\rightleftharpoons$ D has a $\Delta G_{300^{\circ}K}$ of -7000 calories. If we couple or link these two reactions together to produce a new net reaction, A $\rightleftharpoons$ D, the overall process becomes spontaneous. In effect, we have used the free energy change of one favorable reaction to drive an unfavorable reaction.

$$\text{Step 1}\quad A \rightleftharpoons B\quad \Delta G_1\ = +5000\ \text{cal (no)}$$

$$\text{Step 2}\quad B \rightleftharpoons D\quad \Delta G_2\ = -7000\ \text{cal (yes)}$$

$$\text{Net reaction}\quad A \rightleftharpoons D\quad \Delta G_{net} = \Delta G_1 + \Delta G_2$$

$$= -7000 + 5000$$

$$= -2000\ \text{cal (yes)}$$

This process is known as *free energy coupling*, and it is particularly important in biological processes. Most biological processes proceed by a whole series of stepwise reactions with each step having a particular free energy change that can be either favorable or unfavorable. As long as the overall reaction sequence has a negative ΔG_T, the entire sequence of steps (reaction mechanism) can proceed spontaneously even though one or more steps in the sequence are nonspontaneous.

Free energy coupling is the primary means for energy transfer in living cells involving "high energy" phosphates such as adenosine triphosphate (ATP), adenosine diphosphate (ADP), and creatine phosphate. Reaction 11 in Chapter 3 involves the hydrolysis (reaction with water) of ATP to produce ADP plus a phosphate radical (P or P_i), and the reaction has a free energy change, $\Delta G_{310°K}$, of -7300 cal.[2]

Reaction $ATP + H_2O \rightleftharpoons ADP + phosphate$ $\Delta G_{310°K} = -7300$ cal
11

When you use a muscle in your body, a chemical reaction is needed to produce a high yield of energy. This process involves the free energy coupling of Reaction 11 with the following reaction:

---

creatine phosphate $+ ADP \rightleftharpoons ATP +$ creatine $\Delta G_{310°K} = -2800$ cal

---

This provides a net reaction of the hydrolysis of creatine phosphate to yield creatine plus a phosphate radical, which yields a more negative ΔG than either individual reaction and provides the needed energy, as shown below.

Reaction 9: *Free Energy Coupling—Work in Muscle Contraction*

$$ATP + H_2O(1) \rightleftharpoons ADP + phosphate \quad \Delta G_{310°K} = -7300$$

$$creatine\ phosphate + ADP \rightleftharpoons ATP + creatine \quad \Delta G_{310°K} = -2800\ cal$$

Net reaction:[3]   creatine phosphate $+ H_2O(l) \rightleftharpoons$ creatine $+$ phosphate
$$\Delta G_{net} = -10,000\ cal\ at\ 310°K$$

---

[2] This value is at body temperature, 37°C or 310°K, a hydrogen ion, or H^+ concentration of $10^{-7}M$, and in the presence of an excess magnesium ion, Mg^{2+}. The value of ΔG_T for this reaction under actual conditions in a living cell is approximately $-12,500$ cal.

[3] Note that since one mole of ATP and one mole of ADP are on both sides of the equation, they cancel out and do not appear in the net equation.

We can now begin to appreciate the distinction between reaction mechanism (the actual sequence of steps by which a reaction occurs) and the net reaction for an overall process. For example, it is obvious that the net equation for the reaction of water with creatine phosphate does not in any way tell us about the mechanism of the reaction. This is accounted for by the coupling of the two reactions. (Actually the mechanism involves an even greater number of steps, but this crude mechanism serves to illustrate the difference between net reaction and reaction mechanism.)

Study Question 5-6   Given the two reactions

$$S(s) + O_2(g) \rightleftharpoons SO_2(g)$$

$$SO_2(g) + \tfrac{1}{2}O_2(g) \rightleftharpoons SO_3(g)$$

a. Calculate $\Delta G^\circ_{300^\circ K}$ for each reaction (see Section 3-6), and comment on the thermodynamic feasibility of each reaction at 300°K (27°C). In terms of air pollution, would you expect to find more sulfur dioxide (SO_2) or more sulfur trioxide (SO_3) when fuel containing sulfur is burned?
b. Write the net equation obtained.
c. Calculate $\Delta G^\circ_{300^\circ K}$ for this net reaction. Is it thermodynamically feasible?

5-5   Enthalpy Coupling—Hess' Law

We have seen that because ΔG_T and ΔG_T° are additive functions, the separate free energies for a series of reactions can be added to give ΔG_{net} or ΔG_{net}° for the net reaction when the reactions are coupled together. Since ΔH° is also an additive function, we can add, or couple, the ΔH° values for a series of reactions to obtain ΔH_{net}° for the net reaction. This principle of enthalpy coupling is known as *Hess' Law of Constant Heat Summation*.

Hess' Law of Constant Heat Summation

ΔH° depends only on the initial and final states of a reaction, and when two or more chemical equations are added to yield a net equation, the individual reaction enthalpies can be added to yield ΔH_{net}° for the net equation.

This principle can be very useful in obtaining $\Delta H^\circ_{298^\circ K}$ values for reactions that can't be run in the laboratory. Frequently the reaction is too dangerous, chemicals are not available, or the chemist (like all of us) would rather save himself a lot of time and quickly calculate the answer instead of setting up a time-consuming experiment.

For example, suppose we want to calculate $\Delta H^\circ_{298^\circ K}$ for the heat of formation of methane, $CH_4(g)$, from its elements

$$C(s) + 2H_2(g) \longrightarrow CH_4(g).$$

In this case the value, available from Table 3-1, is $-18,000$ cal. However, the value for ΔH_f° could also be obtained by coupling the $\Delta H^\circ_{298^\circ K}$ values for the following three reactions using Hess' Law.

Reaction 1: $C(s) + O_2(g) \longrightarrow CO_2(g)$ $\qquad\qquad\qquad \Delta H^\circ_{298^\circ K} = -94,000$ cal

Reaction 2: $2H_2(g) + O_2(g) \longrightarrow 2H_2O(l)$ $\qquad\qquad \Delta H^\circ_{298^\circ K} = (2)(-68,000)$

$$= -136,000 \text{ cal}$$

Reaction 3: $CO_2(g) + 2H_2O(l) \longrightarrow CH_4(g) + 2O_2(g)$ $\;\; \Delta H^\circ_{298^\circ K} = +212,000$ cal

---

Net Reaction: $C(s) + 2H_2(g) \longrightarrow CH_4(g)$ $\qquad\qquad \Delta H^\circ_{net} = \Delta H_1{}^\circ + \Delta H_2{}^\circ$

$$+ \Delta H_3{}^\circ$$

$$= -94,000$$

$$-136,000$$

$$+212,000$$

$$\Delta H^\circ_{net} = -17,000 \text{ cal}$$

Suppose we want to determine $\Delta H^\circ_{298^\circ K}$ for Reaction 6, the oxidation of ammonia,

$$\text{Net Reaction: } 4NH_3(g) + 3O_2(g) = 6H_2O(g) + 2N_2(g) \qquad\qquad (6)$$

and are given the following:

Reaction 1: $N_2(g) + 3H_2(g) \rightleftharpoons 2NH_3(g)$ $\quad \Delta H^\circ_{298^\circ K} = -22,000$ cal

Reaction 2: $2H_2(g) + O_2(g) \rightleftharpoons 2H_2O(g)$ $\quad \Delta H^\circ_{298^\circ K} = -116,000$ cal.

How can we combine Reactions 1 and 2 so that when added they will yield the net reaction for the oxidation of ammonia? Looking at the net reaction, we see that 4 moles of ammonia are needed on the left-hand side. These four moles of NH_3 could be obtained by reversing Reaction 1 and multiplying it through by two. This means that the sign for $\Delta H^\circ_{298^\circ K}$ would be reversed and the value would be multiplied by two. The net equation has 6 moles of $H_2O(g)$ on the right, which could be obtained by multiplying Reaction 2 by three.

Reaction 1: $(-2\times)$ $4NH_3(g) \rightleftharpoons 2N_2(g) + 6H_2(g)$ $\Delta H^\circ_{298°K} = (2)(+22,000)$

Reaction 2: $(3\times)$ $6H_2(g) + 3O_2(g) \rightleftharpoons 6H_2O(g)$ $\Delta H^\circ_{298°K} = (3)(-116,000)$

Net Reaction: $4NH_3(g) + 3O_2(g) \rightleftharpoons 6H_2O(g) + 2N_2(g)$ $\Delta H^\circ_{net} = (2)(+22,000)$

$$+(3)\times$$

$$(-116,000)$$

$$= -304,000 \text{ cal}$$

Study Question 5-7   Use the reactions listed below to determine $\Delta H^\circ_{298°K}$ for Reaction 3, the burning of octane in gasoline. Compare your answer with the one obtained for Study Question 3-5.

$$\underline{\quad}C_8H_{18}(l) + \underline{\quad}O_2(g) \rightleftharpoons \underline{\quad}CO_2(g) + H_2O(g)$$

Reaction 1: $8C(s) + 9H_2(g) \rightleftharpoons C_8H_{18}(l)$ $\Delta H^\circ_{298°K} = -60,000$ cal

Reaction 2: $2H_2(g) + O_2(g) \rightleftharpoons 2H_2O(g)$ $\Delta H^\circ_{298°K} = -116,000$ cal

Reaction 3: $CO_2(g) \rightleftharpoons C(s) + O_2(g)$ $\Delta H^\circ_{298°K} = +94,000$ cal

5-6 Summary of Chemical Energetics

We have now seen how the four fundamental questions of chemical energetics, or thermodynamics, can be answered.

Four Fundamental Questions of Chemical Energetics

1. Can a reaction occur spontaneously?
2. How much energy can be released or absorbed when a reaction takes place?
3. How far can the reaction go?
4. How can we make a reaction go further?

The answers to these questions are found by using and combining the First and Second Laws of Thermodynamics. The second law alone provides a criterion to answer the first question of spontaneity if both the *system plus surroundings* are taken into account ($\Delta S_{net} > 0$ for any spontaneous process). However, if the first law (energy, or enthalpy, principle) is combined with the second law (entropy, or disorder, principle), a new function, the Gibbs free energy change ($\Delta G_T = \Delta H - T\Delta S$), can serve as a criterion of spontaneity for the *system only* at constant temperature and pressure. For spontaneous reactions at constant temperature and pressure, ΔG_T is $-$, and the value of ΔG_T is a measure of the maximum work or energy that is available.

First and Second Laws of Thermodynamics

First Law: In all macroscopic chemical and physical changes, energy is neither created nor destroyed but merely transformed from one form to another. The total energy of the system plus surroundings remains constant.

$$\Delta H_{system} + \Delta H_{surroundings} = 0$$

Second Law: Any system plus its surroundings tends spontaneously toward a state of increasing entropy, or disorder. The total entropy of the system plus surroundings increases.

$$\Delta S_{total} = \Delta S_{system} + \Delta S_{surroundings} > 0$$
for any spontaneous process

First and Second Laws combined

Gibbs free energy change, or reaction tendency

$$= \boxed{\begin{array}{c} \text{tendency toward} \\ \text{minimum enthalpy} \end{array}} - \boxed{\begin{array}{c} \text{tendency toward} \\ \text{maximum entropy} \end{array}}$$

= maximum work available to drive a reaction

$$\Delta G_{system} = \Delta H_{system} - T\Delta S_{system}$$

or

Chemical reactions at constant temperature and pressure proceed spontaneously to a state of lower Gibbs free energy.

$$\Delta G_{system} < 0$$

for any spontaneous process at constant T and P

A spontaneous process in a closed system continues until it reaches a state of dynamic equilibrium where the free energy of the system has reached a minimum and the entropy of the system plus surroundings has reached a maximum. *At equilibrium* the free energy and entropy are no longer changing, and ΔS_{net} and $\Delta G_{system} = 0$.

The value of the free energy change, ΔG_T, is a measure of the distance of the reaction from its equilibrium position, and the sign of ΔG_T determines whether the reaction will proceed spontaneously to the right (ΔG_T is $-$) or to the right (ΔG_T is $+$) to reach the equilibrium position.

The value of ΔG_T, the standard free energy change, or the value of the equilibrium constant, K_{eq}, provide a measure of how far the reaction can go at a particular temperature. A reaction can be made to go further by changing the temperature, concentrations of reactants and products, or by coupling it with a reaction with a favorable ΔG_T (free energy coupling). These relationships are governed by four simple equations.

Reaction *Not* at Equilibrium

$$\Delta G_T = \Delta H - T\Delta S$$

$$\Delta G_T = \Delta G_T^\circ + 4.6\ T \log Q$$

Reaction at Equilibrium

$$\Delta G_T^\circ = \Delta H_{298^\circ K}^\circ - T\Delta S_{298^\circ K}^\circ$$

$$\Delta G_T^\circ = -4.6\ T \log K_{eq}$$

Thus, with only two laws based on observing everyday phenomena and four simple equations, we can answer the four questions of chemical energetics. Before turning our attention to the three questions of chemical kinetics, we shall take a deeper look at the meaning and significance of entropy.

There is something fascinating about science, one gets such wholesale
returns of conjectures out of such a trifling investment of fact.

Mark Twain

5-7 Review Questions

5-1. What are two general methods for increasing the percent yield of a reaction? Can both approaches be treated from the standpoint of thermodynamics? Explain.

5-2. Distinguish between ΔG_T° and ΔG_T.

5-3. Explain how a spontaneous reaction with a negative ΔG_T° could become nonspontaneous.

5-4. What two variables can be changed to shift the equilibrium position of a reaction?

5-5. What is Le Chatelier's principle?

5-6. Indicate whether the reaction under each of the following sets of conditions is spontaneous or nonspontaneous.
  a. ΔH is $-$, the disorder increases, T = 1000°C.
  b. The enthalpy of the products is lower than that of the reactants, ΔS is $+$, T = 1000°C.
  c. The reaction is endothermic, ΔS is $+$, T = 25°C.
  d. ΔH is $+$, disorder increases, T = 1000°C.
  e. ΔH is $-$, order increases, T = 25°C.
  f. The reaction is exothermic, ΔS is $-$, T = 1000°C.

5-7. How can an endothermic reaction be spontaneous?

5-8. Why is it reasonable to say that at room temperature (25°C) the spontaneity of reactions is determined only by the change in enthalpy (ΔH)?

5-9. What is free energy coupling and why is it important?

5-10. Write an equation for the hydrolysis of ATP. Why is this reaction so important?

5-11. Distinguish between net reaction and reaction mechanism.

5-12. Summary
  a. What are the four fundamental questions of chemical energetics?
  b. State the First and Second Laws of Thermodynamics.
  c. What are the criteria for spontaneity in terms of entropy? in terms of Gibbs free energy?
  d. What are the two equations for calculating ΔG_T° when a reaction is at equilibrium at standard conditions?
  e. What are the two equations for calculating ΔG_T when the system is not at equilibrium and not at standard conditions?

5-13. Why is it useful to plot log K_{eq} vs. 1/T for a reaction?

5-14. What is Hess' law of constant heat summation and how can it be useful?

Chapter 6

Entropy—A Closer Look

**6-1 Entropy, Disorder, and Randomness—
Positional and Motional Entropy**

According to the second law, spontaneous processes in nature tend toward an increase in entropy when both the system and its surroundings are considered. In a rather glib manner we have stated that entropy is a measure of disorder. Actually, the terms "order" and "disorder" are difficult to define clearly, but intuitively we picture "order" as characteristic of a collection of objects arranged in a definite pattern (that is, the pattern is *not* haphazard or random). A neat stack of blocks sitting on the floor of a room would be considered an ordered, or nonrandom, arrangement. A collection of objects not arranged in a definite pattern is considered to be disordered, haphazard, or random. If several young children played with the neat pile of blocks, the blocks would become distributed over the room in a disordered or random manner. We would not expect the blocks to return to their original ordered state spontaneously (a lot of energy, encouragement, and perhaps threats would be required to reverse the process).[1]

There is essentially only one ordered way for the blocks to be stacked, but there are millions and millions of disordered arrangements. When there are a number of different ways of arranging something, this disordered state is more probable. Later in this chapter we shall see that this provides the connection between entropy and probability.

The type of disorder, or randomness, discussed above is *positional disorder*. The difference in the two states for the blocks is one of position, relative to total volume. If objects are lined or stacked up in a small volume, there is a low degree of positional disorder, while if they are randomly distributed over a large volume, there is a high degree of positional disorder.

---

[1] Mothers might sometimes refer to young children as "entropy's little helpers."

There is a second type of disorder based on the relative motion, or average kinetic energy, in a system. This type of randomness is called *motional, or thermal, disorder*. Consider a large number of ducks cooped up in a small cage in the corner of a room. Their positional disorder is very low, since they occupy a small volume. Consider what happens if a wolf enters the room. The positional disorder will remain low because of the confines of the cage, but the ducks will become nervous and begin to move about rather frantically and their *average motional disorder*, or randomness, will increase. In systems of molecules and atoms this increase in average motion, or kinetic energy, occurs when the temperature of the system is raised.

Now suppose the cage door is opened when the wolf is not in the room. The positional disorder will increase drastically as the ducks spread about the room but their average motional disorder will increase only slightly. If the wolf then enters the room, the positional disorder cannot increase significantly, because of the finite size of the room, but motional disorder will increase drastically as the ducks attempt to evade the wolf.

In applying these ideas to aggregates of molecules or atoms, we must, therefore, distinguish between two types of entropy: *positional entropy* and *motional entropy*. *Positional* entropy is basically a function of whether the chemical substances are in the solid, liquid, or gaseous state or dissolved in solution, while *motional* entropy is a function of the temperature of the substance and its molecular complexity. As the temperature is increased, the average thermal motion of the atoms or molecules is increased regardless of whether they are in the solid, liquid, or gaseous state.

Molecular complexity is also involved in motional entropy. Compare the motions for a simple monatomic substance (consisting of one atom) and a more complex diatomic substance (consisting of two atoms linked together, for example, O_2 or O-O). The more complex diatomic molecule has more ways to move (it can vibrate and rotate in a number of different ways while moving through space), and it will therefore have a higher motional entropy. It can be likened to a duck running about (translational motion) flapping its wings (vibrational motion).

6-2 Entropy of Solids, Liquids, and Gases—
Estimating Relative Entropies

The molecules, ions, or atoms in a solid are very close together and their motion is highly restricted. In other words, they have a low positional energy. As the temperature is raised, the particles begin to vibrate and there is a steady rise in motional entropy. Eventually enough energy is put into the system to break the chemical bonds holding the solid together, and it melts to form a liquid. The particles in the liquid state are still relatively close together, but they are freer to move and the overall arrangement is fairly disordered. In going from the solid to the liquid state there is a relatively large increase in positional entropy. (See Figures 2-4 and 6-3.)

If the liquid is heated, the motional entropy again increases until enough energy is put into the system to break the bonds between the particles in the liquid state. At this point the liquid begins to boil and is converted into a gas. The boiling, or massive evaporation, of a liquid involves a drastic increase in positional entropy, or disorder, since the gas occupies a much larger volume. If the gas is then heated, the motional entropy will increase. At any given point the total entropy of the system is the sum of the positional and motional entropies.

The difference between the solid, liquid, and gaseous states is somewhat analogous to a large group of students under several different conditions. The solid state with low entropy is approximated by several hundred students sitting in a lecture hall during a lecture.[2] If the lecturer gives the students a two-minute break but asks them to remain in the lecture hall, this approximates the liquid state. The positional and motional disorder are increased but pockets of order (small groups of students) occur here and there. The "gaseous" state is approximated when the class is over and the students have spread out over the entire campus. This conversion from the "liquid" to the "gaseous" state involves a very large increase in positional entropy.

In actual practice the entropy of melting (solid to liquid) for most solids involves an entropy increase of about 2–6 entropy units. The entropy of vaporization (liquid to gas) involves a much greater relative change in position, or volume, and it typically involves an entropy increase of about 20–26 entropy units.

The sequence of entropy changes that occur when a particular substance is changed from its solid state (below its melting point) to its gaseous state (above its boiling point) is shown in Figure 6-1.

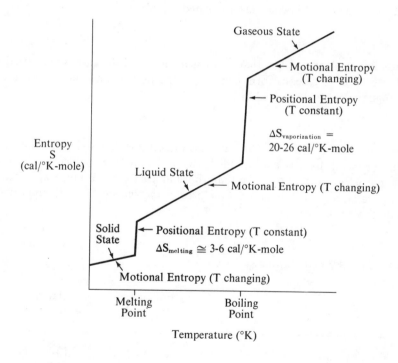

Figure 6-1   Entropy Changes for a Typical Substance Being Heated at Constant Pressure.

[2] If the students are all asleep, the entropy is even lower, and so is the enthalpy.

Study Question 6-1 When a liquid is boiled at constant pressure and converted to a gas, energy is being continually supplied to the system, yet the temperature remains constant during the boiling process. Explain.

Study Question 6-2 For each of the processes below indicate whether you would expect the entropy of the system to increase or decrease and indicate whether the change involves primarily positional or motional entropy.

a. freezing of water,
b. $H_2O(s) \longrightarrow H_2O(g)$,
c. $H_2O(g)$ at $100°C \longrightarrow H_2O(g)$ at $200°C$,
d. heating of a metal pipe,
e. dissolving of salt in water.

Study Question 6-3 Some of the processes in Study Question 6-2 involve a spontaneous decrease in entropy if carried out under the proper conditions. Does this mean that the second law has been violated? Explain.

The enthalpy, or heat, changes that occur when a solid substance is converted into its liquid and then its gaseous form can be plotted in a manner similar to that for entropy (Figure 6-1), as illustrated in Figure 6-2.

Entropy can also be related to the hardness or softness of a solid material. A soft solid will, in general, have weaker bonds between the atoms or molecules, and, as a result, the particles are freer to move than those in a harder substance with stronger bonds. *In general, the softer a material the greater the entropy.* This is supported by the data in Table 6-1.

Table 6-1   Entropy and Hardness.

Element	Symbol	Entropy (cal/°K-mole)	Relative Hardness
Carbon (diamond)	C	0.6	10
Boron	B	1.4	9.5
Silicon	Si	4.5	7
Copper	Cu	8.0	2.7
Calcium	Ca	10.0	1.5
Sodium	Na	12.2	0.5

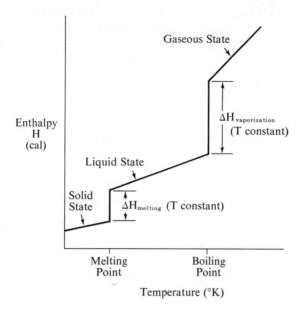

Figure 6-2   Enthalpy Changes for a Typical Substance Being Heated at Constant Pressure.

In Chapter 3 (Section 3-5) you calculated the standard entropy change, $\Delta S^{\circ}_{298^{\circ}K}$, for a number of important reactions. It is possible in many cases to estimate whether a given reaction will have a positive or negative ΔS° without making any calculations. For example, we have seen that a reaction involving a change from solid to liquid to a gas involves an increase in disorder and ΔS° should be positive.

Examples

$$A(s) \longrightarrow B(l) \qquad \Delta S \text{ is } +$$

$$C(l) \longrightarrow D(g) \qquad \Delta S \text{ is } +$$

$$3X(g) \longrightarrow 3Z(s) \qquad \Delta S \text{ is } -$$

If there is a net increase in the total moles of gaseous products over those of the reactants, then ΔS° for the reaction should be positive. If the number of moles of gas decrease, then ΔS° should be negative. Two moles ($2 \times 6 \times 10^{23}$ particles) have a larger number of arrangements than one mole (6×10^{23} particles) of gas. Thus, ΔS should be $-$ for Reaction 1, the conversion of oxygen to ozone.

$$3O_2(g) \rightleftharpoons 2O_3(g)$$

3 moles gas   2 moles gas

ΔS should be $-$.

If the number of moles of gas on each side of the equation is equal, then it is not always possible to predict whether ΔS will be $+$ or $-$, but it will be fairly close to zero (that is, only slightly positive or slightly negative).

$$\text{hydrogen gas} + \text{chlorine gas} \rightleftharpoons \text{hydrogen chloride gas}$$

$$H_2(g) + Cl_2(g) \rightleftharpoons 2HCl(g)$$

$$1 \text{ mole} + 1 \text{ mole} = 2 \text{ moles}$$

$$2 \text{ moles of gas} = 2 \text{ moles of gas}$$

$$\Delta S \text{ is close to 0.}$$

For reactions involving solids or liquids and gases, any net increase in gaseous products should give a positive $\Delta S°$. For example,

$$2A(s) + B(g) \longrightarrow 3C(g) + D(g)$$

$$2 \text{ moles solid} + 1 \text{ mole gas} \longrightarrow 4 \text{ moles gas}$$

$$\Delta S \text{ should be } +.$$

Study Question 6-4   Use the ideas discussed above to predict whether the entropy change for Reactions 2 through 10 (Chapter 3) should be positive $(+)$, negative $(-)$, or close to zero (~ 0). Check your predictions with the values you determined in Study Question 3-6. Explain each answer.

Study Question 6-5

a. Would an endothermic reaction with an increase in the number of moles of gaseous products necessarily be spontaneous? Explain.
b. Under what conditions would it be spontaneous?

Study Question 6-6   Given the hypothetical *exothermic* reaction,

$$6A(g) + 2B(s) \rightleftharpoons 2C(l) + 2D(g),$$

are the following statements true or false?

a. It should be spontaneous at high temperatures.
b. It should be nonspontaneous at a high temperature.
c. $\Delta G°$ should be positive at 25°C.
d. $\Delta S°$ should be positive at 25°C.
e. There is not enough information to make a prediction about spontaneity at high temperatures.
f. There is not enough information to make a prediction about spontaneity at 25°C.

g. It should have a positive enthalpy value.

h. It should involve an increase in order.

i. The equilibrium constant will probably be relatively high at 25°C.

j. Removing D from the equilibrium mixture should shift the equilibrium to the left.

k. Adding A to the equilibrium mixture should increase the amount of C and D present when the system attains equilibrium again.

Study Question 6-7 Which *one* of the following reactions will be nonspontaneous regardless of the temperature? Explain.

a. Reaction 1: $\Delta H = -100$ kcal, $\Delta S > 0$

b. Reaction 2: $\Delta H < 0$, $6A(g) + 2B(s) \rightleftharpoons 5C(g) + 4D(g)$

c. Reaction 3: $\Delta H = 100$ kcal, $\Delta G < 0$

d. Reaction 4: $\Delta H > 0$, $\Delta S > 0$

e. Reaction 5: $\Delta H > 0$, $3X(g) + Y(s) \longrightarrow Z(l) + W(g)$

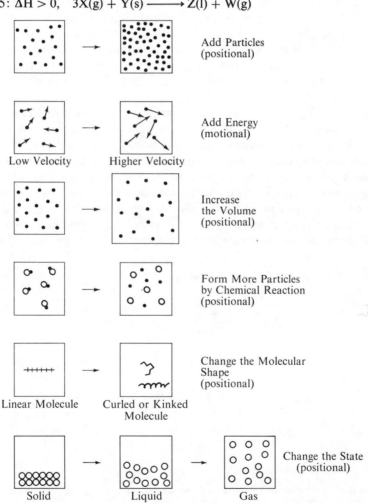

Add Particles
(positional)

Add Energy
(motional)

Low Velocity   Higher Velocity

Increase
the Volume
(positional)

Form More Particles
by Chemical Reaction
(positional)

Change the Molecular
Shape
(positional)

Linear Molecule   Curled or Kinked
Molecule

Change the State
(positional)

Solid   Liquid   Gas

Figure 6-3 Summary of Some of the Ways of Increasing the Entropy. Adapted from Charles Kittel *Thermal Physics,* John Wiley & Sons, Inc., 1969. Used by permission.

6-3 Entropy and Quantization—Statistical Thermodynamics

Actually there are two ways of approaching chemical thermodynamics: the *classical* approach and the *statistical* approach. The classical approach, which is independent of any concepts about the structure of atoms or molecules (quantum mechanics), is based on experimental measurement of the change in macroscopic properties when a system undergoes a change. In other words, classical thermodynamics is based purely on experience—on experiments and observations of bulk matter—and the results are not derived from theory.

Statistical thermodynamics is based on the quantum mechanical theories of atomic and molecular structure, and it represents an attempt to express macroscopic thermodynamic properties through averaging the individual micro states of atoms and molecules. This statistical approach involves the study of trillions times trillions of very small particles, and it is a blend of statistical mechanics and quantum mechanics. When we speak of entropy as a measure of disorder, we refer to a concept from statistical thermodynamics—not classical thermodynamics. In the classical approach, entropy is a quantity that is defined mathematically[3]—it is not given any physical or pictorial meaning.

Prior to 1900 it was believed that all energy states were possible, but this concept was changed with the development of quantum mechanics between 1912 and 1930. The basic principle of quantum mechanics is that energy is quantized. Any atom or molecule can have only certain allowed or *quantized* energy values. For example, the electron in a hydrogen atom (H) can have only certain allowed energy levels, as shown in Figure 6-4. According to the minimum energy principle, the electron would be expected to be found in the lowest possible energy state (its so-called ground state). However, if energy is absorbed from the outside, the electron may make a " quantum jump" from one electronic energy level to another. The energy spacing between levels

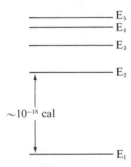

Figure 6-4   Electron Energy Levels in an Atom.

---

[3] Recall that the mathematical definition of entropy change is $\Delta S = \dfrac{q_r}{T}$.

becomes smaller and smaller as we go up the scale. If sufficient energy is put in, the electron is actually separated from the nucleus of the atom and a positively charged ion is formed.

The atom as a whole is continuously moving from one region of space to another. This type of motion is known as *translational motion*. The *translational energy* of an atom is also quantized. However, the difference between translational energy levels is so small that, compared to the electron energy level differences, for all practical purposes, the atom would appear to have a continuous band of translational energies, as shown in Figure 6-5. The relative spacing between electronic energy levels is about 10^{19} times greater than the spacing between translational energy levels.

The total energy of an atom or group of atoms is the sum of their allowed electronic and translational energies and other energies involving interactions of the atoms with their surroundings (for example, attraction to other nearby atoms or molecules and gravity.)

ATOM

$$E_{total} = E_{electronic} + E_{translational} + E_{other}$$

What types of energy levels are found in molecules? Molecules (such as H_2, N_2, O_2, H_2O) also have only certain allowed energy levels for translational motion and for the electrons in the molecule. In a diatomic molecule such as H_2, the two atoms are connected by a chemical bond, which can be crudely visualized as a spring

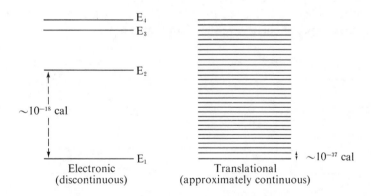

Figure 6-5   Relative Comparison of Electron and Translational Energy Levels in an Atom (drawing not to scale).

(▭). The chemical bond is formed when the electronic energy levels of the two separate atoms merge, or overlap, to form a set of molecular energy levels that have a lower energy than the separate atomic energy levels. Again we see that one of the driving forces for forming a stable molecule is the tendency toward minimum energy.

The two atoms joined by the "spring," or bond, are capable of rotating around a center of gravity and of vibrating toward and away from one another. The frequency of vibration depends on the energy put into the system. Again the energy is quantized— the energies of rotation and vibration cannot be any value but can only assume certain allowed values.

Thus, in molecules, there are two additional types of motion available—*rotational motion* and *vibrational motion*—and the molecule can have *translational, rotational,* and *vibrational* motion, as shown in Figure 6-6.

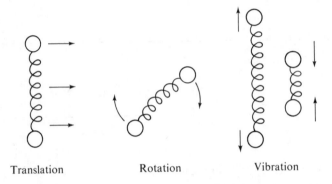

Translation   Rotation   Vibration

Figure 6-6   Types of Motion for a Diatomic Molecule.

The total energy for a molecule is the sum of its allowed vibrational energies plus other outside energies. Each type of energy is quantized, and the relative differences between these types of energy are shown in Figure 6-7.

MOLECULE

$$E_{total} = E_{electronic} + E_{translational} + E_{rotational} + E_{vibrational} + E_{other}$$

Again we see that the differences between translational energy levels in a molecule are so small that, for all practical purposes, translational energy can be considered to be essentially continuous (that is, we can ignore the quantization of translational

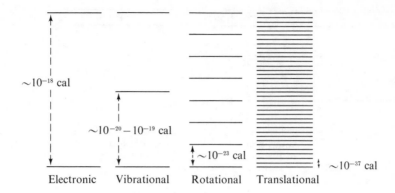

$\sim 10^{-18}$ cal

$\sim 10^{-20} - 10^{-19}$ cal

$\sim 10^{-23}$ cal

$\sim 10^{-37}$ cal

| Electronic | Vibrational | Rotational | Translational |

Figure 6-7 Relative Comparison of Types of Allowed Energy Levels in a Molecule (drawing not to scale).

energy). A particular molecule will have its total energy distributed among these various types of energy. At room temperature or lower the energy will be essentially all translational energy, since there will not be enough energy available to transfer some of the translational energy to the more widely spaced rotational, vibrational, and electronic energy levels.

At low temperatures: $E_{total} \cong E_{translational}$

As the temperature is raised, more of the translational energy will be transferred to rotational energy, and eventually, at high temperatures, to vibrational and electronic energy levels.

Quantization of energy is the basis of the absorption and emission spectra that are used as "fingerprints" to identify different atoms and molecules. Each chemical substance will have a unique set of possible "quantum jumps" between its energy levels. An *absorption spectrum* of allowed "jumps" is obtained by exposing the atoms or molecules to an energy source and measuring the specific set of frequencies that are absorbed. Depending on the relative amount of energy introduced (visible, ultraviolet (UV), infrared (IR), X ray, etc.), the spectrum of absorbed frequencies will be due either to rotational, vibrational, or electronic quantum jumps (or some combination of these). After energy is imparted to a system, the system will return to its original and lower energy state, once the energy input is removed. Again, only certain specific

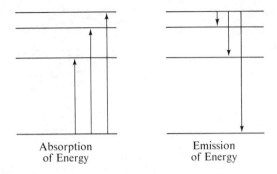

Absorption
of Energy

Emission
of Energy

Figure 6-8   Comparison of Absorption and Emission Spectra.

frequencies of energy will be emitted, and this set of frequencies or energy emissions is known as the *emission spectrum*, as shown in Figure 6-8.

We have seen how energy can be distributed over various types of energy for a *single molecule*. What happens when a *group of molecules* of the same substance are brought together? In a group, each molecule has the same set of molecular energy levels available (as shown in Figure 6-7), but the different molecules do not necessarily have the same energies. Each individual molecule undergoes continual change in the distribution of energy over these levels as a result of its collisions with other molecules. Although the energy of an individual molecule changes, the total energy of all of the molecules remains constant at a constant temperature. This is merely an expression of the First Law of Thermodynamics. The total energy of a group of molecules is the result of a number of different energy distributions for the different molecules, and as the number of molecules in the group is increased, the number of possible arrangements increases. For one mole of any substance, containing 6×10^{23} molecules, the number of possible energy distributions is astronomical.

At room temperature or lower the average energy is due almost solely to the energy distribution in the translational energy levels of the molecules. Since temperature in °K is a measure of the average energy of motion of a group of molecules, the average kinetic energy is dependent only on the Kelvin temperature. The typical energy distribution of a large number of molecules at two different temperatures is given by the Maxwell-Boltzmann distribution curve shown in Figure 6-9.

At a given temperature most of the molecules have energies near the maximum of the curve, and the Kelvin temperature is a measure of the average energy, $E_{average}$.[4] As the temperature is raised, the average energy of the molecules becomes higher.

As the temperature is increased even more, collisions between molecules allow exchange of energy not only between translational levels, but also between rotational energy levels. At even higher temperatures the vibrational (and eventually even the electronic) energy levels contribute to the overall energy distribution. In summary,

---

[4] Actually the energy at the peak of the curve is the most probable energy—which is not quite the same as the average energy—but for our purposes we shall ignore this slight difference.

Molecular Energy Distribution as a Function of
Temperature

Low T: $E_{total} \cong E_{translational} \propto$ temperature in $°K$

Moderate T: $E_{total} \cong E_{translational} + E_{rotational}$

High T: $E_{total} \cong E_{translational} + E_{rotational} + E_{vibrational}$

Very high T: $E_{total} \cong E_{translational} + E_{rotational} + E_{vibrational} + E_{electronic}$

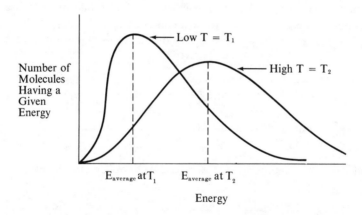

Figure 6-9   Maxwell-Boltzmann Distribution of Molecular Translational Energies.

6-4 Entropy, Probability, and Microstates

What does all of this have to do with entropy? This is the point at which we combine
quantum mechanics and statistical mechanics to account for macroscopic properties.
Quantum mechanics asks, what are the allowed energy levels? Statistical mechanics
asks, how are the particles distributed among the energy levels? Statistical thermodynamics asks, what are the differences in the average energy level distributions between the initial and final states when a system containing trillions times trillions of
atoms or molecules undergoes a physical or chemical change?

Take a jar of white powder and a jar of black powder and mix the two together. The result we would say without hesitation would be a gray powder.[5] Regardless of how much or how long we continue to shake, it is essentially inconceivable that the gray powder will return to separated black and white layers. Why? The statistical answer is that there are billions and billions of ways for the grains of powder to be mixed but far fewer ways for finding them in two layers. The mixed state is more random, or dispersed, and therefore more probable. There is a vanishingly small probability that separation into two layers could occur in billions and billions of years of shaking. There is also a vanishingly small chance that three monkeys typing on three typewriters could eventually produce even one of the works of Shakespeare or that if you hold your hand above a table all of the molecules in the table might leap up spontaneously to meet your hand. In these cases the probability is finite but so extremely improbable that we say it is essentially impossible.

We say that the mixed, or disordered, state of the two powders is more *probable*. This *statistical probability* is a measure of the number of different ways a given situation can occur. The more ways a given state can be realized, the greater its statistical probability, randomness, and entropy.

The same idea can be applied to the number of possible ways that energy can be distributed among the available molecular energy levels in a group of molecules. Consider a simplified system with four possible energy levels, E_0, E_1, E_2, and E_3. Into this system we put 3 hypothetical molecules, x, y, and z, and provide three energy units, or quanta of energy. How can these quanta of energy be shared by these three molecules? A summary of the possible energy distributions is given in Figure 6-10.

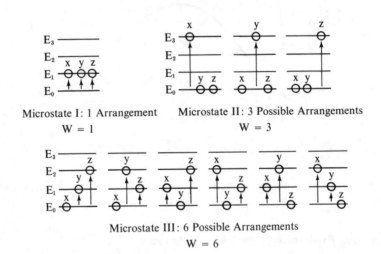

Microstate I: 1 Arrangement
W = 1

Microstate II: 3 Possible Arrangements
W = 3

Microstate III: 6 Possible Arrangements
W = 6

Figure 6-10   Possible Distributions of Three Quanta of Energy for a System of Three Molecules.

[5] Imagine that you are a tiny being even smaller than a speck of the powder; would you agree that the mixture is gray? At this micro level of observation, you would see only black boulders and white boulders—the concept of gray would have no meaning. It is a statistical concept that is meaningful only when viewed at the macro level by a much larger being. What one sees is relative to his scale of observation.

The simplest arrangement is to give each molecule one quantum of energy and move the three molecules from the E_0 to the E_1 level (Microstate I). There is only one way to have this type of energy distribution. We can also give one molecule all three quanta and leave the remaining two at the E_0 energy level (Microstate II). There are three different ways of achieving this distribution. Finally, we can give one molecule one quantum of energy, a second molecule two quanta, and leave the third one at the E_0 level (Microstate III). Figure 6-10 shows the six possible arrangements for this distribution.

Thus, there are 3 general types of distribution and several ways of achieving each type. Each general type is called a *microstate*, or complexion, and the number of ways each type, or microstate, can be achieved is called the *thermodynamic probability* of the microstate and is assigned the symbol W.

Thermodynamic Probability

$W \equiv$ number of ways to achieve a particular microstate or energy distribution type

---

The number of ways, W, in which the energy can be distributed in a particular system can be figured out as done in Figure 6-10. However, as the number of atoms, energy levels, and quanta are increased, this becomes impractical. Fortunately, the number of ways of distributing n particles among n_i different possibilities can be calculated from a relatively simple standard formula obtained from mathematical statistics.

$$W = \frac{n!}{n_0! \, n_1! \, n_2! \dots n_i!},$$

where n is the total number of particles, n_0 is the number of particles in the zero energy level, n_1 is the number in the first energy level, and so on to the i th level. The exclamation points denote factorial numbers; for example, $0! = 1$; $1! = 1$; $2! = 2 \times 1 = 2$; $3! = 3 \times 2 \times 1 = 6$; $4! = 4 \times 3 \times 2 \times 1 = 24$; and so on.

We can use this formula to calculate the thermodynamic probability for each microstate in Figure 6-10. The results are the same as obtained in the figure.

$$\text{Microstate I: } W_I = \frac{3!}{0! \, 3! \, 0! \, 0!} = \frac{3 \times 2 \times 1}{(1)(3 \times 2 \times 1)(1)(1)} = 1$$

$$\text{Microstate II: } W_{II} = \frac{3!}{2!\,0!\,0!\,1!} = \frac{3 \times 2 \times 1}{(2 \times 1)(1)(1)(1)} = \frac{6}{2} = 3$$

$$\text{Microstate III: } W_{III} = \frac{3!}{1!\,1!\,1!\,0!} = \frac{3 \times 2 \times 1}{(1)(1)(1)(1)} = \frac{6}{1} = 6$$

There are a total of 10 different arrangements. Assuming each arrangement to be equally probable, we would expect the molecules to be continually shifting back and forth from one arrangement to another. At any given moment, however, we would expect to find more molecules in Microstate III, because there are 6 possible ways of achieving this type as compared with three ways for Microstate II and only one for Microstate I. Thus, Microstate III is the most random or thermodynamically probable state—it has the highest W.

Study Question 6-8

a. Assume the three quanta of energy are distributed among *four* molecules. How many microstates are there and what is the value of W for each state? Which microstate is the most probable?
b. Assume four quanta of energy are distributed among four molecules with four available energy levels. How many microstates are there and what is the value of W for each state? Which microstate is the most probable?

By looking at the example problem and your answers to Study Question 6-8, you can see that any distribution in which particles are in separate energy levels in ladder-like fashion is the most probable, while the distribution of all particles in the same energy level is the most improbable. In other words, the most probable state is the most random, or disordered, state. This leads us to suspect that the thermodynamic probability, W, must be related in some way to entropy, as we shall see shortly. As the number of particles is increased, the probability of finding the particles in the most "ladderlike" or random distribution becomes much greater.

Consider the number of different five-card poker hands that can be dealt using 52 cards. The total number of different possible poker hands is 2,598,960.[6] Unfortunately the highly ordered arrangement, or microstate, known as a royal straight flush (consisting of an ace, king, queen, jack, and ten in the same suit) has a probability $W = 4$—that is, the probability of getting a royal straight flush is 4 out of 2,598,960. The probability of being dealt four of a kind is 624 out of 2,598,540, and the probability of being dealt a hand of no value in poker (5 random cards) is 1,302,540 out of 2,598,960. (Naturally, in these calculations we are assuming that the dealer does not cheat.) The skillful poker player is one who has some knowledge of the relative probabilities

---

[6] Obtained by $W = \dfrac{52 \times 51 \times 50 \cdots}{5 \times 4 \times 3 \times 2 \times 1} = 2{,}598{,}960$ different hands.

(W values) of getting or drawing certain types of hands along with nerves of steel and a knowledge of psychology.[7]

If we move into chemical reactions where the number of particles may be 10^{23} or higher, the number of possible microstates for distributing energy among these particles is astronomical. For example, in a quart bottle of air the number of possible microstates, or energy distributions, is estimated to be approximately $10^{10^{20}}$, or 10^{100} million million million. It is sobering to remember that there are only about 10^{80} atoms estimated to be in the entire universe.

We indicated earlier that thermodynamic probability, W, and entropy, S, should be related to one another. In order to formulate this relationship we must remember that probability is a multiplicative function, while entropy is an additive function. The total probability is obtained by multiplying the individual probabilities,

$$W_{total} = W_1 \times W_2 \times \cdots W_n,$$

while the total entropy is the sum of the individual entropies,

$$S_{total} = S_1 + S_2 + \cdots + S_n.$$

We saw in Section 4-3 that the logarithm of a multiplicative function (for example, log W) is the sum of the individual logarithms; log W is, therefore, an additive function

$$\log W_{total} = \log W_1 + \log W_2 + \cdots + \log W_n.$$

Thus, we can propose that *entropy*, S, should be proportional to the logarithm of thermodynamic probability, log W. This simple but most important relationship was proposed in the late 1800's by the famous physicist, Ludwig Boltzmann,[8] and it is one of the fundamental equations of statistical thermodynamics.

Boltzmann Equation—Entropy and Thermodynamic Probability

$$S = 2.3 \, k \log W$$

$$S \text{ (in cal/°K-mole)} = 4.6 \log W$$

---

[7] The moral of this story is that if you don't have any of these requirements, don't play poker.

[8] Because of increasing nearsightedness, Boltzmann switched from experimental to theoretical work. His numerous theoretical contributions, including the above equation, are considered to be some of the most important in science. During his lifetime his theoretical ideas were strongly criticized and attacked by his fellow scientists. (Scientists are not nearly so objective as they would have you think.) In 1898, Boltzmann wrote, "I am conscious of being only an individual struggling against the stream of time." He became severely depressed and committed suicide eight years later. A few years after his death his work became widely accepted and acclaimed.

2.3 and k are proportionality constants and the constant, k, is known as the Boltzmann constant. It has an approximate value of 2 cal/°K per mole, and (2) × (2.3) = 4.6. Using this equation we can calculate the entropy for each of the three microstates for distributing three quanta of energy among three molecules (as discussed earlier in this section). These data are summarized in Table 6-2.

Table 6-2   Entropy and Thermodynamic Probability for Various Microstates.

Microstate	Thermodynamic Probability W	Entropy, S S = 4.6 log W in cal/°K-mole
I	1	0
II	3	2.2
III	6	3.6

Study Question 6-9   For a macroscopic system of molecules the thermodynamic probability for certain possible microstates, or energy distributions, is given below. Calculate the entropy for each state. (Remember that log 10^{power} = power.)

Microstate	Thermodynamic Probability W	Entropy, S in cal/°K
I	10^{10}	
II	10^{23}	
III	10^{100}	

In statistical thermodynamics, however, we are not concerned with S but with ΔS, the difference in entropy between the final and initial states. The Boltzmann equation for ΔS can be obtained as follows:

$$S_{final} = 2.3 \text{ k log } W_{final} \quad \text{and} \quad S_{initial} = 2.3 \text{ k log } W_{initial}$$

$$\Delta S = 2.3 \text{ k log } W_{final} - 2.3 \text{ k log } W_{initial}$$

$$\Delta S = 2.3 \text{ k (log } W_{final} - \log W_{initial}).$$

Since $\log A - \log B = \log A/B$, then $\log W_{final} - \log W_{initial}$ can be written as

$$\log \frac{W_{final}}{W_{initial}}.$$

The Botzmann equation for a change in entropy becomes

Boltzmann Equation for ΔS

$$\Delta S = 2.3\ k \log \frac{W_{final}}{W_{initial}}$$

or

$$\Delta S(cal/°K) = 4.6 \log \frac{W_{final}}{W_{initial}}$$

If the final state has a greater probability (W) than the initial state, then $\log W_f/W_i$ will be positive, ΔS will be positive, disorder or randomness increases, and the entropy change is favorable. If the final state has the lower value of W, then $\log W_f/W_i$ will be negative, ΔS will be negative, and the entropy change is unfavorable.

Study Question 6-10   For a chemical reaction $A \rightleftharpoons B$, W_B is 10^{10} and W_A is 10^{16}. Calculate ΔS for the change. Is the entropy factor favorable?

6-5 Enthalpy, Entropy, and Probability

We now have a statistical basis for understanding entropy as a driving force for a spontaneous change. With respect to entropy, *the basic postulate of statistical thermodynamics is that a system will spontaneously try to attain the state with the largest thermodynamic probability, W—that is, the state with the largest number of possible energy distributions among the available energy levels.*

But, this is only one part of the overall driving force for a reaction. Even though this chapter has been concerned primarily with entropy, we must not forget the enthalpy, or ΔH, factor, which is a measure of the tendency toward minimum energy. Indeed, as we saw earlier, at room temperature or lower this is normally the predominating factor.

The counteracting effects of the tendency towards minimum energy and maximum disorder, or thermodynamic probability, can be demonstrated with a simple example. Suppose 1000 small white balls and 1000 small red balls are put in a box and shaken vigorously. The tendency toward maximum disorder, or randomness, insures that the red and white balls will become mixed. There is an extremely large number of ways in which they can be mixed, compared to the number of ways they can form separate red and white layers. Thus, each time we shake the box we would be most surprised if the balls spontaneously formed two separate layers. Suppose, however, that the white balls have small pieces of iron inside and a large magnet is placed below the box. After shaking we would expect to find more white balls than red balls in the bottom layer as a result of the "minimum energy" or enthalpy factor added by the magnet. The system then attains an equilibrium between the tendency toward maximum randomness, or disorder, and the opposing tendency toward a lower energy. In terms of molecules the effect of the enthalpy factor is to give unequal weight to the lower, or ground state, energy level distributions of either the reactants (ΔH is $+$ and unfavorable) or the products (ΔH is $-$ and favorable).

Taking into account the tendency toward minimum energy, or enthalpy, along with the above relationship for maximum entropy or thermodynamic probability, we see that *statistical thermodynamics* yields the following equation for the Gibbs free energy change, ΔG_T. This can be compared with the expression obtained for ΔG from classical thermodynamics.

Equations for ΔG_T

From statistical thermodynamics:[9]

ΔG_T = change in ground state energy $-$ change in thermodynamic probability

$$\Delta G_T = \Delta H_{0°K} - 4.6\, T \log \frac{W_{final}}{W_{initial}}$$

From classical thermodynamics:

$$\Delta G_T = \Delta H_{298°K} - T\Delta S_{298°K}$$

Likewise, the thermodynamic probability can be related to the equilibrium constant, K_{eq}. Suppose an equilibrium exists between A and B as follows:

$$A \rightleftharpoons B.$$

[9] Actually, the statistical thermodynamic equation contains $\Delta H_{0°K}$, the change in energy at the absolute zero of temperature ($0°K$). In this qualitative discussion we can neglect this difference.

Chapter 6

The equilibrium constant is

$$K_{eq} = \frac{[B]}{[A]},$$

which can be expressed in terms of probability as

$$K_{eq} = \frac{\text{probability of finding molecules in energy levels of B}}{\text{probability of finding molecules in energy levels of A}}.$$

We have seen that classical thermodynamics and statistical thermodynamics yield essentially the same results. A very important difference, however, must be pointed out. Classical thermodynamics results exclusively from observation of the macroworld. It is a summary of what we see going on around us, and it does not depend on any theoretical models, or pictures of atomic and molecular properties. Statistical thermodynamics, on the other hand, is based on theory. It starts with the atomic and kinetic molecular theories of the microworld and uses statistical methods to average microstates to predict and explain macroscopic properties. If the atomic and molecular theories were to be changed, this would alter the formulation of statistical thermodynamics, but it would have *no effect on the predictions of the First and Second Laws of Classical Thermodynamics*—just as an alteration in any theory or explanation of gravity would not change the observed law of gravity. Rather, these laws establish a framework within which these models or theories must fit.

6-6 Entropy, Atomic Weight, and Molecular Complexity

A monatomic gas, such as hydrogen (H) or helium (He), has no way to vibrate or rotate around a center of gravity, so that all of its randomness results from the distribution of energy over its translational energy levels. Quantum mechanics shows that the heavier the particle the more closely spaced are its translational energy levels, as illustrated in Figure 6-11.

Thus, for heavy atoms more distributions among the energy levels are possible and the disorder, randomness, or entropy for a large group of heavy atoms should be greater than for the same number of light atoms at the same temperature. This prediction is borne out by the data in Tables 6-3 and 6-4.

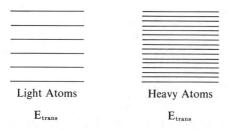

Light Atoms

E_{trans}

Heavy Atoms

E_{trans}

Figure 6-11   Mass and Translational Energy Levels.

Table 6-3 Entropy and Atomic Weight for Monatomic Gases at 1 atm. and 25°C (298°K).

Species	Symbol	Atomic Weight*	Entropy (S°) in cal/°K-mole
Hydrogen	H	1	27
Helium	He	4	30
Argon	A	40	35
Krypton	Kr	84	37
Xenon	Xe	131	41

* Atomic weight is the mass of one mole or 6×10^{23} atoms.

Table 6-4 Entropy and Molecular Weight at 1 atm. and 25°C (298°K).

Species	Formula	Molecular Weight*	Entropy (S°) in cal/°K-mole
		DIATOMIC MOLECULES	
Hydrogen	H_2	2	31
Nitrogen	N_2	28	46
Iodine	I_2	254	63
		TRIATOMIC MOLECULES (X_2Y or Y_2X)	
Water	H_2O	18	45
Hydrogen sulfide	H_2S	34	49
Carbon dioxide	CO_2	44	51

* Molecular weight is the mass of one mole on 6×10^{23} molecules.

We can formulate an approximate rule for predicting the entropy of atoms or molecules with similar shapes.

Entropy and Molecular Weight Rule

The entropy of atoms or molecules with similar shapes normally increases with increasing atomic or molecular weight.

Suppose we compare the entropies of molecules with the same approximate mass, or molecular weight, but with different degrees of molecular complexity. For example, the molecules of water (H_2O), ammonia (NH_3), and methane (CH_4) all have approximately the same molecular weight, but they successively contain a larger number of atoms and are thus more complex. The net result of increased complexity is that with more atoms in a molecule there are more ways the molecule can vibrate and rotate. A simple diatomic molecule like H_2 can vibrate in only one way, but a more complex triatomic molecule like carbon dioxide, CO_2, can vibrate in several ways.

Since more vibrational energy level distributions are possible in more complex molecules, their entropy, or randomness, should be higher provided their masses are approximately equal. This effect would not necessarily become important except at high temperatures where the vibrational energy is a contributing factor. This prediction is borne out by the data in Table 6-5.

Table 6-5   Entropy and Molecular Complexity at 1000°K.

Molecule	Formula	No. of Atoms per Molecule	Molecular Weight	Entropy at 1000°K in cal/°K-mole
Water	H_2O(g)	3	18	56
Ammonia	NH_3(g)	4	17	58
Methane	CH_4(g)	5	16	59

We can formulate another approximate rule for predicting the entropy of molecules with similar molecular weights.

Entropy and Molecular Complexity Rule

At high temperatures the entropy of molecules with similar molecular weights normally increases with increasing molecular complexity.

Study Question 6-11   Which substance in the following pairs would you expect to have the higher entropy?

a. HCl (mol. wt = 36) or HI (mol. wt = 128)
b. PH_3 (mol. wt = 34) or HCl (mol. wt = 36) at 1000°K
c. Ne (mol. wt = 20) or HF (mol. wt = 20) at 1000°K
d. HF (mol. wt = 20) or F_2 (mol. wt = 38)
e. Cl_2 (mol. wt = 71) or CS_2(mol. wt = 76) at 1000°K
f. CH_4 (mol. wt = 16) or CCl_4 (mol. wt = 154)
g. H_2(g) at 300°K or H_2(g) at 1000°K
h. N_2(g) or 2N(g) at the same temperature
i. I_2(s) or I_2(g)

6-7 Summary of Approximate Rules for Predicting Relative Entropy Changes

We have completed a rather detailed look at the very important and useful concept of entropy. It might be useful to summarize the several qualitative rules for predicting relative entropy values and changes.

Rules for Predicting Relative Entropy Values and Changes

Rule 1: A process that leads to a more "mixed-up," disordered, or probable state usually involves an entropy increase.

Rule 2: A process that leads to a larger volume, such as solid to liquid or liquid to gas, results in an entropy increase.

Rule 3: A process that leads to the formation of an increased number of atoms, ions, or molecules usually involves an entropy increase.

Rule 4: In general, the softer a solid, the higher its entropy.

Rule 5: The entropy of substances with similar atomic or molecular shapes normally increases with increasing atomic or molecular weight.

Rule 6: At high temperatures the entropy of molecules with similar molecular weights normally increases with increasing molecular complexity.

Rule 7: As temperature is increased the entropy factor (TΔS) becomes more important. A process in which ΔS is positive will be spontaneous at a sufficiently high temperature.

Although we can use these rules to make a very large number of correct predictions about entropy, we should not get too carried away with the idea that the entropy of most substances and reactions can be predicted qualitatively in this simple manner. Like all rules, these are only approximate ways of classifying experience and there are exceptions. This is what makes science and life interesting.

The law that entropy always increases—the Second Law of Thermodynamics—holds, I think, the supreme position among laws of Nature. If someone points out to you that your pet theory of the universe is in disagreement with Maxwell's equations—then so much the worse for Maxwell's equations. If it is found to be contradicted by observation—well these experimentalists do bungle things sometimes. But if your theory is found to be against the Second Law of Thermodynamics, I can give you no hope; there is nothing to do but to collapse in deepest humiliation.

Arthur S. Eddington
Noted British Astronomer, in the Gifford Lectures of 1927

6-8 Review Questions

6-1. Distinguish between positional entropy, or disorder, and motional entropy, or disorder. Give examples of each.

6-2. Distinguish between the solid, liquid, and gaseous states of a substance in terms of positional entropy.

6-3. Draw a graph showing the entropy changes when a typical solid below its melting point is converted to the liquid state and then to the gaseous state at a temperature above its melting point. Indicate clearly on the diagram the places where the entropy changes are due primarily to changes in positional and motional entropy. Draw and label a similar diagram for the enthalpy changes involved.

6-4. What is the relationship between entropy and the hardness of a substance? Explain why you would expect this relationship.

6-5. Distinguish between the classical and statistical approaches to thermodynamics.

6-6. What is the basic principle of quantum mechanics? What is a "quantum jump"?

6-7. Draw a diagram showing the relative types of energy levels available in an atom or group of atoms.

6-8. Which are farther apart in a given atom: electronic energy levels or translational energy levels?

6-9. Comment on the following statement: Translational energy levels in an atom or molecule are not quantized.

6-10. What additional types of energy levels are available in molecules? Are they quantized? Under what conditions do they contribute significantly to the overall molecular energy?

6-11. At low temperatures, why is it reasonable to say that the total energy is equal only to the average translational energy?

6-12. Distinguish between the absorption and emission spectrum of an atom or molecule.

6-13. Draw Maxwell-Boltzmann distribution curves showing the distribution of energies of a group of molecules at a low temperature and a high temperature.

6-14. What is wrong with the following statement? "The energy of a group of molecules is given by its Kelvin temperature."

6-15. What basic question is asked by quantum mechanics? by statistical mechanics? by statistical thermodynamics?

6-16. What is a microstate, or complexion?

6-17. What is thermodynamic probability and how can it be calculated?

6-18. What is the Boltzmann relationship between entropy (S) and thermodynamic probability (W)? Why isn't $S = kW$?

6-19. What is the relationship between entropy change (ΔS) and thermodynamic probability?

6-20. Compare the equations for Gibbs free energy obtained from the classical and statistical approaches to thermodynamics.

6-21. What is wrong with this statement? "The entropy of atoms and molecules increases as the atomic or molecular weight increases."

Part Two

Chemical Kinetics

Rate and Mechanism of Chemical Reactions

It is not enough that you should understand about applied science in order that your work may increase man's blessings. Concern for man himself and for his fate must always form the chief interest of all technical endeavors, in order that the creations of our mind shall be a blessing not a curse to mankind. Never forget that in the midst of your diagrams and equations.

Albert Einstein

Chapter 7

Is Thermodynamics Enough?
—Chemical Kinetics

7-1 The Questions of Chemical Kinetics

The four fundamental questions of chemical thermodynamics have been considered. We have seen that the free energy change and the equilibrium constant can be used to determine the direction of reaction, the extent to which the reaction can go, and the maximum amount of useful work that can be obtained if it occurs. Thermodynamics provides a description of the equilibrium state of a system, but it does not tell us about the *rate* of approach to equilibrium, nor does it tell us about the path, or *mechanism*, by which the reaction occurs. Several different reactions may have similar values for ΔG_T, but one reaction may occur instantly, another may occur at a slow but measurable rate, and another may show no signs of reaction after centuries. The study of the rate and mechanism of a reaction and the means by which the rate can be altered constitutes the second branch of chemical dynamics, known as *chemical kinetics*. When the answers to kinetic questions are added to those provided by thermodynamics, one has completed the anatomy of a chemical reaction.

Questions of Chemical Kinetics

1. What is the rate of reaction?
2. How can the rate be altered?
3. What is the path, or mechanism, by which the reaction takes place?

Many reactions that are thermodynamically spontaneous proceed so slowly that for all practical purposes no reaction occurs. Consider Reaction 7, the formation of water. In Chapters 3 and 4 (Study Questions 3-5, 3-6, 3-7, and 4-9) you calculated the thermodynamic properties for this reaction, as summarized below.

$$2H_2(g) + O_2(g) \rightleftharpoons 2H_2O(l)$$

	$\Delta H^\circ_{298^\circ K} = -136,000$ cal	(yes)
Reaction	$\Delta S^\circ_{298^\circ K} = -77$ cal/$^\circ$K	(no)
7	$\Delta G^\circ_{300^\circ K} = -112,900$ cal	(yes)
	$K_{eq} = 10^{82}$ ([_____])	

The fact that $\Delta G^\circ_{300^\circ K}$ is very negative and the equilibrium constant, K_{eq}, is very large indicates that the reaction should occur spontaneously at room temperature and pressure and initial concentrations of one molar. Yet, a mixture of these substances, maintained at room temperature and pressure for long periods of time, shows no sign of reaction even after months or years. But if a small match or spark or even a heated platinum wire is introduced, the mixture reacts instantly and explodes violently. How can we explain this odd behavior?

Similarly, Reactions 2 and 3, representing the combustion of coal and n-octane in gasoline, should occur spontaneously from a thermodynamic standpoint. Again, however, when coal and gasoline are exposed to the atmosphere (O_2) at room temperature, they do not burn unless they are ignited by an outside energy source. Why?

The balanced equation for the burning of n-octane, $C_8H_{18}(l)$, also points up the difference between the *net equation* and the *reaction mechanism*. Do 25 molecules of oxygen (O_2) and 2 molecules of octane come together in a 27-way collision to form CO_2 and water? The probability of these 27 molecules coming together in one place at one time in this manner is so inconceivably small that the event would probably not be expected to have occurred in the lifetime of our universe. Even a reaction involving a simultaneous collision between three molecules is highly improbable. Most reactions occur in steps, with each step involving one or two and, very rarely, three molecules at a time. This sequence of steps constitutes the *reaction mechanism*. When all steps are added, with many of the products and reactants canceling out, the final "bookkeeping" tally is the *net equation*. Thermodynamics is concerned only with the initial reactants and final products as tallied in the net equation. Chemical kinetics is concerned with how the reaction gets from the initial to the final state.

The questions of rate and mechanism are crucial for understanding reactions in living systems. From the standpoint of energy, the two most important chemical

$$C(\text{graphite}) + O_2(g) \rightleftharpoons CO_2(g)$$

$$\Delta H^\circ_{298^\circ K} = -94{,}000 \text{ cal} \qquad \text{(yes)}$$

Reaction $\qquad \Delta S^\circ_{298^\circ K} = +2 \text{ cal}/^\circ K \qquad \text{(yes)}$

2 $\qquad \Delta G^\circ_{300^\circ K} = -94{,}600 \text{ cal} \qquad \text{(yes)}$

$$K_{eq} = 10^{68} \;(\boxed{})$$

$$2\,C_8H_{18}(l) + 25\,O_2(g) \rightleftharpoons 16\,CO_2(g) + 18\,H_2O(g)$$

$$\Delta H^\circ_{298^\circ K} = -2{,}428{,}000 \text{ cal} \qquad \text{(yes)}$$

Reaction $\qquad \Delta S^\circ_{298^\circ K} = +229 \text{ cal}/^\circ K \qquad \text{(yes)}$

3 $\qquad \Delta G^\circ_{300^\circ K} = -2{,}496{,}700 \text{ cal} \qquad \text{(yes)}$

$$K_{eq} = 10^{1809} \;(\boxed{})$$

processes in biology are those of *photosynthesis* and *respiration*, as represented by Net Equations 9 and 10.

The net equation for photosynthesis involving the reaction of carbon dioxide and water to form glucose and oxygen is not thermodynamically feasible, yet it does occur in plants. Again, while the net equation is all we need to know to calculate the energy change for photosynthesis, the equation provides no information about the mechanism by which it occurs. Although the complete detailed mechanism of photosynthesis is still unknown, it is known that there are over one hundred sequential steps involved in the conversion of CO_2 and H_2O into a molecule of glucose.

Photosynthesis (net equation)

$$6\,CO_2(g) + 6\,H_2O(l) \rightleftharpoons C_6H_{12}O_6(s) + 6\,O_2(g)$$

$$\Delta H^\circ_{298^\circ K} = +667{,}000 \text{ cal} \qquad \text{(no)}$$

Reaction $\qquad \Delta S^\circ_{298^\circ K} = -63 \text{ cal}/^\circ K \qquad \text{(no)}$

9 $\qquad \Delta G^\circ_{300^\circ K} = +675{,}900 \text{ cal} \qquad \text{(no)}$

$$K_{eq} = 10^{-490} \;(\boxed{})$$

Respiration (net equation)

$$C_6H_{12}O_6(s) + 6\ O_2(g) \rightleftharpoons 6\ CO_2(g) + 6\ H_2O(l)$$

$$\Delta H^\circ_{298^\circ K} = -667{,}000\ \text{cal} \qquad \text{(yes)}$$

Reaction $\qquad \Delta S^\circ_{298^\circ K} = +63\ \text{cal/}^\circ K \qquad \text{(yes)}$

10 $\qquad \Delta G^\circ_{300^\circ K} = -675{,}900\ \text{cal} \qquad \text{(yes)}$

$$K_{eq} = 10^{490}\ (\ \square\ )$$

We see that the net process of respiration involving the oxidation of glucose should be thermodynamically feasible. Yet, if glucose is exposed to oxygen at room temperature and pressure, no appreciable reaction is observed to occur. However, in the cells of living animals this essential reaction does occur. Why? Again, the net equation, while useful in terms of energy predictions, provides no information about the mechanism of respiration in cells. Although the net equation for respiration is the reverse of that for photosynthesis, the two processes have different mechanisms. There are more than seventy individual steps in the respiration process.

It is not surprising that chemical kinetics occupies a position of extreme importance and relevance. The industrial production of any chemical (recall the synthetic food plant in Chapter 1) requires that a rate study be made to find the optimum conditions for achieving a reasonable rate of reaction with minimum cost. In addition, an understanding of the mechanisms for the complex biochemical reactions in man and in other living systems is important for understanding the nature of life and controlling disease.

The field of chemical kinetics is a challenging and important frontier for research. Mechanisms are known for only a relatively small number of reactions. As we shall see in the next section, there are several theories of reaction rates, but no one theory is completely satisfactory.

Kinetics, like any portion of science, can be approached experimentally and theoretically. We shall begin by discussing two major theories of reaction rates—the *collision theory* and the *transition state theory*. This will be followed by a discussion of the experimental approach.

7-2 What Is the Rate?—The Collision Theory

What makes some reactions so slow while others occur very rapidly? What are the factors that determine the reaction rate?

Before discussing these factors it is necessary to define reaction rate. The *rate* of any reaction is defined as the number of molecules reacting in a given period of time. This is normally measured in terms of the *variation in concentration of either reactants or products with time.*

Rate of Reaction

$$\text{Rate of reaction} = \frac{\text{change in concentration}}{\text{change in time}} = \frac{\Delta c}{\Delta t} = \frac{c_2 - c_1}{t_2 - t_1}$$

In one sense, the answer to the question of what determines the rate of reaction seems obvious. The molecules presumably must collide before a reaction can take place, and the more frequently they collide, the faster the rate of reaction. Thus, the rate of reaction should be determined by the *frequency of collision*—the total number of collisions in a given volume in a given time.

A Tentative Hypothesis

Reaction rate $\propto$ total number of collisions per unit time per unit volume

But you are probably beginning to agree with the famous philosopher and mathematician, Alfred North Whitehead, who said, "Seek simplicity, and distrust it." The problem is not quite so simple. Just because two molecules collide does not mean that they will necessarily react. The rate of reaction depends not only on the *collision frequency* but also on the *collision efficiency*. The collision must be an effective one, in that the molecules must collide with sufficient *energy* to bring about reaction. The minimum energy required to bring about a reaction is known as the *activation energy*.

Thus, we have modified our original hypothesis to say that the reaction rate depends on two factors—collision frequency and collision energy.

A Modified Tentative Hypothesis

$$\frac{\text{Reaction}}{\text{rate}} \propto \frac{\text{Collision}}{\text{frequency}} \times \frac{\text{Fraction of collisions}}{\text{having sufficient energy}}$$

Why have we multiplied these two factors instead of adding them? The rate of the reaction depends on the probability of occurrence of two separate events—the collision of the molecules and the fact that they must collide with sufficient energy. The total probability of an event is obtained by multiplying the individual probabilities that make up the event.

We should continue, however, to follow Whitehead's dictum to distrust simplicity. Will a reaction necessarily occur even if the molecules collide with sufficient energy? Consider several ways in which diatomic molecules (for example, H_2 and O_2) could collide, as illustrated in Figure 7-1. Even if these different types of collisions took

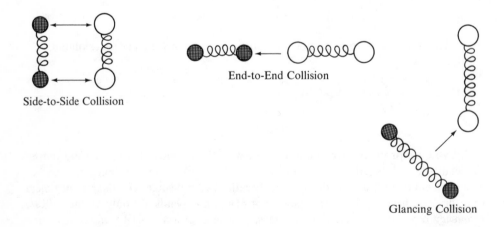

Side-to-Side Collision

End-to-End Collision

Glancing Collision

Figure 7-1   Types of Collisions Between Molecules.

place with the same energy and this energy were sufficient for reaction, this does not mean the reaction would necessarily occur. We see that the molecules must collide with the right *geometry*, or *orientation*. For example, it might be expected (even though this wouldn't necessarily hold true) that the side-to-side collision might offer the most favorable reaction geometry. The colliding species ⬚

could split to form two new molecules ⬚

Thus, we must add another factor affecting collision efficiency—a *collision geometry*, or *orientation*, factor (sometimes called a steric factor).

A Final Hypothesis—The Collision Theory

$$\begin{array}{c}\text{Reaction} \\ \text{rate}\end{array} \propto \begin{array}{c}\text{Collision} \\ \text{frequency}\end{array} \times \begin{array}{c}\text{Fraction of} \\ \text{collisions having} \\ \text{sufficient energy}\end{array} \times \begin{array}{c}\text{Fraction of} \\ \text{collisions with} \\ \text{correct geometry}\end{array}$$

$$\begin{array}{c}\text{Reaction} \\ \text{rate}\end{array} \propto \begin{array}{c}\text{Collision} \\ \text{frequency} \\ \text{factor}\end{array} \times \begin{array}{c}\text{Collision} \\ \text{energy} \\ \text{factor}\end{array} \times \begin{array}{c}\text{Collision} \\ \text{geometry} \\ \text{factor}\end{array}$$

This summarizes a modern version of the theory of reaction rates known as the *collision theory*. If we could somehow determine each of these factors theoretically, it would be relatively easy to calculate the rate of a given reaction. It is possible to use the kinetic molecular theory to get reasonable estimates of the collision frequency factor for simple reacting molecules (which can be assumed to act approximately like hard spheres). However, it is not possible to get estimates of the energy and geometry factors. In other words, the collision theory gives a nice qualitative picture of the factors determining the reaction rate—but that is all. If we want to calculate the rate from first principles, the collision theory provides no means for calculating either the activation energy factor or the geometry factor.

Interestingly enough, chemical thermodynamics has been used to overcome part of this difficulty in the development of the *transition state*, or *activated complex theory of reaction rates*, which will be discussed in the next section. It should be pointed out that a third theory of reaction rates, known as the *theory of absolute reaction rates*, has also been developed. It combines quantum mechanics and statistical mechanics to develop a reaction rate theory. Because of its general similarity to the transition state theory and its mathematical complexity, we shall not deal with it in this book.

7-3 What Is the Rate?—Activation Energy and the Transition State Theory

We have seen that the collision theory predicts that a reaction will occur when molecules collide with at least the minimum energy necessary for reaction (the activation energy) and the right geometry, or orientation. It seems reasonable that when molecules collide under such conditions, they may unite momentarily to produce an unstable species, which can be called the *activated complex*,[1] or *transition state*. This activated complex can then break apart to give the new reaction products, or it can break up to yield the original reactants. It is also assumed that *the activated complex is in equilibrium with the original reactants*.

We can represent the reaction of A_2 molecules with B_2 molecules to give AB molecules as follows:

$$A_2 + B_2 \rightleftharpoons [A_2B_2] \longrightarrow 2\,AB$$

Unstable
Activated
Complex

or

Unstable
Activated
Complex

Assumptions of the Transition State Theory

1. An unstable transition state, called the activated complex, is formed.
2. The activated complex is in equilibrium with the original reactants.

---

[1] The activated complex lasts only a very short time, probably equivalent to one or two vibrations of the molecule (about 10^{-13} seconds). It should be distinguished from a stable intermediate that can form in some reactions.

The overall free energy changes in the reaction are shown in Figure 7-2, and here we can see in one diagram a thermodynamic and kinetic profile of a reaction. The reaction in Figure 7-2, from a thermodynamic standpoint, is *exergonic*, or spontaneous. The free energy of the products is lower than that of the reactants, and ΔG_T for the reaction is negative. However, in order for the molecules to react, they must collide with an energy equal to or greater than the free energy of activation ($\Delta G_{activation}$).

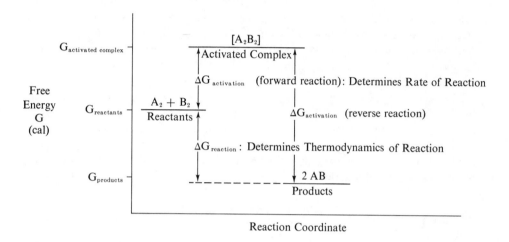

Figure 7-2 Free Energy Diagram for an Exergonic Reaction with a Relatively Slow Rate.

Sometimes this type of diagram is represented as a free energy hill, or barrier. The *free energy of activation* is an energy barrier that must be overcome before reaction can occur, regardless of how favorable the reaction may be thermodynamically. Figure 7-2 shows that there are two free energies of activation—one for the forward reaction ($A_2 + B_2 \rightarrow 2AB$) and one for the reverse reaction ($2AB \rightarrow A_2 + B_2$). We can see that the difference between the forward and reverse activation energies is equal to ΔG_T, the thermodynamic free energy change. This is expected, since the balance between the rates of the forward and reverse reactions determines the rate of approach to the thermodynamic equilibrium state.

$$\Delta G_{reaction} = \Delta G_{act-forward} - \Delta G_{act-reverse}$$

We are concerned with the forward reaction and we shall deal with $\Delta G_{activation}$, or ΔG_{act}, for the forward reaction as the factor determining initial rates of reaction. The higher the free energy of activation, the slower the reaction. The top part of the diagram (height of the activation free energy barrier) determines the *rate of reaction*, or the rate of approach to equilibrium, while the bottom part of the diagram (the free energy of reaction) determines the *thermodynamic spontaneity* of the reaction. Now we can see why H_2 and O_2 gases do not react at room temperature even though the reaction has a large negative free energy of reaction. The free energy of activation is high, and at room temperature the molecules do not collide with the necessary energy for reaction. This explains why a spark will set off an explosive reaction of hydrogen and oxygen gas. The spark causes a number of molecules to attain the necessary free energy of activation. They react and, because the reaction has a large negative $\Delta G_{reaction}$, a great deal of energy is given off. This in turn provides the energy for more molecules to achieve the necessary collision energy with the evolution of even more energy and a sort of chain reaction ensues. Another equivalent way of showing the same information found in Figure 7-2 is given in Figure 7-3.

The free energy of activation, ΔG_{act}, determines *the rate of approach to equilibrium*; the free energy of reaction, ΔG_T°, determines how far the reaction is from equilibrium; and ΔG_T°, the standard free energy change, determines *the position of equilibrium*. The importance of these types of diagrams cannot be overemphasized. They provide in a single diagram an answer to almost all of the major questions of both chemical thermodynamics and kinetics.

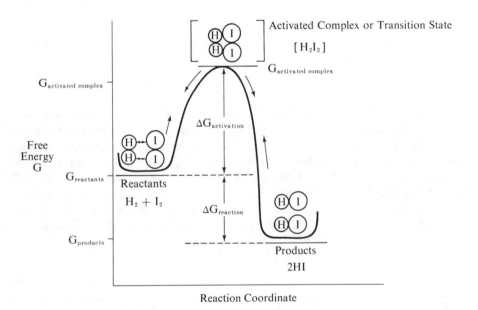

Figure 7-3   Free Energy Diagram for Exergonic Reaction of H_2 and I_2 to Form 2HI.

Transition State Theory

$$A_2 + B_2 \rightleftharpoons [A_2B_2] \longrightarrow 2AB$$

activated
complex

$$\Delta G_{\text{activation}} = G_{\text{activated complex}} - G_{\text{reactants}}$$

Determines rate of reaction or kinetic feasibility

$$\Delta G_{\text{reaction}} = G_{\text{products}} - G_{\text{reactants}}$$

Determines thermodynamic feasibility

A diagram similar to that in Figure 7-2 for an endergonic reaction with a positive $\Delta G_{\text{reaction}}$ is shown in Figure 7-4.

Study Question 7-1   Which *one* of the following is a correct statement about the free energy of activation for a given reaction?

a. The lower it is, the slower the reaction rate.
b. The lower it is, the faster the system reaches the equilibrium position.

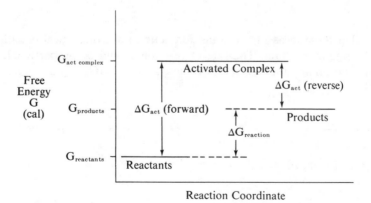

Figure 7-4   Free Energy Diagram for an Endergonic Reaction with a Relatively Slow Rate.

c. The lower it is, the more favorable the thermodynamic spontaneity.
d. The lower it is, the higher the value of the equilibrium constant, K_{eq}.
e. The higher it is, the more the equilibrium position lies to the right.

Study Question 7-2

a. Draw and label a free energy diagram for an *exergonic* reaction ($\Delta G_{reaction}$ is $-$) with a *fast* reaction rate at room temperature.
b. Draw and label a free energy diagram for an *endergonic* reaction ($\Delta G_{reaction}$ is $+$) with a *slow* rate at room temperature.
c. Given the information below on four reactions of the type A → B, draw and label a free energy diagram for each reaction.

	Free Energy of Reaction	Free Energy of Activation (forward)
Reaction 1:	− 100,000 cal	+ 1,000 cal
Reaction 2:	− 100,000 cal	+ 100,000 cal
Reaction 3:	+ 50,000 cal	+ 60,000 cal
Reaction 4:	+ 50,000 cal	+ 100,000 cal

d. Which of these four reactions *will* occur at room temperature? at high temperatures? Explain.

Let us develop these ideas of the transition state into a theoretical equation for describing the rate of reaction. The rate of reaction should be proportional to the frequency of collision and the free energy of activation, or $\Delta G_{activation}$.

Transition State Theory of Reaction Rates

$$\text{Rate} \propto \frac{\text{Collision}}{\text{frequency}} \times \frac{\text{Fraction of collisions with sufficient}}{\text{free energy of activation}}$$

Remember, however, that free energy contains both an energy (enthalpy) term and an entropy term. Thus, the rate could be represented as the product of three factors—a *collision frequency factor, enthalpy of activation factor,* and *entropy of activation factor.*

Transition State Theory of Reaction Rates

$$\text{Rate} \propto \begin{array}{c} \text{Collision} \\ \text{frequency} \\ \text{factor} \end{array} \times \begin{array}{c} \text{Enthalpy of} \\ \text{activation} \\ \text{factor} \end{array} \times \begin{array}{c} \text{Entropy of} \\ \text{activation} \\ \text{factor} \end{array}$$

This looks very similar to the collision theory equation. We can see that the geometry, or orientation, factor is really an entropy factor. For the reaction to occur with the proper orientation is equivalent to saying that there must be a decrease in disorder or an increase in order (that is, $\Delta S_{activation}$ is $-$).

There are many ways for the molecules to collide and not react but only one for them to collide with the proper geometry. The difference between the collision theory and the transition state theory is that the latter provides a possibility for the theoretical calculation of the enthalpy and entropy of activation factors. This results from the assumption that the activated complex is in *equilibrium* with the reactants. This means that there is an equilibrium constant for the formation of the activated complex $[A_2B_2]$.

$$A_2 + B_2 \rightleftharpoons [A_2B_2]_{act}$$

$$K_{eq} = \frac{[A_2B_2]}{[A_2][B_2]}$$

The rate will be proportional to the concentration of the activated complex present at equilibrium. A fast rate means that more activated complex will form.

$$\text{rate} \propto [A_2B_2]$$

The K_{eq} equation can be solved for $[A_2B_2]$:

$$[A_2B_2] = [A_2][B_2]\ K_{eq}$$

Substituting this in the rate equation, we obtain

$$\text{rate} \propto [A_2][B_2]\ K_{eq}.$$

The concentrations of A_2 and B_2 will be proportional to the collision frequency at a given temperature. The more A_2 and B_2 that are present in a given volume (at constant T), the more they will collide with one another. Thus

$$\text{rate} \propto \begin{pmatrix} \text{collision} \\ \text{frequency} \\ \text{factor} \end{pmatrix} \times K_{eq}.$$

From thermodynamics (see Section 4-5) we found the relationship between the standard free energy and the equilibrium constant to be

$$\Delta G^o_{activation} = -4.6\ T \log K_{eq}.$$

Rearranging, we obtain

$$\log K_{eq} = \frac{-\Delta G^o_{activation}}{4.6\ T}.$$

By definition of logarithms, if $\log x = \text{number}$, then $x = 10^{number}$.
Thus,

$$K_{eq} = 10^{-\Delta G^o_{act}/4.6\ T}.$$

A transition state equation is derived by substituting for K_{eq} in the rate equation.

Transition State Theory

$$\text{Rate} \propto \begin{matrix} \text{Collision} \\ \text{frequency} \\ \text{factor} \end{matrix} \times 10^{-\Delta G^o_{act}/4.6\ T}$$

$$\text{Rate} \propto \begin{matrix} \text{Collision} \\ \text{frequency} \\ \text{factor} \end{matrix} \times \begin{matrix} \text{Free energy of} \\ \text{activation} \\ \text{factor} \end{matrix}$$

Since $\Delta G^o_{act} = \Delta H^o_{act} - T\Delta S_{act}$, this can be substituted for ΔG^o_{act} in the above equation.

Transition State Theory

$$\text{Rate} \propto \begin{matrix}\text{Collision}\\\text{frequency}\\\text{factor}\end{matrix} \times 10^{-\Delta H^\circ_{act} - T\Delta S^\circ_{act}/4.6\,T}$$

$$\text{Rate} \propto \begin{matrix}\text{Collision}\\\text{frequency}\\\text{factor}\end{matrix} \times 10^{-\Delta H^\circ_{act}/4.6\,T} + 10^{\Delta S^\circ_{act}/4.6\,T}$$

or

$$\text{Rate} \propto \begin{matrix}\text{Collision}\\\text{frequency}\\\text{factor}\end{matrix} \times \begin{matrix}\text{Enthalpy of}\\\text{activation}\\\text{factor}\end{matrix} \times \begin{matrix}\text{Entropy of}\\\text{activation}\\\text{factor}\end{matrix}$$

If the enthalpy and entropy of activation could be determined, they could then be used in the transition state equation to calculate the reaction rate.[2] The transition state equation indicates that the reaction rate is slow if the free energy of activation is positive—energy must be put into the system to make the collisions efficient. The free energy of activation is a combination of two factors, and they may work together or in opposition to determine the rate. If the enthalpy of activation, ΔH_{act}, is positive and high, the rate should be slow at room temperatures or lower. Only by increasing the temperature can enough of the molecules have sufficient energy for reaction. The entropy of activation factor, however, has a sign opposite that of ΔH_{act}. This indicates that the rate is faster if ΔS_{act} is large. If the activated complex is more disordered than the reactants, the equilibrium constant for the reaction in which it is formed is large and the formation of the activated complex is favored. However, in most bimolecular reactions involving gases, $A_2(g) + B_2(g) \rightleftharpoons [A_2B_2]_{act} \rightarrow 2AB$, the activated complex $[A_2B_2]_{act}$ is more ordered than the reactants $A_2(g)$ and $B_2(g)$, and we would expect ΔS_{act} to be negative and unfavorable. Frequently, the transition state equation is used to calculate the entropy of activation, ΔS_{act}, after the reaction rate has been determined experimentally. The sign of ΔS_{act} provides information about the structure of the activated complex and helps in the postulation of a reasonable reaction mechanism.

What are the differences between the collision theory and the transition state theory? Both are based on the concept of an energy of activation needed before the

---

[2] In actual practice these factors are often difficult to measure or calculate but they are available for some reactions. Another approach, the theory of absolute reaction rates, makes use of quantum mechanics and statistical mechanics to calculate the enthalpy and entropy of activation factors.

reaction rate will be significant. The collision theory focuses on the collision frequency and collision efficiency (energy and geometry), while the transition state focuses on the formation of an activated complex that is in equilibrium with the initial reactants. This assumption of thermodynamic equilibrium allows the introduction of the free energy (ΔG_{act}), enthalpy (ΔH_{act}), and entropy (ΔS_{act}) of activation. These two theoretical approaches may be compared as follows:

The Collision Theory

$$\begin{array}{c} \text{Reaction} \\ \text{rate} \end{array} \propto \begin{array}{c} \text{Collision} \\ \text{frequency} \\ \text{factor} \end{array} \times \begin{array}{c} \text{Collision} \\ \text{energy} \\ \text{factor} \end{array} \times \begin{array}{c} \text{Collision} \\ \text{geometry} \\ \text{factor} \end{array}$$

The Transition State Theory

$$\begin{array}{c} \text{Reaction} \\ \text{rate} \end{array} \propto \begin{array}{c} \text{Collision} \\ \text{frequency} \\ \text{factor} \end{array} \times \begin{array}{c} \text{Enthalpy of} \\ \text{activation} \\ \text{factor} \end{array} \times \begin{array}{c} \text{Entropy of} \\ \text{activation} \\ \text{factor} \end{array}$$

Since the enthalpy, or energy, factor and the geometry, or entropy, factor are constant for a given reaction under fixed conditions, they may be combined to yield a constant, k, called the specific reaction rate constant.

Specific Reaction Rate Constant

$$k = \begin{array}{c} \text{Collision} \\ \text{energy} \\ \text{factor} \end{array} \times \begin{array}{c} \text{Collision} \\ \text{geometry} \\ \text{factor} \end{array} = \begin{array}{c} \text{Enthalpy of} \\ \text{activation} \\ \text{factor} \end{array} \times \begin{array}{c} \text{Entropy of} \\ \text{activation} \\ \text{factor} \end{array}$$

Thus, the reaction rate expression for either theory can be simplified to

---

Reaction Rate Expression

$$\text{Rate} = \frac{\Delta c}{\Delta t} = k \times \text{collision frequency factor}$$

or

$$\text{Rate} = \frac{\Delta c}{\Delta t} = k \times \text{concentration of reactants factor}$$

Since the frequency of collision will be directly dependent on the concentration of reacting species, this factor is essentially a concentration factor. Since the rate is dependent on the probability of collision, the concentration factor will be related to the product of the concentrations—not their sum.

7-4 What Is the Rate?—The Experimental Approach

In order to answer the question "What is the rate of a reaction?" we can take a theoretical approach, using one of the theories of reaction rates, such as the transition state theory, or we can go into the laboratory and measure the rate experimentally. Before we get carried away with our ability to calculate reaction rates while sitting comfortably in a chair, we should realize that these calculations are usually very complex and the data needed for calculating rates of most reactions are not available. In other words, reaction rate theory is still in its infant stage, and the rates of only a few simple reactions have been calculated successfully from theory. This is why theoretical kinetics remains as an important frontier in science.

Thus, *most reaction rates are obtained experimentally—not theoretically*. Recall that reaction rate is the change in concentration per unit time in a fixed volume, rate = $\Delta c / \Delta t$. To measure the rate we need to follow the change in concentration of one or more of the reactants or products as a function of time. If we follow one of the reactants, its concentration will decrease with time. Similarly monitoring one of the products should show an increase in concentration as the reaction proceeds. In actual practice, it is good technique to follow at least two species independently to provide a crosscheck on the results.

The concentration changes can be followed by measuring some chemical or physical property that changes as the reaction proceeds—and an array of approaches is possible.

Examples include measuring the pressure change in a gaseous reaction (where the number of molecules changes as the reaction proceeds), measuring refractive index (bending of light), color intensity, conductivity (where ions are used up or formed), the absorption spectrum, and so on. Any technique may be used that distinguishes one species from another and allows the amount present to be determined with reasonable accuracy.

For example, the rate of Reaction 4, involving the oxides of nitrogen, can be followed by measuring the change in pressure as the reaction proceeds.

$$2NO(g) + O_2(g) \rightleftharpoons 2NO_2(g)$$

$$3 \text{ moles gas} \longrightarrow 2 \text{ moles gas}$$

$$\text{colorless} \longrightarrow \text{brownish-yellow}$$

Since 3 moles of gas are converted into 2 moles of gas, the pressure should drop as the reaction proceeds. Since nitrogen dioxide, $NO_2(g)$, is brownish-yellow and $NO(g)$ is colorless, the rate could also be determined by following the change in color intensity or by measuring the ultraviolet or visible absorption spectrum.

Many of the more recent experimental approaches have been devoted to measuring the rates of very fast reactions occurring anywhere from 1 second to 10^{-12} seconds. Reaction rates as low as $1/100$ (10^{-2}) seconds can be measured by the *flow method* in which a specially designed injection system mixes the reactants in times as short as $1/1000$ (10^{-3}) seconds. The concentration of one or more reacting species is measured by a rapid method (such as spectroscopy) as the chemicals flow down a tube. Even more rapid reactions can be measured by the *relaxation method*. The reaction is first allowed to attain equilibrium. The equilibrium is then very suddenly disturbed. A high-speed electronic device can then measure the rate at which the system returns to equilibrium—known as the relaxation time. Many other intriguing experimental approaches have been developed to measure these extremely rapid reactions.

7-5 How Can the Reaction Rate Be Altered?

Many reactions occur either too slowly or too rapidly, and much of man's efforts are spent in trying to find effective (and economical) ways of changing the rates of various reactions. We have seen that the rate of reaction depends on three factors—the frequency of collision, the energy of collision, and the geometry of collision. Any factor that changes one or more of these factors will change the rate of reaction by changing either the collision frequency, collision efficiency (energy and geometry), or both.

It has been found that six principal factors can affect the rates of chemical reactions: (1) the *nature* of the reacting species and the reaction environment, (2) the *concentrations* of reacting species, (3) the *temperature* of the reacting system, (4) the *degree of subdivision* of reacting particles (for some reactions), (5) *light* (for some reactions), and (6) *catalysts*. Let us examine each of these factors.

(1.) Nature of the Reacting Species and the Reaction Environment

The reaction rate obviously depends upon the particular chemicals undergoing reaction. Most reactions take place in the atmosphere or in an aqueous solution. One way of altering the rate of some reactions is to change the reaction environment. Gases can be bubbled into a solution for reaction, or a reaction that normally takes place in aqueous solution can be changed to a nonaqueous solvent or system. When there is an interaction or attraction between reacting molecules and solvent molecules, this can drastically alter the rate and in some cases it can change the mechanism by which the reaction occurs.

(2.) Effect of Concentration of Reactants—Rate Laws

Concentration changes affect the *frequency of collision*. Suppose we have the reaction $A + B \rightleftharpoons AB$ going on in a closed container at a given temperature. Then the reaction step determining the rate involves the collision of A and B molecules. If we increase the concentrations of A or B or both A and B in the container, the number of collisions per unit time will be increased and the rate will increase. The rate of the reaction will be proportional to the product (not the sum) of the concentrations of these reacting species. This is due to the fact that the rate depends on the probability of collisions of both A and B, and an increase in the concentration of one or both reactants increases the probability of collision. The total probability is obtained by multiplying individual probabilities.

Rate Expression, or Law

$$A + B \rightleftharpoons AB$$

$$\text{Rate} \propto [A][B]$$

or

$$\text{Rate} = k[A][B]$$

---

where k is the *specific rate constant* for a particular reaction. As we saw in Section 7-3, the specific rate constant, k, can be thought of as the product of the energy, or enthalpy, factor and the geometry, or entropy, factor.

Similarly, the rate expression, or law, for the bimolecular collision of 2 molecules of the same species (2A $\longrightarrow$ A$_2$) is

$$2A \longrightarrow A_2 \text{ or } A + A \longrightarrow A_2$$

$$\text{rate} = k[A][A] = k[A]^2.$$

For reactions that occur as a simultaneous collision of 3 molecules (a relatively improbable and rare occurrence), we can have the following rate laws:

Trimolecular Collision		Rate Law
(1) A + B + C $\longrightarrow$ products		rate = $k[A][B][C]$
(2) 2A + B $\longrightarrow$ products		rate = $k[A]^2[B]$
(3) 3A $\longrightarrow$ products		rate = $k[A]^3$

The rate expression obtained is known as the *rate law*. It shows us how the rate of a reaction varies with the concentration of reactants. It should be noted that increasing the partial pressure of one or more reactants will also increase the rate by increasing the frequency of collision.

Study Question 7-3   Suppose the concentrations of both A and B are doubled in the reaction A + B $\rightleftharpoons$ C. By what factor will the reaction rate be increased?

(3.)   Effect of Temperature

Temperature has a significant effect on reaction rate, and it is one of the main methods used for altering and controlling the rates of chemical reactions. An increase in temperature normally[3] causes a large increase in reaction rate, while decreasing the temperature normally reduces the rate. Increasing the temperature increases the average kinetic energy of motion, or velocity, of the molecules. Increasing the average velocity does two things: it increases the *collision frequency* and the *energy of collision* so that more molecules will collide having energies equal to or greater than the activation energy. Thus, the effect of temperature on reaction rate is very important, because it increases both *collision frequency* and *collision energy*.

(4.)   Effect of Degree of Subdivision of Reactant Particles

If the reactants are in two different states (for example, a solid reacting with a gas or a solid reacting with a liquid), then we say that the reaction is *heterogeneous*. For

---

[3] There are some exceptions to this general rule of thumb. The rate of some reactions, such as Reaction 4, $2NO(g) + O_2(g) \rightleftharpoons 2NO_2(g)$, and Reaction 1, $3O_2(g) \rightleftharpoons 2O_3(g)$, actually decreases with increasing temperature.

Chapter 7

heterogeneous reactions, the state of subdivision, or amount of surface area available for reaction, has an important effect on reaction rate. We know that a large log burns slowly, but if the same log is converted to sawdust the reaction occurs very rapidly. Breaking the reactant into smaller aggregates creates more surface area, and more molecules are available for reaction. Increasing the surface area, therefore, increases the *frequency of collision*. In living cells the cell membranes provide large surface areas and are important in increasing the rate of reaction of some cellular processes.

(5.)   Effect of Light

The energy of activation for *some* reactions can be obtained by the addition of energy in forms other than heat. For example, a mixture of hydrogen gas (H_2) and chlorine gas (Cl_2) will not react at room temperature, if kept in the *dark*. However, when exposed to light, particularly sunlight, the mixture will explode even at room temperature to form hydrogen chloride gas.

$$\text{Dark} \qquad H_2(g) + Cl_2(g) \longleftarrow 2HCl(g) \qquad \text{No reaction}$$

$$\text{Sunlight} \qquad H_2(g) + Cl_2(g) \longrightarrow 2HCl(g) \qquad \text{Explosion}$$

Reactions whose rates are increased by light are called *photochemical reactions*, and the added light energy increases both the frequency and energy of collision. The mechanism for this explosive reaction involves the absorption of a quantum of light energy by the Cl_2 molecule. This quantum of energy can cause the dissociation of a Cl_2 molecule.

$$\text{Step 1:} \qquad Cl_2 + light \longrightarrow Cl + Cl$$

If this is all that happened, then a single quantum of light would result in the formation of two molecules of hydrogen chloride, but actually, *a single quantum* of light can cause the formation of anywhere from 10,000 to 1,000,000 molecules of hydrogen chloride. This is the result of a *chain reaction* initiated by the single quantum of light.

Chain Reaction Mechanism

$$\text{Step 1:} \qquad Cl_2 + light \longrightarrow \textcircled{Cl} + Cl$$

$$\text{Step 2:} \qquad \textcircled{Cl} + H_2 \longrightarrow HCl + \textcircled{H}$$

$$\text{Step 3:} \qquad \textcircled{H} + Cl_2 \longrightarrow HCl + \textcircled{Cl}$$

$$\text{Step 4:} \qquad \textcircled{Cl} + H_2 \longrightarrow HCl + \textcircled{H}$$

$$\text{Step 5:} \qquad \textcircled{H} + Cl_2 \longrightarrow HCl + \textcircled{Cl}$$

and so on and on

In each step one of the atoms produced (enclosed in a circle) reacts with another atom in the next step to produce another atom which reacts in the next reaction and so on.

The function of the single quantum of energy is merely to initiate this sequence, or chain, of reactions. The chain does not go on forever, and it can be terminated—for example by an H reacting with another H or Cl reacting with another Cl.

(6.) Effect of Catalysts

So far we have seen that the energy of activation can be supplied in two ways, through heat and light. What is used to provide the activation energy for reactions occurring inside the human body? Light is not available. Living tissue never rises above about 37°C, and this temperature is not high enough to provide the energy of activation needed for many reactions that do occur in the body.

Many reactions that do not proceed at all or have slow rates when mixed alone can be made to take place much more rapidly by the introduction of an outside substance. These substances that increase the rate of reaction without themselves appearing in the overall net equation are called *catalysts*.

A hypothetical catalyzed reaction is represented below, where X is the catalyst added to speed up the reaction.

$$A + B + X \rightleftharpoons C + D + X$$

Note that the catalyst is somehow used to speed up the reaction but it is not used up. This means that a very tiny amount of catalyst can be used over and over again. Thus, the net reaction may be written as follows:

$$A + B \rightleftharpoons C + D$$

Another example is found in the reaction of the deadly air pollutant sulfur dioxide gas (SO_2), and oxygen gas (O_2) to produce sulfur trioxide gas (SO_3), which can then react with water to produce sulfuric acid. Like most things, this reaction is both helpful and harmful. On the positive side it is an important reaction in the industrial production of sulfuric acid—a very essential and useful chemical for making valuable products. Unfortunately, this same reaction can occur to produce sulfuric acid on our delicate lung tissue when we breathe air polluted with sulfur dioxide.

$$2SO_2(g) + O_2(g) \rightleftharpoons 2SO_3(g)$$

Although the reaction has a negative ΔG_T and is favored thermodynamically, the energy of activation is so high that the rate is very slow at room temperature. Increasing the reaction temperature will increase the rate of reaction and make it more kinetically feasible. However, this increase in temperature will make the reaction less feasible from a thermodynamic standpoint.

Study Question 7-4   Explain why the reaction of $SO_2(g)$ and $O_2(g)$ will be less feasible thermodynamically as the temperature is raised.

This changes dramatically if nitrogen dioxide, $NO_2(g)$, is added as a catalyst. If a little $NO_2(g)$ is added, the reaction proceeds rapidly. A crude mechanism for the catalytic activity of NO_2 is that it reacts with SO_2 to produce the SO_3 plus NO. The NO then reacts with the O_2 to become NO_2 again.

Step 1: $\quad 2SO_2(g) + 2NO_2(g) \longrightarrow 2SO_3(g) + 2NO(g)$

Step 2: $\quad 2NO(g) + O_2(g) \longrightarrow 2NO_2(g)$

Net reaction: $\quad 2SO_2(g) + O_2(g) \longrightarrow 2SO_3(g)$

You may remember from our discussion of air pollution reactions in Section 3-2 that the internal combustion engine very conveniently provides the nitrogen dioxide that helps to catalyze this reaction. The art of kinetics and survival is to provide $NO_2(g)$ for the commercial manufacture of sulfuric acid, while eliminating or minimizing its presence in the air we breathe.

Since the catalyst is the same before and after the reaction, its free energy has not changed and it therefore does not contribute to the overall free energy change for the reaction. Thus, *a catalyst does not alter either the value of the equilibrium constant, K_{eq}, or the free energy of reaction.* A catalyst cannot make a thermodynamically unfavorable reaction occur—it can only make a difficult but possible reaction approach equilibrium faster. The catalyst does not alter either the equilibrium position or equilibrium constant. Instead, *a catalyst merely speeds up the rate of approach to the equilibrium position.*

How does a catalyst speed up the rate of attaining equilibrium? Although the detailed mechanisms by which many catalysts work are not known, it is generally agreed that a catalyst increases the rate of reaction by *lowering the free energy of activation for the reaction*, as illustrated in Figure 7-5.

The catalyst lowers the free energy of activation by changing the mechanism of

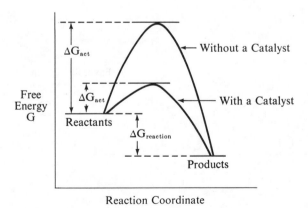

Figure 7-5   Comparison of Free Energy Changes for a Catalyzed and an Uncatalyzed Reaction.

the reaction. This new mechanism, or path, can be provided in different ways depending on the particular reaction and catalyst involved. For example, the catalyst can react to form a new activated complex or intermediate that requires less energy of activation than the old activated complex. When the complex breaks up to form products, the catalyst is released and can be used again. In other cases where the catalysts are solids, they act as surfaces on which reacting molecules can be adsorbed and then react (a kind of reaction landing field). In this case the adsorption of the reactants might change their energy and allow them to react at a lower energy, or the catalyst surface can promote a more favorable reaction geometry and thus lower the free energy of activation by making the entropy of activation more favorable. For example, if two molecules must collide like this in order to react,

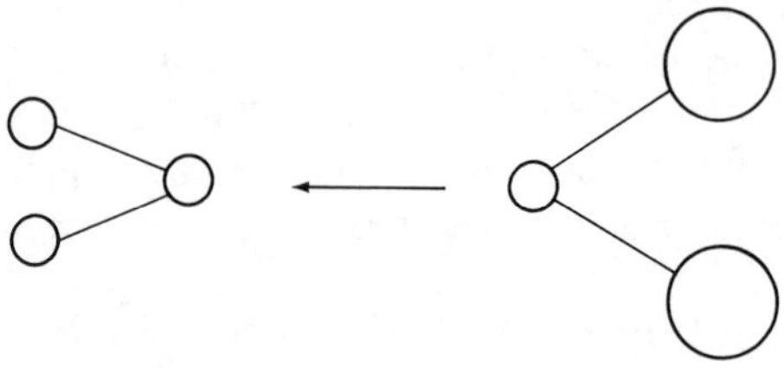

the frequency of fruitful collisions would be enhanced if one of the reaction molecules were adsorbed on a catalytic surface, as follows:

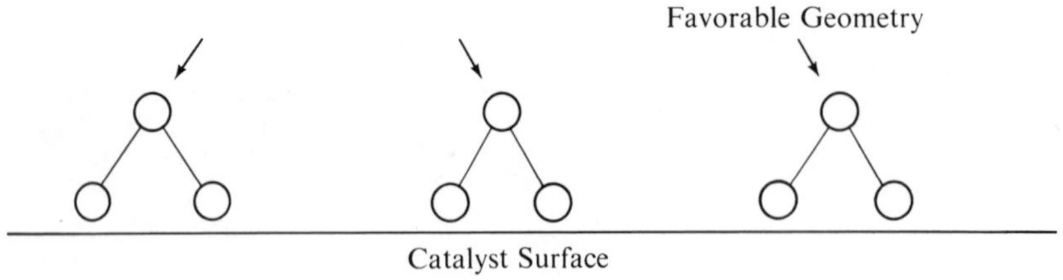

Favorable Geometry

Catalyst Surface

Consider Reaction 8, the Haber synthesis of ammonia. In Section 5-3 we found that $\Delta G_{\text{reaction}}$ becomes less favorable as the temperature is increased. At the same time the rate of reaction at room temperature is so slow that for all practical purposes no reaction occurs. To make this industrial process thermodynamically feasible, a low temperature is needed; but to make it kinetically feasible, a high temperature is needed.

The Haber Process

$$N_2(g) + 3H_2(g) \rightleftharpoons 2NH_3(g)$$

Thermodynamic Data

$$\Delta H^\circ_{298^\circ K} = -22,000 \text{ cal} \quad \text{(yes)}$$

$$\Delta S^\circ_{298^\circ K} = -47 \text{ cal/}^\circ K \quad \text{(no)}$$

$$\Delta G^\circ_{100^\circ K} = -17,300 \text{ cal} \quad \text{(yes)}$$

$$\Delta G^\circ_{300^\circ K} = -8,000 \text{ cal} \quad \text{(yes)}$$

$$\Delta G^\circ_{500^\circ K} = +1,500 \text{ cal} \quad \text{(no)}$$

$$\Delta G^\circ_{700^\circ K} = +10,900 \text{ cal} \quad \text{(no)}$$

Thermodynamically spontaneous at low T

Kinetic Data

Rate very slow at low T and rapid at high T

The way out of this dilemma is to find a suitable catalyst. In the Haber process iron is normally used as a catalyst.

$$N_2(g) + 3H_2(g) \xrightarrow[\text{catalyst}]{\text{Iron}} 2NH_3(g)$$

This increases the rate so that a compromise temperature around 400°–600°C is used in the actual process. The reaction is thermodynamically nonspontaneous at these temperatures, but the cost for introducing relatively small amounts of energy to make the reaction occur still makes the reaction economically feasible.

The iron catalyst serves as a solid surface, or "landing field," upon which the reaction can occur with a lower activation energy. The following sequence may[4] represent the overall reaction mechanism, where the subscript "a" denotes a species adsorbed on the surface of the particles of solid iron catalyst.

$$\text{Step 1:} \quad N_2 \longrightarrow 2N_a$$

$$\text{Step 2:} \quad H_2 \longrightarrow 2H_a$$

$$\text{Step 3:} \quad N_a + H_a \longrightarrow (NH)_a$$

$$\text{Step 4:} \quad (NH)_a + H_a \longrightarrow (NH_2)_a$$

$$\text{Step 5:} \quad (NH_2)_a + H_a \longrightarrow (NH_3)_a$$

$$\text{Step 6:} \quad (NH_3)_a \longrightarrow NH_3$$

A catalyst is not some magical substance that increases the rate of reaction by its mere presence. It participates in the reaction, and in this way provides a new mechanism with a lower free energy of activation, so that the overall reaction may occur more rapidly.

It must be emphasized, however, that the detailed mechanisms for many, if not most, catalytic reactions are poorly understood. Practically all major chemical industries need cheap and effective catalysts, and an enormous amount of time and money goes into catalysis research.

With the Haber process we have a good example of a blend of chemical kinetics and thermodynamics. We see why it is essential to answer all of the seven questions of chemical dynamics before investing heavily in a chemical plant. If possible, we need to find even cheaper and more efficient ways of producing ammonia by the Haber process. Ammonia is the key chemical needed to make nitrogen fertilizers. At present, ammonia is made on a large scale only in developed countries. Unless ammonia for fertilizers can also be produced on a large scale in underdeveloped countries, the present famine rate of 10–20 million per year may increase drastically. The primary lesson of thermodynamics is that we must consider both the system and its surroundings, or environment. Unfortunately, the impact of building fertilizer plants all over the world could also threaten survival through the ensuing air pollution and the pollution of water by nitrates when part of the fertilizer is washed into rivers and streams.

Study Question 7-5   Draw and label a free energy diagram for an endergonic reaction occurring with and without the presence of a catalyst.

Study Question 7-6   Catalysts speed up the rate of reactions. *Inhibitors* are negative catalysts that slow down the rate of a reaction that is too fast. For example, chemicals are added as inhibitors, or preservatives, to most commercial foods to cut down the rate of reactions that lead to spoilage.

---

[4] This is only a hypothesis. The Haber process has been used for over 50 years and it has been studied by hundreds of chemists, but its exact mechanism is still not known.

Draw and label a free energy diagram similar to Figure 7-5, comparing a reaction with and without an inhibitor.

Study Question 7-7 Which one of the following statements is correct for a catalyst?

a. It raises the free energy of activation.
b. It decreases the free energy of reaction.
c. It decreases the value of K_{eq}.
d. It slows down the rate.
e. None of the above is correct.

Study Question 7-8 Consider two gases A and B in a closed container at room temperature. Determine the correct answer(s). The rate of reaction between A and B to form AB will

a. be decreased if the initial number of molecules of A in the container is doubled.
b. be increased if the initial number of molecules of B in the container is increased.
c. be decreased if the partial pressure of A is decreased.
d. be increased if the temperature is raised.
e. be decreased if an inhibitor is added.
f. be increased if a catalyst is added.

7-6 Enzymes—The Catalysts of Life

One group of catalysts deserves special study. These are the *enzymes*—the catalysts of life. A living cell can be likened to a chemical factory or engine. Thousands of reactions occur continuously in a carefully integrated network—large molecules are being broken down and others are being synthesized. Essentially none of these reactions proceeds spontaneously at a rate demanded by the processes in the living cell. Each reaction is catalyzed by a specific enzyme, and as a result cells can function because they contain enzymes. Some scientists therefore describe life as a system of integrated cooperating enzyme reactions. What are enzymes, what are their characteristic properties, and how do they function?

Enzymes are specialized protein molecules that increase the rate of a particular chemical reaction. Like other catalysts, they lower the free energy of activation and thus increase the rate. The rate of an enzyme-catalyzed reaction is much more rapid than the same reaction without an enzyme. Indeed, reactions catalyzed by enzymes frequently proceed from 100,000 to 10,000,000 times faster. In a thousandth of a second they can catalyze complex sequences of reactions that might require weeks or months in the laboratory. Enzymes have certain distinct characteristics.

1. *They are highly specialized proteins, made by cells from amino acids.* This does not mean however, that all proteins are enzymes. Over 1,000 enzymes are known, and each can catalyze only one specific type of chemical reaction or group of reactions. The activity of

some enzymes depend only on their structure as a protein, but others require an additional nonprotein substance, called a *cofactor*. The cofactor is either a divalent (+2 charge) metal ion of substances such as magnesium, calcium, iron, copper, zinc, cobalt, or manganese, or some nonprotein organic molecule. Certain (but not necessarily all) vitamins are coenzymes. If a diet lacks certain vitamins or if the system cannot use certain vitamins, then diseases such as scurvy, beri-beri, and pernicious anemia can result.

2. *They are inactivated by heat.* Since they are proteins, their molecular structure can be disrupted at temperatures above normal body conditions.

3. *They work most efficiently at certain optimum conditions of temperature and pH[5] (acidity).* For example, human body temperature is maintained around 37°C, or 98.6°F, and body enzymes are usually more efficient near this temperature, as indicated in Figure 7-6.

    In a similar manner various fluid systems in the body are maintained at a relatively constant pH, or acidity, and enzyme efficiency drops when the acidity optimum is not maintained. *Metabolism* is the total series of chemical reactions occurring in a living cell, each reaction requiring a particular enzyme, all operating at a certain optimum efficiency. The *metabolic rate* can thus increase or decrease when enzyme efficiency is changed by altering body temperature[6] or pH. Some animals, like man, maintain a fairly constant body temperature regardless of the external temperature. Their metabolic rate is generally not dependent on environmental temperature changes. On the other hand, the body temperature and therefore the enzyme efficiency of some animals, such as reptiles, change with the external temperature. Animals with this type of fluctuating metabolic activity may die if the external temperature becomes too high or low. In order to survive, they must become inactive or move out of the sun during periods of the year or day when temperatures are high and increase their activity or move into the sun when temperatures are low.

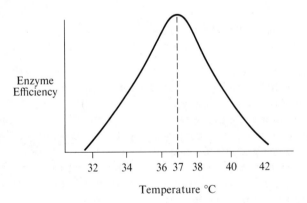

Figure 7-6   Effect of Temperature on Enzyme Efficiency.

[5] pH, which is a measure of acidity, or hydrogen ion concentration, is defined as minus the logarithm of the hydrogen ion concentration in moles per liter.

$$pH = -\log [H^+]$$

A solution with a pH of 7 is neutral. If the pH is less than 7, the solution is acidic, and a pH greater than 7 indicates a basic solution.

[6] For example, one can determine the approximate temperature by noting the calling frequency of crickets and certain types of frogs.

4. *Enzyme activity can be slowed down or blocked by certain specific inhibitors.* Some anesthetics (for example, chloroform) slow down enzyme action by temporarily decreasing or inhibiting enzyme activity and thus slowing down the reactions which they catalyze.

Most of the deadly *poisons* work by blocking the action of key enzymes. *Cyanide* causes the cessation of aerobic (O_2) respiration by blocking the action of the enzyme which catalyzes the consumption of oxygen in respiration. *Arsenic compounds* also inhibit enzyme activity, and the deadly nerve gases and liquids (including some of the commonly used home garden insecticides) act by attacking enzymes in the nervous system. The arrival of a nerve impulse at one nerve cell stimulates the formation of a very small amount of the chemical acetylcholine along the length of the nerve cell. This compound then reacts with a receptor protein on the next cell, and the signal is passed on to control certain specific body functions or actions. Once the signal has been passed on, the acetylcholine must hydrolyze, or react with water, so that the cell can receive and transmit the next impulse. Different nerve poisons block this entire process either by preventing the formation of acetylcholine or by preventing its subsequent hydrolysis. Unfortunately, it only takes a few drops or breaths for a fatal dose. When excess acetylcholine accumulates, there is a rapid loss of control over all body activity and the result is convulsions, choking, paralysis, and death within a matter of minutes after exposure.

On the more optimistic side, enzyme inhibitors can also be used beneficially in some types of *chemotherapy*—the use of chemicals or drugs to destroy infectious organisms. Some of these agents inhibit the growth of a microbe, some destroy the tissue, and others block enzyme activity. By the addition of a particular competitive inhibitor, a particular bacterial enzyme can be put out of action. One important example is found in the sulfa drugs. Some types of bacteria need a chemical called *para*-aminobenzoic acid to complete part of their system of enzyme reactions. A molecule of a sulfa drug, such as sulfanilamide, resembles *para*-aminobenzoic acid closely enough (see Figure 7-7) that the bacteria cannot discriminate between the two molecules. The enzyme process is thus inhibited, an important metabolic process is not initiated, the growth of the microbe is inhibited, and it may die.

The survival of the animal, not just the cell, is the ultimate goal. Thus, the body must have a means of coordinating the functions of the various parts of the body. Much of this "communication" occurs through the nervous system. But a large number of cellular activities are regulated by a special group of messenger chemicals called *hormones* (from

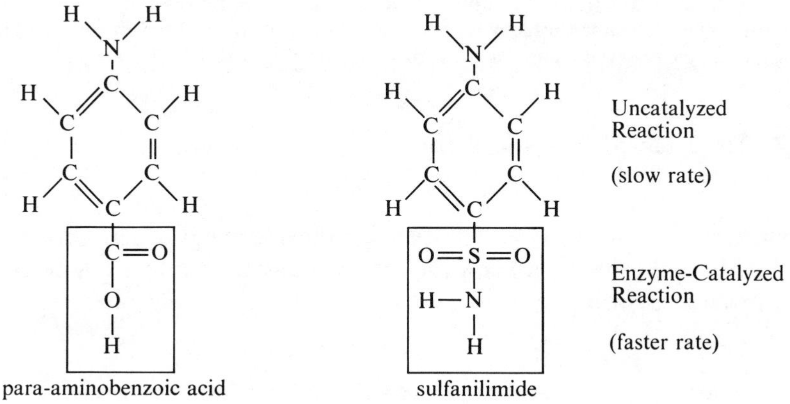

para-aminobenzoic acid   sulfanilimide

Uncatalyzed Reaction

(slow rate)

Enzyme-Catalyzed Reaction

(faster rate)

Figure 7-7   Comparison of the Similar Structures of Para-aminobenzoic Acid and Sulfanilamide.

the Greek word meaning to arouse or excite). They are secreted by endocrine glands scattered throughout the body and carried through the bloodstream to alter and control certain metabolic functions.

For example, the testes and ovaries produce their respective sex hormones, testosterone and estrogen. Most birth control pills contain two types of hormones—an estrogen and a progestrogen. They act in a number of ways to prevent pregnancy. But primarily they act on the pituitary gland to interfere with the normal production of two hormones that act together to release an egg from the woman's ovary—so that ovulation does not occur and pregnancy is prevented.

With several years of intensive research and development activity[7] it may be possible by 1980 to have a greatly improved chemical method for birth control. Although social acceptance would still be a major function in its adoption—as is always the case in complex matters involving human emotion, customs, and ignorance—it might aid in defusing the population bomb. Present means of birth control have the disadvantages of too high an incidence of failure or undesirable side effects in some women, or they rely on the woman remembering to take a pill each day. What is needed is a long-term reversible birth control method that gives a couple both the freedom to have a child and the freedom not to. Humoral research may provide such an improved control method. For example, hormones in a capsule could be impregnated under the skin of the female.[8] Minute amounts of hormone would be released into the bloodstream over a long period of time to inhibit or alter the normal sexual hormone and egg reproduction cycle so that fertilization of the egg would not be possible. When the couple make a conscious decision to have a child, the capsule could be removed or its effect reversed by another chemical.

In addition to the benefits from hormones, a deficiency or excess of certain hormones can lead to disorders such as goiter, diabetes, giantism, and dwarfism. In spite of an enormous amount of research activity in the field of hormones, a detailed biochemical picture of how a hormone works is still not available. Different hormones probably have different mechanisms, and a number of theories have been advanced.

One theory, however, offers the idea that some hormones may control processes by affecting enzyme activity. Some scientists feel that research along these lines could lead to the control of rapidly growing cancer cells, by regulating the transport of nutrients through the cell membranes.

5. *Enzymes are highly specific in their action.* Enzymes usually catalyze only one specific reaction or group of reactions. Although this specificity is far from being understood, it is believed to be related to the particular sequence of the amino acids that link together to form the apoenzyme, or protein portion of an enzyme. Apparently, if a particular protein is to function as an apoenzyme, it must have the right sequence of amino acids and a specific shape that leaves certain chemical groups exposed in a particular pattern so that they can serve as active sites for catalyzing the reaction.

7-7 Theories of Enzyme Action

How do enzymes work? How do they lower the free energy of activation for a particular reaction? We don't yet know the complete answers to these questions, but intensive research is in progress and several theories have been proposed. The answer

---

[7] In 1970 funding for research in reproduction exclusive of drug-firm expenditures was only about $35 million. It is estimated that $150–200 million per year is needed to support this crucial field of research. For a good nontechnical review of birth control research, see O. Harkavy and J. Maier, "Research in Reproductive Biology and Contraceptive Technology: Present Status and Needs for the Future," *Perspectives in Family Planning*, 2 (1970), 5.

[8] One type of capsule is already being tested in clinical trials with over 400 women.

seems to lie in the specific sequence of amino acids in the enzyme protein and their shape or three-dimensional structure as mentioned above. A useful but approximate theory for explaining enzyme action is known as the *"lock and key theory."*

When an enzyme (given the symbol E) catalyzes a specific reaction, it first must attach itself to the surface of a particular reactant molecule, called the *substrate* (given the symbol S). This produces an unstable intermediate species (activated complex) known as the *enzyme-substrate complex* (given the symbol E–S). It has been proposed that there is a "lock and key" fit (much like fitting a piece in a jigsaw puzzle) between the substrate and a particular part of the surface of the large enzyme molecule. This specific portion of the enzyme molecule is called the *active site*, and it apparently contains specific chemical groups that can interact with certain types of substrate molecules. This bond between the enzyme active site and the substrate molecule apparently stretches or strains the substrate molecule and changes its energy so that it might be more susceptible to chemical attack by other reactant molecules. The substrate molecule can thus more readily be converted to the products (given the symbol P) which break away from the active site. The enzyme molecule is regenerated and is ready to repeat the cycle by reacting with another substrate molecule, as summarized below and in Figure 7-8. In one minute a single enzyme molecule may go through this cycle as many as one million times.

Enzyme Reaction

$$S \;+\; E \;\rightleftharpoons\; [E\text{–}S^*] \;\rightleftharpoons\; E \;+\; P$$

substrate + enzyme = enzyme–substrate = enzyme + product
complex

According to this theory the specificity of enzyme action is due to a specific required geometrical fit between enzyme and substrate. Enzyme activity can thus be blocked by the addition of an inhibitor molecule with a similar molecular shape, or key, that also fits into the "lock." Here we must distinguish between the "poisoning" of the enzyme and the temporary blocking of active sites. "Poisoning" involves a permanent blocking of active sites while "competitive inhibition" involves temporary blockage. In addition, enzyme activity can be disrupted when the geometry of the enzyme active site is changed so that the key no longer fits the lock. For example, the mutation of a gene or the incorrect synthesis (amino acids out of sequence) of the enzyme by the organism could alter the shape of the active site.

How can the "lock and key" theory account for the fact that most enzymes have an optimum temperature for operation and become completely deactivated beyond a certain temperature? It has been hypothesized that high temperature somehow damages

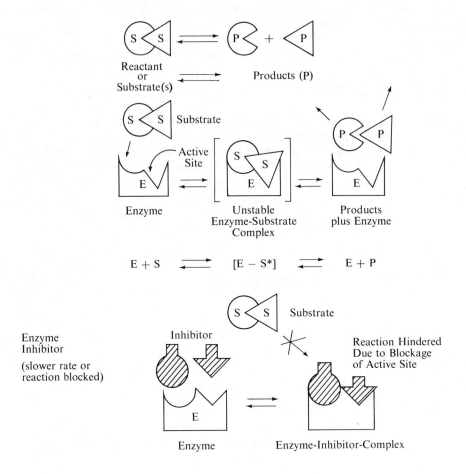

Figure 7-8   The Crude "Lock and Key" Theory of Enzyme Action and Inhibition.

or changes the enzyme's active site. Enzymes are proteins, which consist of poly-peptide chains linked to one another by hydrogen bonds. High temperatures might break these bonds and change the surface geometry of the enzyme by allowing the alpha-helix structures of the proteins to unravel.

The "lock and key" model is useful but crude. It accounts for many but not all of the properties of the thousand or so different enzymes. In addition, it is a somewhat vague model from a chemical standpoint. What is happening to the electrons in the enzyme and substrate at the active sites? What sort of chemical bonding is involved? Newer hypotheses for enzyme activity focus on these aspects. According to Koshland's "induced fit" hypothesis or model, the enzyme is "flexible" rather than rigid. The substrate may bring about changes in the structure of the enzyme so that a fit is possible. Koshland has proposed an "orbital steering" hypothesis to explain how enzymes can increase rates by such large factors. He suggests that enzymes align the outermost electrons of the substrate molecules so that they can easily overlap or be shared with other atoms. This electron, or orbital, steering would greatly increase the rate of formation of chemical bonds.

In any case, the overall effect of the enzyme is to lower the free energy of activation for a particular reaction. Consider Reaction No.10, involving respiration in living cells.

$$C_6H_{12}O_6(s) + 6O_2(g) \rightleftharpoons 6CO_2(g) + 6H_2O(l)$$

$$\text{At } 300°K \ \Delta G_{reaction} = -759,000 \text{ cal}$$

$$\text{At } 300°K \ \Delta G_{activation} \text{ is high}$$

We can see that in terms of the net equation the reaction is thermodynamically feasible but kinetically unfeasible, as shown in Fig. 7-9. In biological systems the net

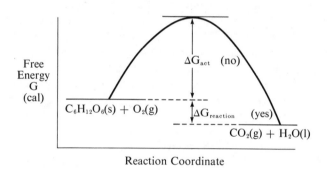

Figure 7-9   Free Energy Diagram for NET Respiration Reaction.

reaction is broken down into a sequence of steps with a specific enzyme catalyzing each step—in effect catalysts help convert a high energy barrier or "mountain" into a series of small hills.

$$A \xrightarrow{E_1} B \xrightarrow{E_2} C \xrightarrow{E_3} D \xrightarrow{E_4} E \quad \text{etc.}$$

In the case of the respiration reaction there are at least one hundred steps, or "hills," and the effect of catalysts is shown symbolically in Figure 7-10. Note also that if only one enzyme in the sequence is poisoned or hindered from its specific task, the

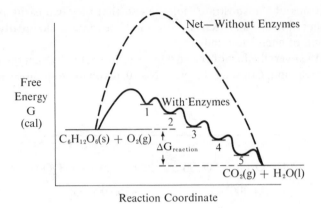

Figure 7-10   Free Energy Diagram for Respiration Showing Effect of Enzymes in Lowering Activation Energy.

entire reaction sequence is disrupted. Certain diseases, disorders, or death can result from metabolic disruptions of only one enzyme.

Obviously enzyme chemistry is one of the more important areas of research. A more complete understanding of the chemistry of enzymes would be a major step forward in controlling and favorably altering the major processes in life, including those of birth control, disease, and the prolongation of life. It is even conceivable that such an understanding could help us deal with the undesirable physiological effects of certain types of pollutants which we are ingesting and inhaling in ever increasing amounts. If specific enzyme actions were understood, perhaps it would be possible to design "antidotes" or inhibitors that would prevent or slow down the physiological damage caused by specific pollutants.

7-8   What Is the Reaction Mechanism?

We have discussed the first two questions of chemical kinetics: "What is the rate of a reaction?" and "How can it be altered?" It is now time to turn to the third question: "What is the *mechanism*, or sequence of steps, by which a particular reaction proceeds?"

Chemists would like to know the mechanism of a reaction not only because of the intellectual challenge it provides, but also because a knowledge of the mechanism can have far-reaching practical results. If we understand a particular mechanism, we may see how it can be manipulated or altered to obtain a higher yield for a desirable product or to eliminate or minimize undesirable side products. Many drugs can cure a particular ill, but they frequently have undesirable side effects which prevent their use. If the reaction mechanism is known, the chemist may be able to alter the molecular structure of the drug or alter the reaction sequence so as to preserve the good effects and eliminate undesirable effects.

The overall plan for determining the reaction mechanism can be summarized as follows.

Plan for Determining Reaction Mechanism

Step 1: Devise an experimental setup and measure the rate of reaction.

Step 2: Use the experimental data to determine the rate law, or expression—that is, what species affect the rate and what is the mathematical relationship between their concentrations and the rate?

$$\text{For example, A + B} \rightarrow \text{AB}$$

$$\text{rate} = k[A][B]$$

Step 3: Try to postulate the rate-determining step or steps from the rate equation. In general, a sequence of reactions will have one slow step, and this step will determine the overall rate. The rate equation normally (but not always) involves the concentrations of the species in this rate-determining step, and it can frequently be used to deduce the slow step in the overall sequence of reactions.

Step 4: Formulate one or more overall mechanisms consistent with the rate-determining step and the net reaction.

Step 5: Design further experiments that will support a particular mechanism or eliminate certain mechanisms. If a relatively stable intermediate compound is postulated in one mechanism, try to detect its presence experimentally.

As we can see, the determination of reaction mechanism is a kind of intellectual and experimental "chess" game, and it is not surprising that it attracts many outstanding scientists who seek challenging problems.

In Section 7-1 we indicated that it is highly improbable for a reaction to occur by a mechanism involving the simultaneous collision of even as many as three molecules. Thus, *most reactions proceed by a mechanism or series of sequential steps with each step involving either the dissociation of a single molecule* ($A_2 \rightarrow A + A$) *or the bimolecular collision of two atoms or molecules* ($A + B \rightarrow AB$).

Let's begin by writing the mechanism for a large group of people inside a theater moving to the street after the show is over. Let us assume that the theater has been

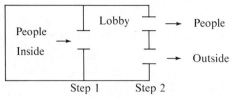

Step 1: People Inside → People in Lobby (slow rate-determining step)

Step 2: *People in Lobby → People Outside (fast)*

NET: People Inside → People Outside

Figure 7-11   Illustration of Rate-Determining Step (RDS), or Bottleneck, Principle.

poorly (and illegally[9]) designed, so that the people must first pass through one narrow door to reach the lobby but have a choice of two doors leading from the lobby to the street. This involves a two-step mechanism, and the first step will be the slowest. It is the "bottleneck," or *rate-determining step* (RDS), for the entire sequence, as shown in Figure 7-11.

Consider the net equation for the hypothetical reaction

$$A + B + E \longrightarrow G + H + I$$

It is highly improbable that the net equation represents the mechanism by which the reaction takes place, since it would involve the low probability of three different molecules colliding simultaneously with the right energy and geometry. A possible mechanism, or reaction sequence, involving the formation and subsequent reaction of various intermediates is given below. Note that the intermediates cancel out to give the overall net equation.

Step 1: $A + B \longrightarrow C + D$ (slow rate) (RDS)

Step 2: $C + E \longrightarrow F + G$ (fast rate)

Step 3: $D + F \longrightarrow H + I$ (fast rate)

Net reaction: $A + B + E \longrightarrow G + H + I$

Again in this mechanism, the first step is the slow, or rate-determining step (RDS). Since A and B are the only species involved in this step, we might expect that the rate would depend only on the concentrations of A and B. The experimentally determined rate equation would be expected to have the form

$$\text{rate} = k[A][B].$$

We can compare the free energy diagrams for the *net reaction* and for the sequence of reactions making up the net reaction, as shown in Figures 7-12 and 7-13.

---

[9] Fire codes require multiple exits in any public building.

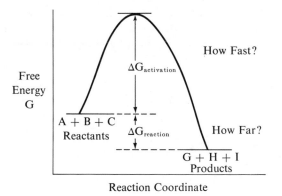

Figure 7-12   Free Energy Diagram for *Net Reaction*.

Note that the free energy of activation for the first step (ΔG_1) in Figure 7-13 is the same as the overall free energy of activation (ΔG_{act}) for the net reaction shown in Figure 7-11. *Again we see that the overall reaction rate of a sequence of reactions is determined by the rate of the slowest step in the system—that is, the step with the highest activation energy.*

This is a most important point. If we want to change the overall rate of a sequence of reactions or events, it does no good whatsoever to change the rate of the fast steps. Efforts at rate control should be directed to the *rate-determining step*.

This "bottleneck," or rate-determining step, principle is crucial in most life processes, which occur by the coupling of a long sequence of reactions in a *metabolic pathway*. The product of one step is a reactant for the next, and each step requires a specific enzyme catalyst. The speed of the overall sequence cannot be any faster than the speed of the slowest step. Since reactions in the body normally occur in the body at constant temperature (37°C), this means that *the key to changing the rate of most*

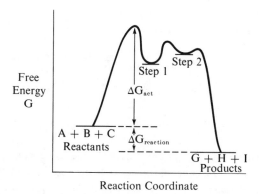

Figure 7-13   Free Energy Diagram for *Reaction Mechanism*.

biological processes lies in changing the enzyme activity for the slowest, or rate-determining, step. Again we can see how an understanding of reaction mechanism might allow one to cure a disease or promote a desirable physiological change.

Study Question 7-9   A hypothetical mechanism for a reaction is given by

$$\text{Step 1:} \qquad A + B \longrightarrow C + D \quad \text{(fast)}$$

$$\text{Step 2:} \qquad B + D \longrightarrow 2E \qquad \text{(fast)}$$

$$\text{Step 3:} \qquad C + E \longrightarrow 2F \qquad \text{(slow)}$$

Net:

a. Write the net reaction for the process.
b. Draw and label a free energy diagram for the same reaction assuming that $\Delta G_{\text{reaction}}$ is *positive*.
c. Draw and label a free energy diagram for the same reaction showing the reaction mechanism free energy profile.
d. Explain how the diagram in c would be changed if a catalyst were added to slow down the reaction but Step 3 still had the slowest rate.
e. Write the expected experimental rate law, or equation, for this reaction.

The mechanism for Reaction 4, involving the oxides of nitrogen, has been studied and two mechanisms have been proposed.

Possible Mechanism No. 1

$$\text{Step 1:} \qquad 2NO(g) \rightleftharpoons (NO)_2 \text{ or } N_2O_2 \quad \text{(slow)}$$

$$\text{Step 2:} \qquad (NO)_2 + O_2(g) \longrightarrow 2NO_2(g) \quad \text{(fast)}$$

$$\text{Net:} \qquad 2NO(g) + O_2(g) \longrightarrow 2NO_2(g)$$

Possible Mechanism No. 2

$$\text{Step 1:} \qquad NO(g) + O_2(g) \rightleftharpoons NO_3(g) \quad \text{(slow)}$$

$$\text{Step 2:} \qquad NO_3(g) + NO(g) \longrightarrow 2NO_2(g) \quad \text{(fast)}$$

$$\text{Net:} \qquad 2NO(g) + O_2(g) \longrightarrow 2NO_2(g)$$

Which (if either) of these mechanisms is correct? One approach would be to try to detect experimentally the presence of the two proposed intermediates $(NO)_2$ and $NO_3(g)$. In this case, however, they are short-lived and difficult to detect. Another approach would involve the use of the concepts of chemical bonding (quantum mechanics) to predict whether either intermediate represents a feasible chemical structure. Using this approach scientists have ruled out mechanism No. 1 on structural grounds. Mechanism No. 2 does seem feasible—but it may not be the actual mechanism. Can you formulate other possible mechanisms?

When sunlight is added to this reaction, $2NO(g) + O_2(g) \rightleftharpoons 2NO_2(g)$, in the presence of organic hydrocarbons, an extremely complex series of reactions occurs to yield the basic ingredients of Los Angeles, or brown air, smog—known scientifically as photochemical smog. The initial effect of the sunlight is to provide the energy to dissociate NO_2.

$$NO_2 + light \rightleftharpoons NO + O$$

The highly reactive O can then react with hydrocarbons in a long and complicated mechanism to produce numerous compounds that irritate the eyes and lungs and cause millions of dollars of damage to crops. Unfortunately, the presence of trace amounts of other air pollutants, sulfur dioxide, $SO_2(g)$, or ozone, $O_3(g)$, greatly increases the rate of formation of photochemical smog. If NO_2 could be removed or very drastically reduced from automobile exhaust, we would not have much, if any, photochemical smog. Unfortunately this is one of the pollutants that is not being reduced significantly by the new emission reducers on automobiles.

7-9 Kinetic Analysis in Everyday Life

The kinetic idea that the overall rate of a process consisting of a series of events is determined by the slowest step in the sequence has considerable application, *by analogy*,[10] in everyday life—in the operation of a business, the assembly production of any item, the performing of household chores, the procedure for studying, and any process in life whose end result depends on a sequence of connected events.

Efficiency experts in any business or assembly line production look at the entire operation to determine the sequence of events and the rate at which each occurs. In this way they can find the slow step or steps in the manufacturing process and increase production or efficiency by trying to increase the rate of these bottleneck steps. For example, either additional or faster machines or more workers might be provided for a particular step, while excess workers at a noncritical fast step might be laid off or moved to a slower step in the process. It is amazing to see how many processes, office or secretarial pools, and other businesses, are operating inefficiently with a great waste of time and money. Indeed, in many cases extra workers or very expensive machines are often found at a noncritical step. The simple application of the rate-determining step principle can save enormous amounts of money and increase profits.[11]

The process by which you study provides another chance to apply kinetic analysis. How do you study for daily classes, prepare for tests, and write term papers? Are you a model of efficiency, or are your study habits so grossly disorganized that most

---

[10] Thermodynamics and kinetics apply only to reacting systems of molecules. The ideas in this section are developed *only* by analogy and do not represent scientific fact or theory.

[11] It should be noted that even if a kinetic analysis is made, it needs to be repeated frequently. The Second Law of Thermodynamics insures that nicely worked out production systems tend to become disordered with time.

efficiency experts would have a heart attack if asked to analyze your procedures? Are your study habits well thought out or do they represent programmed "academic brinkmanship" (known as cramming or flunking, depending on the outcome)? Do you study the right subject at the wrong time or in the wrong sequence? We all know that because of the cyclic nature of the academic process, tests and papers will all occur in quantity at easily predictable intervals. Do you plan ahead for these periods or do you put everything off until the last moment so that you have four or five "mountains," or activation energy barriers, to conquer in only one or two days?

If you have experienced "all nighters" and other such academic crises, you may need to apply kinetic analysis to your study habits. In the body most net reactions involve a free energy barrier, or "mountain," that is impossible to climb in one step. In the cell these "mountains" are not climbed in one step but in a careful sequence of enzyme catalyzed reactions, or steps, so that the entire process is accomplished by climbing a series of smaller energy hills instead of scaling Mt. Olympus.

Perhaps we should borrow this simple but powerful concept and apply it to the process of studying and of accomplishing other tasks.

A Useful Kinetic Idea

Make hills out of mountains.

In other words, study so many pages a day, work so many problems a day or do part of the research for the term paper each week. Most of us know this obvious solution to our difficulties, but we lack the necessary fortitude or desire to fight the Second Law of Thermodynamics that never ceases to erode our attempts to create order.

Study Question 7-10 An assembly line for the production of "crackerjacks" is set up as follows:

Step 1:	Toasting of corn kernels	150 lb/hr
Step 2:	Cooling the popped corn	140 lb/hr
Step 3:	Packaging	145 lb/hr
Step 4:	Insertion of prize in each package	15 lb/hr
Step 5:	Sealing the package	300 lb/hr

a. What is the rate-determining step?
b. As manager of this company what things would you do (be specific) to increase the efficiency of this production line?

Study Question 7-11   Make a kinetic analysis of your school cafeteria line, starting with standing in line and ending with being seated at the table. List the individual steps and estimate the relative rate (very fast, fast, slow, very slow) for each step. Indicate the rate-determining step or steps and give a detailed plan for improving the efficiency of the entire process.

7-10 Summary

In this chapter we have found that the rate of any single-step reaction is determined by three factors—the *collision frequency factor, the activation energy factor,* and *the geometry, or entropy, factor.*

The collision frequency is related to temperature and concentration. Increasing the temperature or the concentration of reactants will increase the total number of collisions per unit of volume.

Collision efficiency (energy and geometry) is related to the free energy of activation and temperature. The free energy of activation consists of two factors, the enthalpy of activation and the entropy of activation. The latter factor to some degree is related to the geometry involved when the reactants collide to form the activated complex.

The free energy of activation determines the kinetic feasibility of the reaction, while the free energy of reaction determines the thermodynamic feasibility. A high value of the free energy of activation means that the reaction will be slow because the reactants either do not have enough energy to react, the reaction geometry is unfavorable, or both. An increase in temperature results in more particles having sufficient energy to overcome the free energy barrier, and the rate is increased.

Catalysts provide another way of increasing reaction rate. A catalyst provides a different mechanism with a lower free energy of activation. Enzymes are particularly important catalysts in living systems.

Most reactions proceed by a *mechanism* involving a sequence of reactions, with each step typically involving a collision between two molecules. The slowest step in a sequence of reactions is the *rate-determining step,* and this principle can also, by analogy, be applied to improving the efficiency of many everyday processes.

We have now completed our study of the seven fundamental questions of *chemical dynamics.* By combining the answers of the four fundamental questions of *chemical thermodynamics* (Chapters 2, 3, 4, and 5) with the three questions of *chemical kinetics* presented in this chapter, it is possible to obtain a profile of a chemical reaction and thus determine overall reaction feasibility. By understanding how the equilibrium position and rate can be altered, it is possible in many cases to increase reaction feasibility and yield.

The fundamental theories of chemical dynamics are still poorly developed and poorly understood, and this field represents an important and exciting research frontier. In the next section we shall see how thermodynamics and kinetics can be applied to an even wider range of man's processes and activities.

> Scientific research consists in seeing what everyone else has seen, but
> thinking what no one else has thought.
>
> *A. Szent-Györgyi*
> *Nobel-laureate*

7-11   Review Questions

7-1.   Distinguish between chemical thermodynamics and chemical kinetics.

7-2.   What are the three fundamental questions of chemical kinetics?

7-3.   Distinguish between the net equation for a reaction and the reaction mechanism.

7-4.   Criticize the following statement: The reaction mechanism is more important than the free energy change in determining whether a given reaction will occur.

7-5.   What is the rate of a reaction?

7-6.   List the three factors that determine the rate of a reaction according to (1) the collision theory and (2) the transition state theory.

7-7.   Distinguish between collision frequency and collision efficiency. What two factors determine collision efficiency?

7-8.   Give two ways for decreasing the collision frequency for a reaction.

7-9.   Give two ways for increasing the collision efficiency for a reaction.

7-10. Is it possible to increase the collision frequency while at the same time decreasing the collision efficiency? If so, how? If not, why not?

7-11. Distinguish between the collision theory and the transition state theory of reaction rates. What are the advantages or disadvantages of each?

7-12. What is an activated complex? How does it differ from a stable intermediate?

7-13. What are the two assumptions of the transition state theory, and which assumption allows the use of thermodynamics?

7-14. Distinguish between the free energy of activation and free energy of reaction, and explain what each tells us about reaction feasibility.

7-15. The following hypothetical reactions are both exergonic:

$$1 \quad AC + D_2 \longrightarrow AD + CD$$
$$2 \quad G_2 + B_2 \longrightarrow 2GB$$

Reaction 1 has a higher free energy of activation and a more negative free energy of reaction than Reaction 2. On the same graph sketch free energy diagrams for both reactions.

7-16. Criticize the following statement: The free energy of activation determines the position of equilibrium.

7-17. Explain how you might follow the rate of the following reaction experimentally: $3O_2(g) \rightleftharpoons 2O_3(g)$.

7-18. What is the flow method for determining the rate of fast reactions? What is the relaxation method?

7-19. List the five major steps usually followed in determining the mechanism of a reaction.

7-20. Explain (at the molecular level) why an increase of concentration of a reactant can cause an increase in reaction rate.

7-21. Given the reaction A + B $\rightleftharpoons$ AB, explain why tripling the concentration of both A and B would be expected to give a ninefold increase in reaction rate.

7-22. Explain (at the molecular level) why an increase in temperature usually causes a significant increase in reaction rate.

7-23. Account for the following observations.
   a. Oxygen and hydrogen apparently do not react to form water at room temperature even though the free energy of reaction has a large negative value at this temperature.
   b. An increase in temperature increases the rate at which oxygen reacts with hydrogen.
   c. A spark will cause hydrogen and oxygen to react explosively.
   d. The addition of finely divided platinum powder will also cause the two gases to react explosively.

7-24. Explain (at the molecular level) why smoking in a coal mine will set off an explosion.

7-25. When hydrogen gas, $H_2(g)$, and bromine, $Br_2(g)$, are mixed in a brown glass bottle, no reaction occurs, but if mixed in a colorless bottle, they may react explosively. Explain (at the molecular level).

7-26. What is a photochemical reaction?

7-27. Explain how the absorption of just *one* quantum of light energy can in some reactions produce as many as a million product molecules.

7-28. What is a rate law, or expression?

7-29. What will happen to the rate of a reaction that involves the collision of two molecules in each of the following cases?
   a. The concentration of one of the reactants is decreased (at constant temperature).
   b. The temperature is lowered.
   c. A catalyst is added.
   d. An inhibitor is added.

7-30. Distinguish between a catalyst and an inhibitor. Which would probably be present as an additive in some foods? Why? Explain in terms of kinetics why many foods should be stored in a refrigerator.

7-31. Explain in words and with a free energy diagram how a catalyst increases the rate of a reaction.

7-32. Catalysts decrease the free energy of activation for a reaction. What effect, if any, does this have on the overall free energy of reaction? Explain. What effect does it have on the equilibrium position? Explain.

7-33. What are enzymes?

7-34. How do most poisons work?

7-35. What are hormones? What is their possible relationship to enzyme activity?

7-36. How could hormone research possibly lead to a means for helping us deal with the population explosion?

7-37. Explain the specificity of enzymes.

7-38. Outline in words and diagrams the "lock and key" theory of enzyme activity. Use the same theory to explain how an inhibitor or poison can block enzyme activity. Describe the more modern "induced fit" theory.

7-39. Draw and label a free energy diagram for
   a. an exergonic one-step reaction with a fast rate,
   b. a slow one-step endergonic reaction,
   c. an exergonic two-step reaction in which the first step is fast and the second step is slow.

7-40. A lump of sugar enclosed in the hand so that it attains body temperature will not decompose. Yet, if we swallow the sugar, it is "burned," or decomposed, at body temperature. Explain. On the same graph draw two net free energy diagrams illustrating the difference between the two situations.

7-41. A hypothetical reaction occurs in the following steps:

Step 1: $A + B \longrightarrow C + D$ (fast)
Step 2: $C + D \longrightarrow E + F$ (slow)
Step 3: $E + B \longrightarrow D$ (fast)

   a. What is the net equation?
   b. What is the rate-determining step?
   c. Draw a mechanism free energy diagram for this reaction sequence. Assume the net reaction is endergonic.
   d. Which of the following statements is true?
      (1) The overall rate will depend on the concentrations of A and B.
      (2) The overall rate will depend on the concentrations of C and D.
      (3) The overall rate will depend on the concentrations of E and B.
      (4) The first step will have the highest activation energy.
      (5) The last step will have the lowest activation energy.

7-42. List at least five processes in everyday life (other than those discussed in the text) that could be studied by use of kinetic analysis.

7-43. A group of students is preparing a ten-page underground newspaper. The pages have been printed and are stacked in piles page-by-page. The pages must be (1) assembled in order, (2) straightened, and (3) stapled in sets.

   a. If three students work together, each performing a different operation, which step will probably be rate-determining?
   b. What would be the overall effect if nine eager volunteers are added to the third step? What would be the effect if these volunteers worked on the first step?
   c. What would be the effect if these volunteers worked on the first step?
   d. As supervisor, you decide to be democratic (after all, it is an underground newspaper) and add three volunteers to each step. From a kinetic viewpoint was this a wise decision? Explain.

Part Three

Thermodynamics, Kinetics, and Life

In science as in other human activities, the speed of progress is less
important than its direction. Ideally, knowledge should serve
understanding, freedom, and happiness rather than power,
regimentation, and technological development for the
sake of economic growth.
Scientists must give greater prominence to large human concerns when
choosing their problems and formulating their results. In addition to
the science of things, they must create a science of humanity.

René Dubos

Chapter 8

Entropy and Information

8-1 Cybernetics, Homeostasis, and Feedback

Most people in industrialized countries are aware that we are no longer in the "atomic age." Instead, we have entered a new era—the *"age of cybernetics."* To many this new field offers hope for understanding and dealing with many of the complex problems facing modern man.

Cybernetics is the field of control and communication and it can be applied to machines or to living organisms. Although the ideas of control and communication have a long history, the fusion of these ideas into a new theoretical field is relatively new. In 1947 Norbert Wiener, a mathematician, formulated the basic ideas and coined the word *cybernetics* (from the Greek word kybernes, meaning pilot or steersman). Cybernetics is the science of controls. But control of a system requires an input of *information* in the form of *feedback*. Thus, the communication that allows control of a system is based upon the concept of information feedback.

Cybernetics is a general term that applies to living and nonliving systems. The term *homeostasis* is used to describe cybernetic systems in living organisms—the

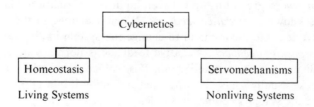

Figure 8-1 Branches of Cybernetics.

self-regulating information feedback systems that maintain the internal environment of living organisms within certain physiological limits. The term *servomechanisms* is used to describe the control mechanisms for nonliving instruments or machines, as summarized in Figure 8-1.

A system with information feedback control has definite advantages because it can correct for errors. *Feedback* means that part of the output of the system is fed back into the system to give information that will allow the system to change in order to maintain a particular state. Information feedback tends to minimize uncertainty. For example, driving a car with a blindfold eliminates the information feedback necessary for controlling the system, and uncertainty is extreme. Information can minimize uncertainty and decrease randomness, or disorder. As we shall see later, this concept provides the key to the relationship between entropy and information. One of the simpler examples of a feedback system is that of a thermostat, as shown in Figure 8-2.

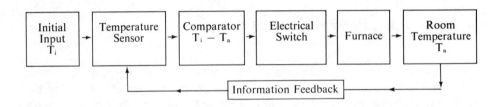

Figure 8-2   Cybernetic Diagram for a Thermostat.

The desired room temperature, T_i, is the initial input of information. It is fed to a temperature receptor, or sensing device, which senses the actual room temperature, T_a, and compares it with the desired temperature. If the actual temperature, T_a, is less than desired temperature, T_i, the switch cuts the furnace on; and if T_a is greater than T_i, the furnace is cut off. Information on the actual temperature, T_a, is fed back continuously to maintain the temperature near the desired value.

Notice that a rise in room temperature causes the thermostatic mechanism to decrease the heat, while a fall in room temperature causes more heat to be supplied. This negative, or reverse, relationship between input of information and the response of the system is known as *negative feedback*. When a change in the system in one direction is converted into a command to change the system in the same direction, this is known as *positive feedback*. Negative feedback is normally necessary to keep a living or nonliving system in equilibrium. Positive feedback, which is often called runaway feedback, usually[1] has the opposite and undesirable effect of disrupting equilibrium, and the system becomes unstable.

[1] There are some systems in which positive feedback is important—those in which it is necessary to reinforce a weak signal. We will be concerned only with negative feedback.

For example, if the wires on the thermostat were hooked up backward, a rise in temperature would cause the furnace to cut on, and the higher the temperature, the hotter the house.

Study Question 8-1   Draw a cybernetic diagram similar to Figure 8-2 to show what would happen when an electrician hooks up the wires on a thermostat backward, so as to provide positive instead of negative feedback.

A living organism has thousands of these communication and control systems. The cybernetic system is a stress-response system designed to maintain or control some variable (temperature, fluid movement, acidity, etc.) around some optimum value. Such control requires a continuous flow of information among the parts of the system.

Living cybernetic, or control, systems generally operate in a common manner. An initial information input, or *stimulus*, activates a *receptor*, or sensing device (perhaps the eyes). The receptor then transmits a signal over a sensory pathway (sensory nerve system) to a response selector, usually called a comparator, or *modulator* (the brain). It "selects" the appropriate response and sends a signal along the correct motor pathway (motor nerve system) to an *effector* (muscles or glands) that carries out the command. This produces a *response* (perhaps a hand movement or secretion of a substance) and this new information is then fed back (negative feedback) to the *receptor* or sensors. From there it goes again to the comparator, or modulator, and so on, in order to maintain some optimal or steady value for a particular variable.

This sequence can be expressed in a general diagram for a homeostatic control system, as shown in Figure 8-3.

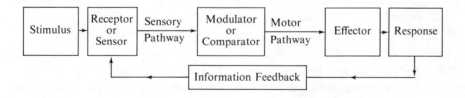

Figure 8-3   Generalized Diagram for Homeostatic Control.

A good example of cybernetic control, or homeostasis, in living organisms is the method by which the body temperature of warm-blooded animals is kept constant through sweating. As we saw earlier, the human body may be considered as a complex system of enzyme controlled reactions, with each enzyme having an optimum operating temperature range. The human body must, therefore, have a mechanism for maintaining body temperature within a few degrees of its normal body temperature of 37°C or 98.6°F.

For mammals such as man, called *endotherms*, an internal physiological mechanism is used to control temperature. Other organisms, such as lizards, have no internal mechanism for temperature control, and they must control body temperature by moving into or away from the sun or some other heat source. Such organisms are called *ectotherms.*

An endothermic organism, such as man, can control body temperature through sweating. The excreted water evaporates and removes heat from the skin. To maintain this temperature with the normal energy-producing metabolic processes, the body must dissipate heat at a rate comparable to the heat flow from a 100-watt electric light bulb. If the air temperature gets too high with humidity approaching 100 per cent, the rate of heat loss may not be sufficient to maintain body temperature, and the result is "heat stroke." On the other hand, if the environment is so cold that it causes too rapid a heat loss, sweating decreases, blood flow decreases, hands and feet get cold, and eventually shivering (an energy-producing mechanism) may occur.

When the temperature of the environment rises, sensory nerve impulses send signals to a special region of the brain known as the hypothalamus, which in turn sends out signals that cause the sweat glands to secrete water. The increased sweating cools the body because, as the water molecules evaporate from the skin, they remove heat from the body. As the temperature falls, the stimulation of the hypothalamus is decreased and sweating is reduced. More complicated cybernetic systems will have an array of numerous comparators, activators, and various information feedback loops.

Study Question 8-2   Draw a cybernetic, or homeostatic, diagram for temperature control in man.

Study Question 8-3   Draw a cybernetic diagram for temperature control in an ectothermic organism, such as a lizard.   .

Study Question 8-4   Draw a cybernetic diagram indicating what happens when you decide to put on the brakes while driving a car.

Most homeostatic systems have working limits. If the external conditions exceed these operating limits, the system goes out of control and runaway feedback takes over. If human body temperature exceeds 42°C, or 107.6°F, the negative feedback mechanism for temperature control breaks down and is replaced with a positive, or runaway, feedback mechanism. The body can't get rid of the heat fast enough and at a high enough temperature the "thermostatic mechanism" fails and the metabolic rate (rate of the chemical reactions necessary for life) increases. This produces more body heat, which causes the chemical reactions to go even faster, giving more body heat and so on—ultimately resulting in death. Similarly, the thermostatic mechanism can break down when the temperature becomes too low and chemical reactions do not proceed at a rate fast enough to provide the necessary energy to maintain life.

Life must be maintained by having certain variables such as body temperature, blood pressure, body fluid content, respiration, hormone activity, and other processes controlled so that they fluctuate around values on a *homeostatic plateau,* as indicated in Figure 8-4. It is important to realize that the principle of maintaining a homeostatic plateau is essential to maintaining life at all levels.

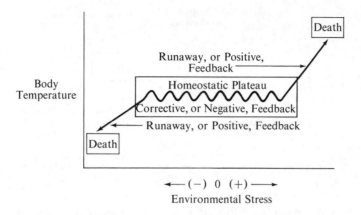

Figure 8-4   The Homeostatic Plateau for Body Temperature.

Another example of a cybernetic system is seen in the mechanism of population control. When overpopulation of a species occurs, relative to the resources available for maintaining life, then the population is decreased by such things as starvation, disease, or—in the case of man—warfare. Some animals (apparently not man) have an automatic physiological mechanism that decreases fertility under conditions of crowding and stress. Man, of course, as a "rational" species can decide to limit population in order to prevent undesirable death rate control through war, famine, disease, or ecological catastrophe.

Man has another advantage. Through technological changes and revolutions he has been able to raise the carrying capacity, or homeostatic plateau. Some examples of technological innovations that provide increased carrying capacity are the bow and arrow, domestication of animals, cultivation of plants, the industrial revolution, and atomic energy. With each innovation the human population has risen sharply, as shown in Figure 8-5. We now appear to be in a very critical transition period moving rapidly on a J-curve toward a new maximum, or homeostatic plateau (S-curve), with respect to resources and the disruption of our environment through too many people using too much energy. Unless we decide rationally to convert the J-curve to an S-curve by a simultaneous decrease in birth rate and the rate of consumption, the increased death-rate mechanism will prevail. At this moment we apparently have no technological miracle on the horizon to raise the plateau, and we will be in a critical transition state until population and resources are brought into a state of equilibrium—one way or the other.

Study Question 8-5   Draw a cybernetic diagram for

a. the five-step scheme for determining the mechanism of a reaction (see Section 7-8),
b. any system of your own choosing (for example, driving a car, buying something, studying, and so on).

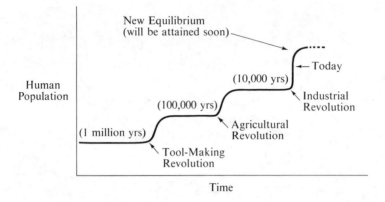

Figure 8-5   Man's Ability to Extend Carrying Capacity through Technological Innovation. The Periodic Transition from a J-Curve to an S-Curve.

We can see that cybernetics, or homeostasis, in living systems is crucial to stability and survival. Homeostatic controls are found in all systems of the body, and these physiological controls involve negative feedback to maintain life. The endocrine glands, the kidney, the rate of respiration, and all other critical functions respond to changing levels of necessary chemicals. The rather simple and perhaps obvious examples used in this section are deceiving. *The real value of cybernetics is its application to exceedingly complex systems.* A good example of a complex cybernetic system would be the entire information and control system for the space program— including initial planning, the building and assembly of all the parts and subsystems throughout the country, the checking and control of all systems before launch and during the launch, trip and return phases.

Cybernetics, or homeostasis, is a theme or characteristic of life from the molecular level to the population level. Cybernetic theory is now being applied in our attempt to understand the complex interactions in our biosphere and in society as well as the interactions between man and machine and machine and machine. It can and is beginning to be applied to economic and political systems,[2] to mass transportation systems, to developing cybernetic models of cities, countries and eventually the globe in order to enable us to predict the impact of man's intrusion into the biosphere and to suggest what we should and should not do. Just as the industrial revolution freed some (but still not the majority) of mankind from routine tasks based on muscular energy, the cybernetic revolution could free man from routine "intellectual" tasks of monitoring, predicting, and controlling (or showing us why we can't or shouldn't attempt to control) many of the simultaneous and exceedingly complex processes upon which life depends. Instead of spending most of our lives in routine and "dehumanizing" tasks, a "cybernetic revolution" offers us the possibility of being freer to explore a wide range of creative and purposeful activities.

[2] An interesting application of cybernetics to city government is found in E. S. Savas, "Cybernetics in City Hall," *Science*, *168*, 1066 (1970).

The development of cybernetics during the past two decades may be one of our most important discoveries. At the same time that our capacity for disruption of the global ecosystem threatens our survival, we have developed a technological concept that potentially enables us to deal with exceedingly complex systems. Like all scientific ideas, cybernetics can be used for good or evil—to guide missiles or to guide us in living in harmony with nature. The choice is ours.

8-2 Information, Order, and Negentropy

Since a cybernetic system involves control through an input of information, usually in the form of negative feedback, the theory of cybernetics is based on information theory. An important part of this theory is that the concept of *information* is closely linked to the concept of *entropy*, as defined on a statistical basis.

What is the relationship between entropy and information? We constantly receive information via devices such as clocks, books, stop lights, and so on. Messages represent a form of pattern and organization. The words in a dictionary convey a certain degree of order and information. If an author such as Shakespeare takes some of these words and creates a play, the pattern of words in the play represents a different pattern of order or entropy.

Receiving a message involves an increase in information. A positive information change, ΔI, means that there has been an increase in order. This is equivalent to a negative change in entropy, or disorder (that is, $-\Delta S$). Brillouin, one of the founders of information theory, suggests that the information change brought about by a message can be interpreted as the negative of the entropy change in the passage between two states. He suggests that information can be considered as *negentropy* (the negative of entropy).

Information and Negentropy

$$\text{Entropy} \propto \frac{\text{Disorder or}}{\text{disorganization}} \propto \frac{\text{Ignorance or}}{\text{missing information}}$$

$$\text{Information} \propto \frac{\text{Order or}}{\text{organization}} \propto \frac{\text{Negative of entropy}}{\text{or negentropy}}$$

Change in information = Negative of the entropy change
$$\Delta I = -\Delta S$$

The information change, ΔI, may be expressed in entropy units of calories per °K, but this is not a very useful unit for expressing the quantity of information in a message. The basic unit for information content is the *binary digit*, abbreviated as *bit*. A *bit* is defined as the quantity of information necessary to specify a single state in a system whose states may be specified by a binary choice (yes or no, 1 or 2, etc.) between two alternative events of equal probability. It is important to note that a gain of information (increase in order) does not mean that the second law is violated. An observer can obtain a bit of information only as a result of an increase in the entropy of the surroundings. For example, a stop light that provides information by blinking on and off requires electrical energy, and the production of this electrical energy increases the disorder of the surroundings to a greater degree than the order increase due to the information received.

The concept of entropy or negentropy as a measure of information provides a means for measuring information quantitatively. This approach can then be used to analyze information input and feedback in cybernetic systems.

8-3 Entropy, Poetry, and Literature[3]

The concept of entropy and the Second Law of Thermodynamics are important in poetry and literature in at least two ways. First, the idea of the never-ceasing forces leading to decay and disorder, as embodied in the second law, are important in describing the human condition. As a result, the idea of the second law appears again and again in poetry and literature.

One of the oldest expressions of this law is in the nursery rhyme

> Humpty Dumpty sat on a wall
> Humpty Dumpty had a great fall!
> All the King's horses
> And all the King's men
> Couldn't put Humpty Dumpty together again.

Proverbs such as "don't cry over spilt milk" and "don't expect a second chance" can be regarded as expressions of the second law. Man continually engages in attempts to create order, but only at the expense of greater disorder in the surroundings. Carl Sandberg very eloquently noted man's attempts to create order from disorder, but

---

[3] For more information on this topic, see Chapter 9 of the superb elementary introduction to thermodynamics by S. W. Angrist and L. G. Hepler entitled *Order and Chaos* (Basic Books, Inc., 1967). This book is highly recommended as a readable and interesting introduction to thermodynamics.

he recognized that in the end man would lose the battle. His poem "Under"[4] describes this struggle.

> I am the undertow
> Washing tides of power
> Battering the pillars
> Under your things of high law.
>
> I am a sleepless
> Slowfaring eater
> Maker of rust and rot
> In your bastioned fastenings,
> Caissons deep.
>
> I am the Law
> Older than you
> And your builders proud.
>
> I am deaf
> In all days
> Whether you
> Say "Yes" or "No."
>
> I am the crumbler:
> Tomorrow.

Another poem by Carl Sandberg, entitled "Limited,"[4] provides an expression of tendency toward increasing disorder.

> I am riding on a limited express, one
>   of the crack trains of the nation,
> Hurtling across the prairie into the
>   blue blaze and dark air go fifteen
>   all-steel coaches holding a thousand people.
> (All the coaches shall be scrap and rust
>   and all the men and women laughing
>   in the diners and sleepers shall
>   pass to ashes)
> I ask a man in the smoker where
>   he is going and he answers: "Omaha."

[4] From *Chicago Poems* by Carl Sandburg, copyright 1916 by Holt, Rinehart & Winston, Inc. Copyright 1944 by Carl Sandburg, reprinted by permission of Holt, Rinehart & Winston, Inc.

Note Franz Kafka's statement: "In the fight between you and the world, back the world." Shakespeare in Cymbeline said

> Golden lads and girls all must
> As chimney-sweepers, come to dust.

Study Question 8-6   From your own reading try to identify other poems, books, passages, songs, or works of art that seem to express the idea embodied in the Second Law of Thermodynamics. Can you identify some that express man's desire to fight or to overcome the second law?

A second use of entropy is in the statistical analysis of language. It is frequently desirable to analyze a poem or a piece of literature on the basis of its linguistic structure. A given poem, novel, or other work is divided into structural elements—for example, the number of letters in a word, the number of syllables in a word, the syntactical character of a word, or the use of certain phrases. Since this analysis involves various types of information, it can be expressed in terms of negentropy. Different textual materials can then be analyzed for frequency of occurrence of a particular linguistic characteristic.

With the advent of high speed computers this rather tedious form of linguistic analysis can be done rapidly and accurately. The uses of this technique are only beginning to be explored. Some obvious uses would be to aid in establishing authorship. For example, a comparison of the works of Marlowe and Shakespeare could be made to obtain data to support or deny the allegation that Shakespeare plagiarized[5] some of Marlowe's works. The writings of a particular author over a period of time could be analyzed to see if he made any characteristic structural changes as he went through different "periods" of writing. In addition, linguistic characteristics of various languages could be compared to see which languages carry the most information per syllable. This type of analysis could be used to design a universal language.

Wilhelm Fucks[6] of Germany has used this technique to examine a number of works of literature. Table 8-1 summarizes some of his work on the analysis of the number of syllables per word and negentropy per word for works in several languages.

In analyzing these results, remember that a high value of negentropy indicates a high information content. We can see that English appears to have the lowest number of syllables per word. This means that it takes simpler words on the average to express messages in English as compared with the other languages. On the other hand, the average negentropy per word is lowest for English and greatest for Latin. This can be

---

[5] Wilson Mizner said, "when you steal from one author, it's plagiarism; if you steal from many, it's research."

[6] Wilhelm Fucks, "Mathematische Analyse des literaischen Stils," in *Studium Generale*, **6**, 506 (1953).

Table 8-1   Average Number of Syllables and Negentropy per Word for Various Works.

Author	Work	Language	Average Number of Syllables per Word	Average Negentropy per Word
Shakespeare	Othello	English	1.29	0.29
Galsworthy	Forsythe Saga	English	1.34	0.33
Huxley	Brave New World	English	1.40	0.37
Thomas Mann	Buddenbrooks	German	1.74	0.51
Karl Jaspers	Der Philosophische Glaub	German	1.89	0.50
Euripides	Orestes	Greek	2.13	0.59
Sallust	Epistulla II	Latin	2.48	0.64

interpreted to indicate that the amount of information per word in English, on an average, is less than that in other languages. It would be interesting to analyze a passage in the Bible or other work that has been translated into numerous languages to see how much information is lost or added in translation from one language to another.

One important point must be made about calculations of negentropy, or information content. It represents a structural linguistic analysis. Although it is certainly a useful quantitative technique for comparing certain aspects of various works, it provides no analysis of the value or beauty or wisdom of the information received.

Information theory has also been applied to analyses of art, music, esthetic perception and to psychology.[7] Indeed, the use of the entropy concept in cybernetics has only begun to be explored.[8]

8-4   Entropy and the Genetic Code

The most amazing property of all living cells is their ability to reproduce themselves almost exactly, not just once but over and over again, for thousands and thousands of generations. In the past two decades we have learned a great deal about this

---

[7] Further applications of entropy, information and cybernetics are found in A. Moles, *Information Theory and Esthetic Perception*, University of Illinois Press, 1968; J. R. Pierce, *Symbols, Signals and Noise*, Harper and Row, Publishers, 1961 (Chapters 12 and 13); *Modern Systems Research for the Behavioral Scientist*, W. Buckley, ed. Aldine Publishing Co., 1968, and C. G. Jung, On Psychic Energy. In *Collected Works*, Vol. 8, Princeton University Press, 1959.

[8] For a readable review of the impact of information theory on communications and engineering see E. N. Gilbert, "Information Theory After 18 Years," *Science*, **152**, 320 (1966).

transmission of genetic information—summed up by the term genetic code. What is the genetic code, and how is it related to entropy? Let us begin by reviewing some of our present knowledge of molecular genetics.

Nucleic Acids and Proteins—The Molecules of Life

The cells in most organisms contain a control center known as the cell nucleus. Inside the nucleus we find the chromosomes, which in turn contain the *genes*—basic units used to transfer information from parent to offspring. One of the major triumphs of modern molecular biology was the discovery that genes are composed of a class of molecules called *deoxyribonucleic acids*, or *DNA*, and that genetic information is contained in the particular sequence of building blocks, or nucleotide units, in the DNA molecule. Thus, the basic idea of modern molecular genetics is that *all usable genetic information consists of nucleotide sequences in DNA molecules.*

But this is only part of the story. Life processes require that a vast number of specific *protein* molecules be manufactured continuously in the cell in the right amount, at the right time, and in the right place. Proteins are long-chain three-dimensional molecules made by linking together subunits called amino acids. *How is the genetic information contained in the sequence of nucleotide units in DNA translated into a sequence of amino acids in protein molecules?* This is the fundamental question of the genetic code. This synthesis of proteins, using DNA as a master template, is now believed to involve another important class of nucleic acids, called the *ribose nucleic acids*, or *RNA*. *Thus, the key molecules of life appear to be the nucleic acids (DNA and RNA) and the proteins.* To understand their role in the transmission of genetic information we must first examine their structures.

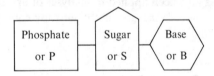

Figure 8-6   A Nucleotide Unit.

DNA and RNA. Nucleic acids are long-chain molecules, or polymers, of units called *nucleotides.* Each nucleotide unit contains three parts: a phosphate group (P or P_i), a pentose sugar (S), and a nitrogenous base (B). A nucleotide can be represented schematically, as shown in Figure 8-6. These single nucleotide units can be joined to form a long, single-strand polynucleotide, or nucleic acid, with the linking, or bonding, of units occurring between the sugar group (S) of one unit and the phosphate group of another unit, as shown in Figure 8-7.

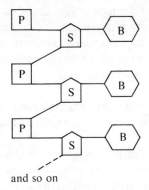

and so on

Figure 8-7 A Polynucleotide, or Nucleic Acid, Formed by Linking Nucleotide Units.

The phosphate group consists of a phosphorus atom and four oxygen atoms joined to form the following structure:

$$
\begin{array}{c}
① \\
\uparrow \\
\vdots \\
O \\
| \\
O{=}P{-}O\cdots\rightarrow② \\
| \\
O
\end{array}
$$

At position ① the phosphate is linked to a sugar unit in one nucleotide, and at position ② it is linked to a sugar unit in a second nucleotide unit.

$$
\text{Ribose Sugar Unit}
$$

Ribose Sugar Unit   Deoxyribose Sugar Unit

Figure 8-8 The Two Pentose Sugar Units in a Nucleotide.

A pentose sugar (S) is a carbon–hydrogen–oxygen compound containing 5 carbon atoms. There are two kinds of pentose sugars found in nucleotide units: ribose and deoxyribose. The deoxyribose sugar contains one oxygen less than the ribose sugar, as shown in Figure 8-8. Dotted lines show bonds to phosphate and nitrogen base groups in one nucleotide and to the phosphate in another nucleotide unit in the chain. Points ① and ② are linked to phosphate groups, and the bond at point ③ is to the nitrogen base.

In the nucleic acid, RNA, the sugar unit is ribose, as the name *ribo*nucleic acid indicates. In DNA, deoxyribose is the sugar unit, as the name *deoxy*ribonucleic acid implies.

There are two general classes of nitrogenous bases found in nucleic acids—the *purines* and the *pyrimidines*. These are carbon–hydrogen–oxygen–nitrogen compounds existing in pentagonal or hexagonal ring structures. The purines have two connected rings while the pyrimidines consist of a single ring. There are a number of different purine and pyrimidine bases, but only five different ones are commonly[9] found in the nucleic acids RNA and DNA, as shown in Figure 8-9.

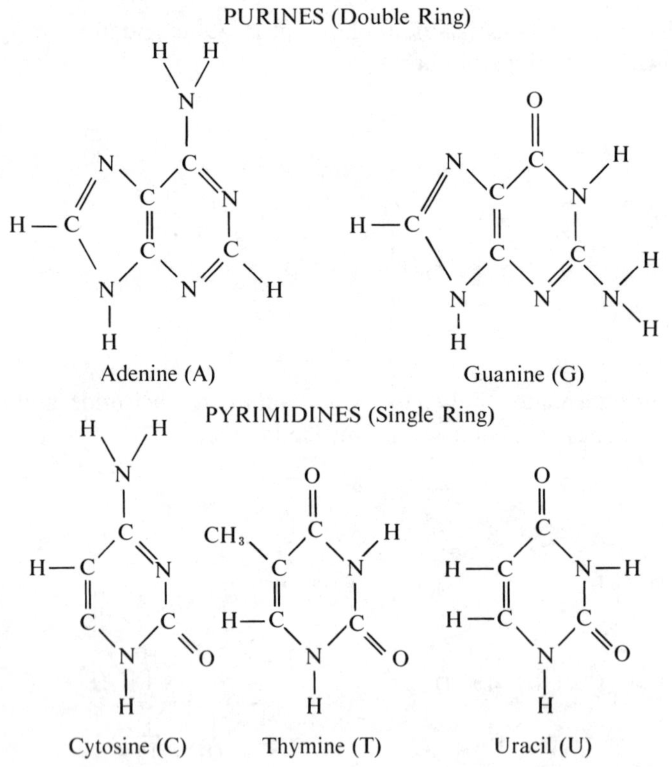

PURINES (Double Ring)

Adenine (A)   Guanine (G)

PYRIMIDINES (Single Ring)

Cytosine (C)   Thymine (T)   Uracil (U)

Figure 8-9   The Five Nitrogen Bases Most Commonly Found in Nucleic Acids.

[9] Other bases are found in some nucleic acids, but these are by far the most common.

Only four of these five bases are commonly found in DNA and RNA. Uracil (U) is not found in DNA and thymine (T) is not found in RNA, as shown in Table 8-2.

Table 8-2   Nitrogen Bases Commonly Found in Nucleic Acids.

Name	Symbol	Commonly Found in
Adenine	A	DNA and RNA
Guanine	G	DNA and RNA
Cytosine	C	DNA and RNA
Thymine	T	DNA only
Uracil	U	RNA only

We can now put the three units—phosphate, deoxyribose sugar, and the four nitrogenous bases (A, G, C, and T)—together to form the nucleic acid, DNA, as shown in Figures 8-10 and 8-11.

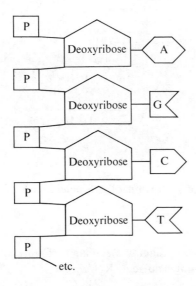

Figure 8-10   Schematic Diagram of a Portion of a Single Chain of Deoxyribonucleic Acid, DNA.

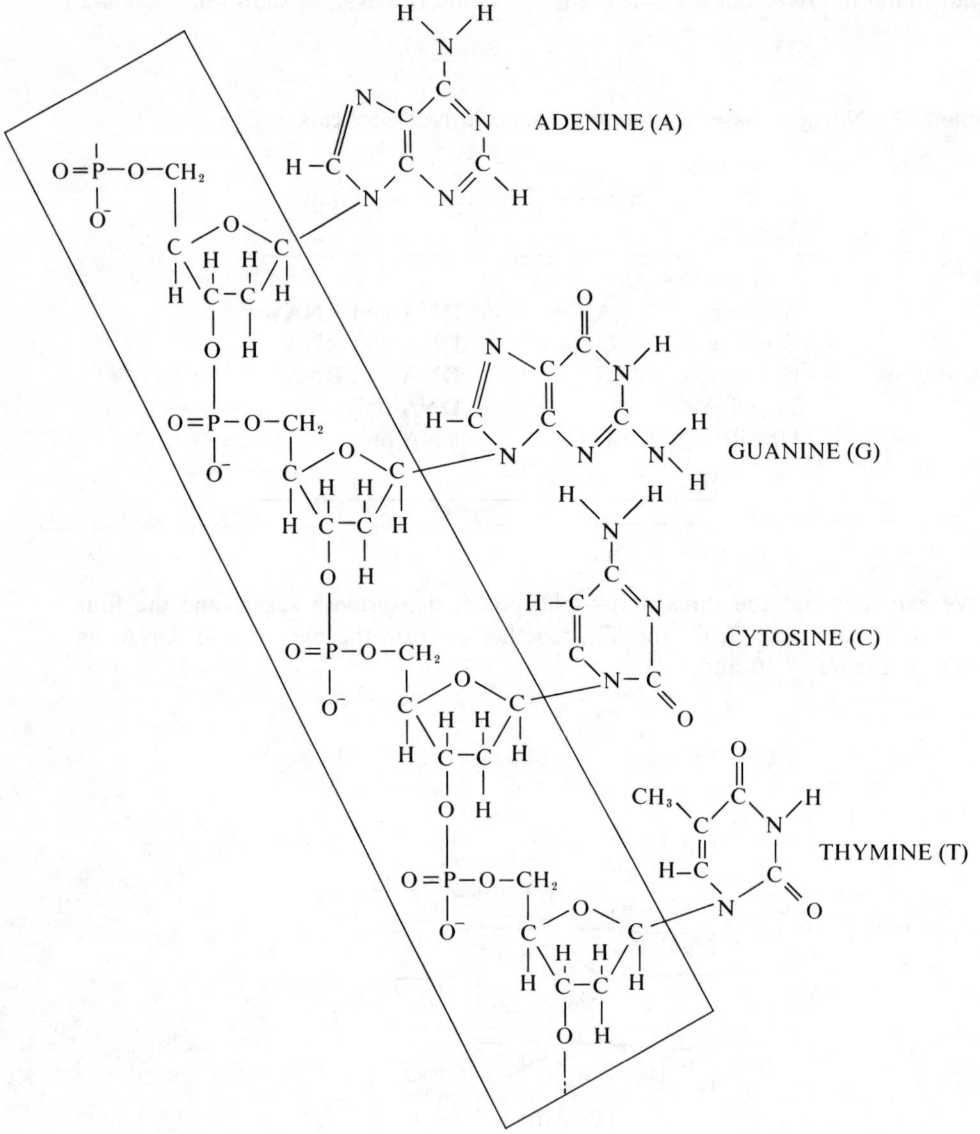

Figure 8-11 Diagram of a Portion of a Single Chain of Deoxyribonucleic Acid, DNA.

Study Question 8-7 Draw diagrams similar to Figures 8-10 and 8-11 showing a portion of a strand of ribonucleic acid, RNA.

Note that the phosphate (P) and deoxyribose sugar units form a sort of "backbone" for the molecule. This phosphate-sugar backbone remains the same, but the four nitrogenous bases may occur in many different orders in the long chain, as shown below.

A G C T T C G G C T A G T T C A A G

The variation, or specific nature, of a particular strand of DNA is determined by the sequence of the purine and pyrimidine bases attached to the sugar-phosphate backbone. A DNA chain with only 100 rungs could be arranged in over 10^{130} different ways—a number 10^{50} times larger than all of the atoms estimated to be in the universe.

In 1953, M. H. F. Wilkins, J. D. Watson, and F. H. C. Crick proposed that DNA is not usually in the form of a single chain, or strand, as shown in Figures 8-10 and 8-11. Instead, it is normally found in two very long chains twisted about one another in the form of a regular double helix, as shown in Figure 8-12.

It can be seen that the molecule forms a sort of spiral staircase with the sugar-phosphate backbone forming the outside handrails and with pairs of bases from each strand linking together by weak hydrogen bonds to form the individual steps, or rungs, in the staircase.

Watson and Crick analyzed X-ray diffraction data on the DNA molecules and, from this and other data, deduced that only certain base pairings occur. Adenine (A) is always paired with thymine (T), and cytosine (C) is always paired with guanine (G), as summarized in Table 8-3.

The DNA molecule is a carrier of information; its information is contained in four possible pairings of the nitrogen bases that form the "rungs" connecting the two

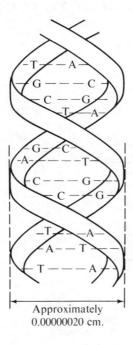

Approximately
0.00000020 cm.

Figure 8-12   A Portion of the DNA Molecule as a Regular Double Helix—the Watson-Crick Model.

Table 8-3 Possible Base Pairings in DNA.

Name of Pair	Symbol	Abbreviated Letter
Adenine-Thymine	A-T	A
Thymine-Adenine	T-A	T
Guanine-Cytosine	G-C	G
Cytosine-Guanine	C-G	C

threads of the DNA molecule, as summarized in Figure 8-13, where D stands for deoxyribose sugar.

Letters can be used to designate the particular sequence of rungs. Since adenine (A) is always paired with thymine (T) and guanine (G) is always paired with cytosine (C), a single letter representing the base on the left of the rung can be used as a letter to describe which of the four types of rungs is found. Thus, the message in a DNA molecule consists of four letters, represented by A, G, C, and T.

A typical DNA molecule might have 30,000 rungs or steps in the staircase and different DNA molecules would have different sequences of information. The message spelled out in such a molecule could be expressed as

A C G T G C A C C A C T G A C C T T A G C . . . .

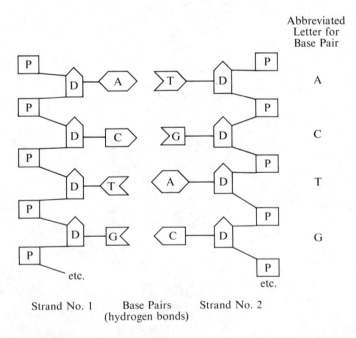

Figure 8-13 Base Pairing between Strands in a Portion of a DNA Molecule According to the Watson-Crick Model.

Remember that genes are segments of DNA molecules. Thus, all genes consist of arrays of the four possible base pairings, but they differ from one another in the particular sequence of the base pairs. In this sense the DNA molecule contains *information* in the form of the particular sequence of basic pairs. This is the genetic information that can be passed on. DNA is found in all species, and the essential genetic difference between you and a horse lies in the coded sequence of the base pairs. Just as a sentence in English contains information according to the particular sequence of the 26 possible letters of the alphabet, the DNA molecule passes on genetic information according to the sequence of the 4 possible letters in the genetic alphabet. A huge DNA molecule can contain many genes, or messages, just as a sentence may contain many words.

The Architecture of Proteins

Proteins carry out many important roles in the cell, including functioning as enzymes in regulating the rate of reactions in the cell and serving as structural fibers in muscle, hair, nails, and skin.

They are very large molecules or polymers formed by linking together a chain of basic building block units derived from *amino acids*. The basic repeating unit in a protein contains four essential elements: carbon, hydrogen, nitrogen, and oxygen arranged as shown in Figure 8-14. These basic units are repeated many times and in

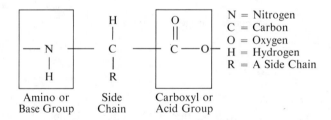

Figure 8-14   The Basic Repeating Unit in a Protein. Different Proteins Have Different R or Side-Chain Units.

different sequences to form a particular protein molecule. All units contain the same amino (base) group and carboxyl (acid) group, but they differ in the nature of the characteristic "R or side-chain group." The number, kind, and arrangement of the atoms in the "R group" distinguish one protein unit from another. The proteins of life are constructed from about 20 different units, each derived from a particular amino acid.

The repeating units are derived from one of 20 different amino acids, which have different R groups (R, R′, R″, etc.) and contain an additional hydrogen (−H) attached to the nitrogen and a hydroxyl group (−OH) attached to the carbon, as indicated in Figure 8-15. Figure 8-16 shows the actual structure of five of the 20 common amino acids. Proteins are formed by linking together various combinations of the twenty common amino acids in chains of 50 to over 1,000 amino acid units. The extremely

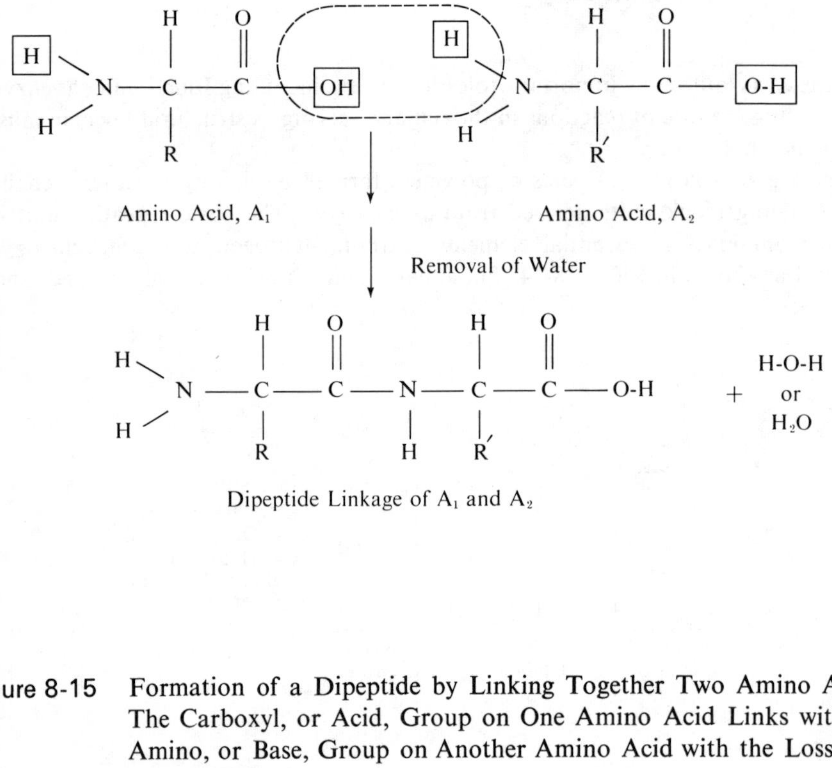

Figure 8-15   Formation of a Dipeptide by Linking Together Two Amino Acids. The Carboxyl, or Acid, Group on One Amino Acid Links with the Amino, or Base, Group on Another Amino Acid with the Loss of a Water Molecule.

large number of possible combinations results in a vast number of different proteins with different sizes, shapes, and specialized properties. As we saw in Chapter 7, this chemical specificity is a key property for proteins that serve as enzymes.

The molecular key to the uniqueness of each type of protein is in the particular sequence of amino acids—known as the *primary structure*. The 20 amino acids serve as the alphabet to build words of varying length and composition. Changing the sequence alters the properties or "meaning" of a protein just as changing a letter in

Name	Structural Formula	Abbreviation

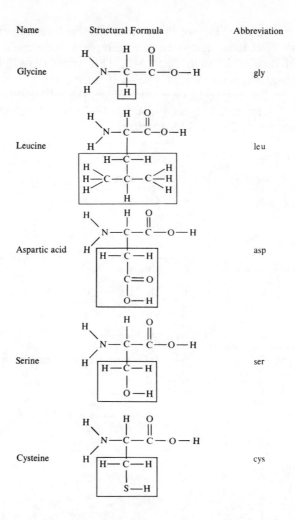

Figure 8-16   Structures of Five of the Common Amino Acids. Characteristic R
              Groups (shaded portion) Are Attached to a Common Backbone of
              Amino and Carboxyl Groups.

a normal word can make a word useless or nonsensical or give it another meaning. For
example, changing one letter in the word *slip* can give it different and meaningful
forms, such as sli*m* and sli*t*, or nonsense forms such as sli*h* or sli*x*.

Similarly a change in the amino acid or letter sequence in a protein can result in a
new protein that is not able to carry out its function. For example, people with sickle-
cell anemia have an abnormal hemoglobin protein molecule that differs from normal
hemoglobin by only one out of 300 amino acids. This minute change at the molecular
level alters the ability of the hemoglobin to combine with oxygen.

Thus, the primary message in a protein is based on the sequence of amino acids.

. . . - gly - val - asp - leu - cys - gly - ser - his - leu - . . .

But this is not the entire picture. Proteins have higher levels of structure, or organization that also affect their properties. For example, a linear chain may fold back and cross-link with itself as shown in Figure 8-17. This type of cross linking can occur with the presence of amino acids such as cysteine (Figure 8-16). If two cysteine molecules occur relatively close together in a peptide chain, the sulfhydryl (—S—H) groups on each cysteine can interact to form a disulfide (—S—S—) bond that links two parts of a chain together to form a folded structure (Figure 8-17).

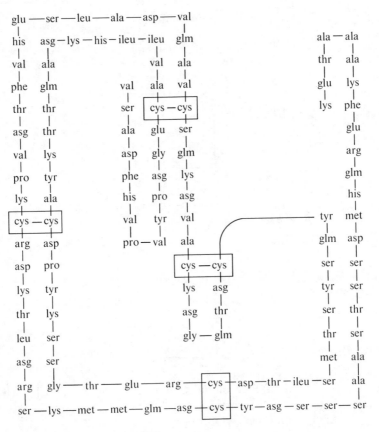

Figure 8-17   Folded Structure of a Single Chain of the Enzyme Ribonuclease as a Result of S—S Bonds Between Two Cysteine Units.

This type of bonding is important in the "permanent wave" treatment for human hair, since hair is a protein containing cysteine. At an early stage of growth the S—S bonds do not form and the hair is straight, but later exposure to air and other conditions causes the S—S bonds to form, and the hair stiffens and becomes coiled or wavy. To remove and then reform this waviness or stiffness in a particular "hairdo," chemicals are added to break the —S—S— bonds. The hair becomes more flexible and can be shaped into a desired pattern. When heated, the —S—S— bonds reform and hold the hair in the new shape—a "permanent wave" that really only lasts for a short time.

The sequence of amino acid units and any disulfide (—S—S—) cross-links in a chain constitute the primary structure of a protein. However, more complex shapes, known as the secondary, tertiary, and quaternary structures of a protein, are possible. Few proteins exist as a linear chain. Most are coiled or twisted in a number of ways through the formation of relatively weak *hydrogen bonds* that form between the carboxyl, or —C=O, group of one amino acid and the amino, or —N—H, group of a nearby amino acid, as shown in Figure 8-18.

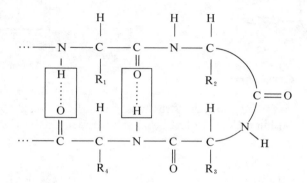

Figure 8-18   Cross-Linking in a Protein Chain by Weak Hydrogen Bonding between the Amino Group (—N—H) in One Amino Acid and the Carboxyl (—C=O) Group in Another Amino Acid Unit.

When hydrogen bonding occurs between amino acid units in the same chain, the protein usually coils in the form of an *alpha helix* while hydrogen bonding between two separate polypeptide strands leads to the side-to-side linking of the chains in an untwisted form known as a *beta configuration*, found, for example, in silk and other protein fibers. Although a single hydrogen bond is weak, the formation of many hydrogen bonds in a long-chain protein act together to produce stable and more complex shapes, known as *secondary structures*, as shown in Figure 8-19.

Many of the properties of proteins result from the breaking and reforming of their weak hydrogen bonds in the secondary structure of the proteins. Proteins with the alpha helix structure, which includes most enzymes, are usually soluble and reactive chemically in the cells, while proteins in the beta configuration are less reactive and are frequently found in structural parts of an organism, such as bone and cartilage. The properties resulting from hydrogen bonding can be changed or destroyed by changing conditions such as acidity or temperature. In a relatively acidic solution the hydrogen bonds are broken and the helical structure disappears. Gentle heating can also break the hydrogen bonds and allow the helix to uncoil. When cooled, the helices can form again, but if the heating is vigorous as in the cooking of an egg, the helical structure is irreversibly destroyed, and the protein is said to be *denatured*. This is somewhat analogous to pulling a coiled door spring. If not stretched too far, it springs back into shape, but if pulled too far it kinks and does not return to its former shape.

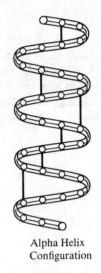
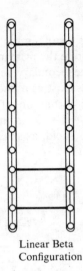

Alpha Helix
Configuration

Linear Beta
Configuration

Figure 8-19 Secondary Protein Structures Formed by Weak Hydrogen Bonding within a Chain (alpha helix) and between Chains (beta configuration).

These secondary structures can fold, roll up, or twist together in threads or ropes to form even more complex geometrical shapes, known as *tertiary structures*, as represented in Figure 8-20.

Even more complex or *quaternary protein structures* can form when several protein molecules join together. The simple case in Figure 8-21 shows several helices joined together to form a "rope, or cable," structure.

It is not surprising, therefore, that there are so many different proteins with different

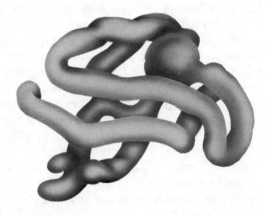

Figure 8-20 The Tertiary Structure Formed by the Crumpling or Folding of an Alpha Helix. This is a Crude Representation of a Molecule of Myoglobin, with the Off-Colored Shape in the Middle Being an Iron-Containing Granule. Myoglobin Is Found in Muscle Tissue and Is Active in the Process of Supplying Oxygen to the Muscle.

Figure 8-21   A Quaternary Protein Structure Formed by the Joining of Several Helices to Form a Rope, or Cable.

shapes, sizes, and specific properties. It is important to remember, however, that the *specific sequence and type of amino acids in the primary structure is the key factor leading to these more complex shapes and properties.*

The Genetic Code

Now that we understand some of the structural characteristics of the molecules of life—the nucleic acids and the proteins—we can discuss the process through which this genetic information is transferred. If DNA is the primary informational molecule, how is this information duplicated and transmitted to the millions and millions of other molecules—particularly proteins—being formed in the cell every few minutes? Typical bacterial cells *replicate*, or produce copies of themselves, within 20 minutes. For the cell to replicate, a large number of new, specific chemical compounds must be synthesized or assembled from the surrounding medium and linked to form a new cell.

If DNA is to guide this process, it must have general functions: (1) it must somehow provide directions for its own duplication as the cell divides so that its information is transmitted to the daughter cells, and (2) it must specify the structures of the proteins to be made continually during the lifetimes of these cells. The key questions are: How does DNA replicate itself and what is the code mechanism by which it translates

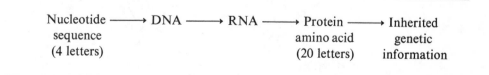

Figure 8-22   A Simplified Version of the Watson-Crick Central Dogma for the One-Way Flow of Genetic Information from DNA to RNA to Protein.

its four-letter nucleotide-base sequence (A, C, G, T) into an amino acid sequence using 20 letters?

In 1953 Watson and Crick postulated their double helical structure for DNA (Figure 8-12) which accounted for its x-ray diffraction pattern. They also suggested a simple mechanism for the transfer of genetic information—known as the Watson-Crick hypothesis. As it was used and extended it became known as the *central dogma* in the transmission of genetic information. It states that genetic information flows in one direction from DNA to RNA to proteins, as summarized in Figure 8-22.

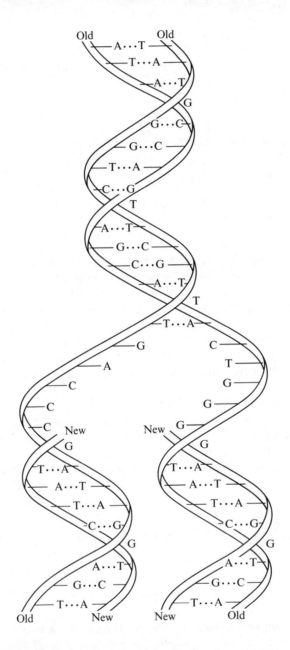

Figure 8-23   Early, Crude Hypothesis for Replication of DNA.

According to central dogma, three major processes occur in the transmission of genetic information: *replication*, or copying, of the DNA; *transcription*, or transcribing the message, from DNA to a form of RNA; and *translation*, in which the genetic message is decoded and translated into the 20-letter alphabet of the proteins.

An early hypothesis of the *replication* process proposed that the two strands of the DNA molecule separate, or "unzip," much like an opening zipper, with each separate strand then serving as a template for the formation of a new daughter strand. Once the molecule completely "unzips," one parent DNA can yield two daughter-DNA molecules that are exact replicas of the parent, as represented in Figure 8-23.

Nobel laureate Arthur Kornberg and Okazaki modified this earlier version of DNA replication in what is now called the Kornberg-Okazaki hypothesis. In the earlier model both new strands are being built in the same direction at the same time, but the Kornberg-Okazaki proposal, which is supported by an increasing amount of experimental evidence, states that as one strand is being built in a continuous strand in one direction, the other strand is built in short segments in the opposite direction, with these segments being joined later into a continuous strand by a joining enzyme, as shown in Figure 8-24.

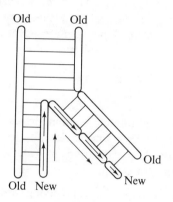

Figure 8-24   Kornberg-Okazaki Hypothesis for DNA Replication. One Strand Is Copied Backwards in Short Segments that Are Joined Later by an Enzyme.

The next two processes of transcription and translation are still not understood in any detailed sense, and there are a number of hotly debated proposals. A crude mechanism can, however, be outlined. Not only can DNA replicate itself, but in a similar manner the two strands can uncoil and bond to a form of RNA, known as messenger-RNA or m-RNH, as symbolized in Figure 8-25. Recall that RNA contains the sugar ribose instead of deoxyribose and that it contains the nitrogen base uracil, or U (Figure 8-9), in place of thymine, or T. Thus, the four-letter A, G, C, T alphabet in DNA is transcribed into a four-letter A, G, C, U alphabet in messenger-RNA.

What is the mechanism for translating the four-letter DNA or RNA language into the 20-letter protein language? How can a sequential structure of four different units be arranged in messenger-RNA so as to recognize 20 different units in a protein molecule? The answer to this was provided about 15 years ago by the physicist George Gamow and confirmed experimentally in 1962 by Francis Crick, Sidney

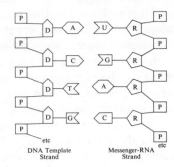

DNA Template Strand   Messenger-RNA Strand

Figure 8-25   Transcription of the Genetic Message from DNA to Messenger-RNA. Uracil (U) Takes the Place of Thymine (T) in DNA.

Brenner, and their co-workers. The question of how many nitrogen bases are needed to select one specific amino acid is (at least in retrospect) a relatively simple mathematical problem. Could it be a doublet code involving two bases for each amino acid? With two bases the number of possible arrangements is 4^2, or 16. Since there are 20 known amino acids in protein synthesis, a doublet code apparently would not work.

Study Question 8-8

a. Verify that there are only 16 different combinations of the four letters A, U, G, C using a doublet code of two letters per code unit, or codon.
b. Make a list of the 64 possible combinations of A, U, G, C using a triplet code with three letters per code unit, or codon.

However, if the code were a triplet code, involving three bases for each amino acid, then there would be 4^3 or 64 possible amino acids, more than enough.

Thus, in order to transcribe the message in the messenger-RNA and put one particular amino acid on the end of a growing protein chain, it is proposed that three base pairs, or rungs, of the m-RNA chain are needed. An RNA chain of 900 bases would be needed to translate the "genetic message" to a protein chain of 300 amino acids. These three rungs, or triplets of base pairs, which provide the instructions, or code, for assembling each particular amino acid, are called *codons*, and the genetic code reduces to a triplet code.

Triplet Genetic Code

codon ≡ three bases ≡ one amino acid
m-RNA   protein

Experimental evidence indicates that this assumption is reasonable, and the particular sequence of three bases for each of the 20 common amino acids has been determined experimentally. For example the code or three bases needed for the five amino acids in Figure 8-16 are glycine (UGG), leucine (UUC, UUG, or UUA), aspartic acid (UAG), serine (UUC), and cysteine (UUG). It was particularly surprising to find that for at least three of the 20 amino acids (asparine, leucine, and threonine), the code is *degenerate* with more than one triplet of bases yielding the same amino acid. It was also discovered that the same triplet of bases can yield different amino acids. For example, UUG can give both leucine and cysteine. Some scientists believe that this degeneracy might allow a doublet rather than a triplet code.

But how is this code transfer process carried out in the cell? After the messenger-RNA (or m-RNA) template is formed from one of the unraveled DNA strands, it is believed that the m-RNA travels to a portion of the cell known as a ribosome, which is about 70 percent ribosomal-RNA or r-RNA, a third type of RNA. The ribosome serves as a kind of "factory," or assembly line, for translating the four-letter RNA code (A, U, G, C) into the 20-letter amino acid code, with one codon, or base-pair triplet, needed for each amino acid. This translation process is carried out by a second type of RNA, known as *transfer-RNA* (t-RNA).[10] Each transfer-RNA molecule has a particular codon, or three-base pair, sequence on one end; and on the other end it has bonds that are highly specific so that one and only one amino acid molecule can attach there, as shown in Figure 8-26.

As the messenger-RNA slides over the ribosomal-RNA, the codons on one end of the transfer-RNA are matched to the codons on the messenger-RNA template originally formed from the DNA strand. A particular triplet of base pairs on one end of the transfer-RNA "recognizes" and aligns with a particular amino acid on the other end

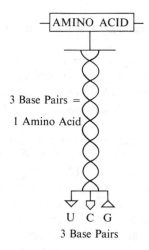

Figure 8-26   Schematic Diagram of Transfer-RNA.

[10] Sometimes the designation soluble-RNA, or s-RNA, is also used.

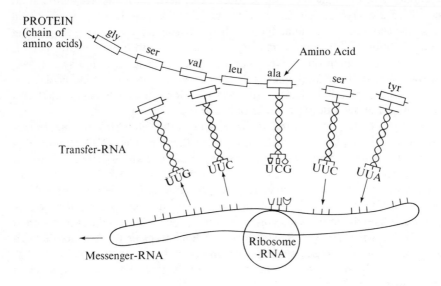

PROTEIN
(chain of
amino acids)

Amino Acid

Transfer-RNA

UUG UUC UCG UUC UUA

Ribosome
-RNA

Messenger-RNA

Figure 8-27   Crude Representation of Protein Synthesis.

of the transfer-RNA molecule, and by repetition of this process amino acids are linked together in a particular sequence to form a specific protein, as symbolized in Figure 8-27.

The overall process, or central dogma, for transmitting the *genetic code* can be summarized as follows:

Step 1:   Replication   DNA ⟶ another DNA
Step 2:   Transcription   DNA ⟶ messenger-RNA
Step 3:   Translation   messenger-RNA + transfer-RNA + specific amino acid
$\xrightarrow{\text{ribosome}}$ RNA peptides ⟶ Complete protein

Watson has estimated that under optimal conditions a protein chain of about 280 amino acids can be linked together in about 10 seconds. A typical cell contains some 15,000 ribosomes, so that in the cell a total of (280 × 15,000), or about 1/2 million, amino acids are being linked together every 10 seconds. In this way a single fertilized egg cell can multiply into an embryo and eventually grow into a human being whose body contains about 10^{14}, or 100 trillion, cells—again with each of these cells linking together about 1/2 million amino acids every few seconds during its lifetime. This amazing and incredibly complex process is guided by the information in a single original DNA molecule. If the total number of DNA molecules in your body could be stretched out in a straight line, they would form a thread running back and forth from the Earth to the sun (a distance of 93 million miles) some 33 times.

How valid are these ideas about the genetic code and the processes of genetic transfer as reflected by the central dogma? Many have assumed that the problem has been solved when indeed the work and ideas to date merely represent an extremely important beginning. We don't really know the details of the genetic transfer process, and indeed there has recently been an increasing amount of evidence to suggest that the central dogma is not necessarily always followed—it may not be a dogma after all. A growing number of experiments indicate that the flow of genetic information may not always be from DNA to RNA, but that in some cases the flow may be the reverse from RNA to DNA, with RNA acting as a template to produce DNA—a kind of chicken and egg problem has arisen to intrigue scientists and spur further research into the mystery of life. The search has merely begun, and many exciting events are to come.

Entropy and the Genetic Code

Now that we have a crude summary of the transmission of genetic information, what does all of this have to do with entropy? We are dealing with the transmission of information, and we have seen that negentropy can be considered as a measure of information. Negentropy could provide a means for calculating, or quantifying, the bioinformational content in living organisms.

If we could calculate the information content of DNA, it might be possible to calculate the information content of a single cell and then, by knowing the approximate number of cells in an organism (10^{14} for man) to calculate the information content of an organism. This is a relatively new and exciting application of information theory, and various calculations have been made using different sets of assumptions.[11] At this stage these calculations are incomplete, and indicate only some relative comparisons. One calculation indicates that a typical DNA molecule with a molecular weight of one million would contain about 8000 bits of information. Based on this, a typical cell would contain about 10^{12}, or one trillion, bits of information and a typical human body, which contains approximately 10^{14} cells, would contain the astronomical number of 10^{26} bits of information.

A single DNA molecule carries the equivalent of about 120,000 English words, or about two copies of this book. The DNA molecules in a single cell are equivalent to the information in about 50,000 books equal in length to this book—that is, about one trillion words for every man on earth today. The human body, which contains about 10^{14} cells, would contain the information equivalent of about 10^{23} books of this size. If the books containing the information equivalent to that in the cells of only one man were lined up, they would reach around the earth 10 million times.

But one must be cautious about overextending these rather simple calculations based on information theory. Information theory assumes that the choices are binary—that is, between two possibilities of *equal probability*. All of the information in DNA cannot necessarily be expressed in simple binary probability, and all of the possible links between two molecules are not necessarily equally probable. Furthermore, the idea of " information " in the DNA molecules is somewhat ambiguous. The

[11] D. H. Andrews and M. L. Boss have done some fascinating work connecting entropy and genetic information. For a description of their work, see Chapter 25 in D. H. Andrews, *Introductory Physical Chemistry*, McGraw-Hill Book Co., (1970).

amino acid sequence has several different functions, and thus there may be several types, or levels, of "information" in the DNA molecule. Much more research is needed on specifying the different types of "information" in the DNA molecule as it functions in a cell.

We have only scratched the surface in applying entropy and information theory to biological systems. Although information theory was formulated almost two decades ago, its applications to biology to date have been disappointing.[12] It remains a challenging frontier for those who can learn to use it or to modify the theory so that it can be applied more readily toward understanding the great beauty and complexity of life and toward enhancing rather than degrading human dignity and freedom.

> Each human being is unique, unprecedented, unrepeatable. The species
> *Homo sapiens* can be described in the lifeless words of physics and
> chemistry but not the man of flesh and bone.

> *René Dubos*

8-5 Review Questions

8-1. What is cybernetics?

8-2. Distinguish between cybernetics and homeostasis; between homeostasis and servo-mechanisms.

8-3. What is feedback?

8-4. Distinguish between positive and negative feedback.

8-5. What type of feedback is necessary for a system to remain stable? Explain.

8-6. Draw a general diagram showing the six components normally found in a simple homeostatic system.

8-7. Distinguish between ectothermic and endothermic organisms and give examples of each.

8-8. Draw cybernetic diagrams for three everyday systems not discussed in this chapter.

8-9. What is a homeostatic plateau? Draw a diagram indicating its meaning.

8-10. Draw a cybernetic diagram for population control.

8-11. What two major things distinguish man from other mammals in terms of population control?

8-12. Distinguish among the changes in entropy, negentropy, and information.

8-13. What is a bit of information?

---

[12] For a summary of some of the applications and failures of information theory in biology, along with some suggestions for modifying the original theory, see H. A. Johnson, *Science*, *168*, 1545 (1970).

8-14. Explain how negentropy, or information theory, can be used to carry out structural linguistic analysis of poems, novels, or other literary works. Give some examples of how this type of analysis can be used. What are some of its limitations?

8-15. What are some of the possible relationships between information theory and the genetic code? What are some of the problems or limitations?

8-16. Distinguish between deoxyribonucleic acid and ribonucleic acid.

8-17. Explain in words and with a diagram how the code in a DNA molecule can be expressed as a four-letter code.

8-18. What are nucleotides? nucleic acid? polynucleotides?

8-19. What is a gene?

8-20. What is a codon?

8-21. Explain briefly in words and in a diagram the proposed three-step process for the transfer of the genetic information from DNA to protein molecules. How is this central dogma of molecular genetics now being questioned.

8-22. What is a protein?

8-23. Distinguish between messenger-RNA and transfer-RNA.

8-24. What is the value of entropy, or information theory, to biology? What are some of its limitations at present?

Chapter 9

Thermodynamics, Kinetics, and Evolution

9-1 Does Life Violate the Second Law?

The previous chapter offered a glimpse of how entropy is related to information theory and cybernetics. And we saw that this relationship provides a basis for understanding some of the processes by which the functions of life are maintained through homeostatic control and through the transfer of genetic information.

In this chapter we shall explore the relationship between entropy and free energy and the origin and evolution of life on this planet. But first we must explore more deeply the relationships between life and the second law.

One of the first questions that arises is whether living organisms violate the second law. The formation and maintenance of a living organism give a vivid demonstration of the transformation of disorder into order. An array of atoms and molecules is synthesized to form a highly organized living cell. These cells then multiply and arrange themselves into even more highly organized forms. This process appears to contradict the tendency toward disorder imposed by the second law. Life appears to be an "island of decreasing entropy in a sea of increasing entropy."

Life Feeds on Negentropy

Until the 1940's, there was considerable speculation and fuzzy thinking about the "apparent" violation of the second law by living organisms. We now know that there are at least two reasons why life is *not* a violation of the second law. The first reason was stated in 1944 by one of the founders of quantum mechanics, Erwin Schrödinger.[1]

---

[1] Erwin Schrödinger, *What Is Life?* Doubleday & Co., Inc. (1956, paperback edition).

He indicated that a living organism feeds on the "negentropy" of its surroundings—it survives by extracting order from its surroundings. This results in an increase in disorder, or entropy, in the surroundings and when *both the system plus the surroundings* are taken into account, there is a net increase in disorder, or entropy, as called for by the second law.

Most apparent contradictions in the second law result when the observer forgets to take the surroundings into account. Human beings eat food and support themselves by burning fossil fuels and using other energy sources. Living cells take in food from their environment in the form of complex and highly ordered foodstuff molecules, such as glucose and amino acids. In the process of respiration these molecules are converted to smaller molecules, such as CO_2 and H_2O, which have a higher entropy than the starting molecules. When the smaller molecules are released back into the environment, there is a net increase in entropy greater than the order created in the cell.

Study Question 9-1   Trace the entropy increases and decreases in the system and surroundings that result when coal is formed, extracted from the earth, and then burned to heat a house. Show why this process does not violate the second law.

Study Question 9-2   Which one of the following changes represents a violation of the second law?

a. freezing of a liquid,
b. evolution of a living organism,
c. $CO_2(g) \longrightarrow CO_2(s)$,
d. $2H_2(g) + O_2(g) \longrightarrow 2H_2O(g)$,
e. a chemical reaction with a negative change in entropy,
f. none of these.

Photosynthesis also seems to contradict the second law, because it results in the organizing of small nutrient molecules into complex and organized forms of plant life. But here we run into the larger entropy increases in our solar system. Photosynthesis depends on solar energy. Nuclear fusion reactions in the sun produce astronomical amounts of disorder—a massive entropy increase that swamps the relatively tiny entropy decrease represented by life on Earth.

Open, Closed, and Isolated Systems

There is a second reason why life processes do not contradict the second law. The thermodynamic principles presented in this book are those of *equilibrium thermodynamics*. They apply strictly to *isolated* and *closed* systems. An *isolated* system is one in which there is no exchange of either matter or energy between the system and surroundings. It is totally isolated—neither matter nor energy ever leaves or enters the system. Strictly speaking, it is an ideal state which can never be achieved—only approached. Since life depends on a continuous input of solar energy, life as we know it is not possible in an isolated system. The type of system to which equilibrium thermodynamics is normally applied is a *closed system*. It is a system in which *energy, but not matter*, can be exchanged between the system and its surroundings. Consider a closed test tube containing chemicals. It can be heated or cooled and thus gain or lose energy, but the amount of matter in the system remains the same even though the chemicals could react to form different substances. As we have seen earlier, a *closed system* tends toward a state of *dynamic equilibrium* where the free energy of the system decreases toward a minimum and ΔG_{system} equals zero. In addition, the entropy of the system plus surroundings increases toward a maximum value, so that ΔS_{net} approaches zero.

A man living in a space capsule is another example of a closed system. His entropy is maintained at a low value, but for him to stay alive the entropy of the space capsule must increase as the complex and highly ordered food molecules are used and broken down into the smaller and more disordered molecules, such as CO_2, H_2O, and human waste. There will be a net increase in entropy in the capsule with time, as required by the second law.

There is a third and very important type of system—the *open system*. In this type, both matter and energy are exchanged between the system and surroundings. Consider an open container of boiling water. Both energy and matter are transferred between the system and surroundings. The growth, maintenance, and decay of all living plants and animals constitute open systems. Thus, *life involves an open, or flowing, system* in which the system is maintained by a balanced exchange of matter and energy with the environment.

Life on the spaceship, mentioned earlier, demands a continuous supply of energy and other essential materials. To stay alive the spaceman must convert his closed system to an open system or somehow maintain a steady state of life in the closed system. One possibility is to return to earth periodically and open the capsule to remove waste materials and take on a new supply of fuel, food, and other essential materials—thereby periodically converting his closed system to an open system.

Alternatively, if he cannot come back for refueling and new supplies—as is the case

---

Types of Systems

Isolated: No exchange of matter nor energy between system and surroundings. (Life not supportable)

Closed: Energy but not matter can be exchanged between system and surroundings. (Life possible only through outside energy and recycling of essential chemicals)

Open: Energy and matter can be exchanged between system and surroundings. (Life maintained by balanced exchange of matter and energy between the organism and its environment)

---

for us on spaceship earth—a recycling system must be set up and maintained at the proper efficiency and rate. Waste products could be fed to algae, which, with a constant input of solar energy, could synthesize food molecules through photosynthesis— providing algaeburgers for breakfast, lunch, and supper. These recycling systems would maintain the spaceman as long as energy from the sun is available and as long as the recycling systems are not destroyed or damaged.

Life Is a Steady State

A closed or isolated system can attain a state of dynamic equilibrium and thus can be treated from the standpoint of equilibrium thermodynamics. An open system, how- ever, cannot attain a state of dynamic or thermodynamic equilibrium. Instead it can attain what is called a *steady state*. The steady state is maintained only by a continual flow of free energy and matter into and out of the system.

A *closed system* tends spontaneously toward a state of *minimum free energy and maximum entropy*. By contrast, an *open system* is maintained on a plateau of higher free energy and lower entropy, as summarized in Table 9-1.

Table 9-1 Comparison of Open and Closed Systems.

System	Attains	Free Energy(G)	Entropy(S)	Exchange with Surroundings
Closed	Dynamic equilibrium	Low	High	Energy only
Open	Steady state	High	Low	Matter and Energy

Life is an open, or flowing, system maintained in a steady state. This requires that the system be maintained on a higher plateau, or pocket of free energy, in a delicate tension between the tendency toward minimum energy on one side and the tendency toward maximum disorder, or randomness, on the other hand. As long as there is a constant input of matter and free energy and a constant drainoff of entropy in the form of heat and waste material, the system maintains its steady state of life. If the necessary matter or energy inputs are cut off, the entropy of the system begins to rise, free energy decreases, and the living system becomes a closed system and eventually dies as shown in Figure 9-1.

Living organisms maintain their steady state within relatively narrow physiological limits in spite of wide fluctuations in environmental conditions. This is accomplished

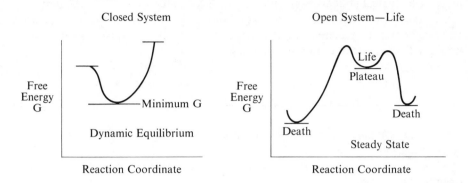

Figure 9-1 Comparison of a Closed System in Dynamic Equilibrium and an Open System in a Steady State.

through the cybernetic mechanism of *dynamic homeostasis*, involving information and negative feedback to alter the nature and rate of the chemical reactions necessary to maintain life. Death can occur by excessive ordering with limited adaptability to environmental stress or by excessive disorder where any critical part of the life processes fail. Man struggles during his lifetime to maintain a small island, or pocket, of order in an ocean of disorder. Within minutes after death the universe is more disordered than it was before we were born. Man can create temporary domains of order, but attempting to create order on a universal scale is like trying to freeze the Atlantic ocean with a small ice cube.

Nonequilibrium, or Steady-State, Thermodynamics

The steady state represents life's way of temporarily escaping the increasing entropy imposed by the second law. If living systems deviate from the steady state, they become inefficient and produce entropy at a higher rate. They survive by producing entropy at a minimal rate in the steady state.

This means that life cannot be described by the equilibrium thermodynamics developed in the earlier sections of this book. Instead, life and other open systems are described by a relatively new branch of thermodynamics known as *nonequilibrium thermodynamics*, or *steady-state thermodynamics* (also called irreversible thermodynamics). The basic principles of nonequilibrium thermodynamics still involve energy and entropy, but the emphasis is on energy flow through the steady-state system. An open system is constantly subjected to forces tending to move it toward thermodynamic equilibrium. To prevent this drift toward equilibrium—which means death—the organism must constantly perform work to maintain its steady state. The system must therefore be connected to an outside energy source so that energy can flow through the system. Thus, free energy flow, rather than entropy, becomes the focus in steady-state thermodynamics. Uses of this field are growing and, as it is applied more and more to biological systems, some important developments will be forthcoming.[2]

This does not mean, however, that equilibrium thermodynamics cannot be applied to living systems. It merely means that in applying it, we must take into account both the living system and its surroundings.

Equilibrium Thermodynamics: Can be applied to living, or open, systems by also considering the surroundings.
Nonequilibrium, or Steady-State, Thermodynamics: Can be applied exclusively to open, or living, systems.

Study Question 9-3   Which of the following are in a state of dynamic equilibrium and which are in a steady state?

a. a flame burning at a constant temperature,
b. a theater ticket booth that constantly has a line of five persons,
c. a growing tree,
d. a decaying tree,
e. the liquid water and water vapor inside a sealed glass container at constant temperature,
f. a dam with water flowing over it, with the water level in the lake above the dam being maintained at a constant level.

9-2 Heat Death of the Universe—Fact or Fiction?

Most books on thermodynamics contain the statement of the second law given by Clausius: "Die Entropie der Welt strebt einem Maximum Zu," or "the entropy of the universe is increasing toward a maximum." Using this version of the second law, many philosophers and scientists have envisioned the so-called "heat death," or "entropy death," of the universe at some point, probably billions of years into the future.

As we convert solar energy to other forms, it is degraded into less and less useful forms. Eventually all energy ends up at the lowest rung on the ladder of energy—heat energy. This addition of heat represents an entropy increase. Of course we can make

---

[2] Most books on irreversible, or nonequilibrium, thermodynamics require considerable mathematical sophistication. One important and readable application of nonequilibrium thermodynamics is H. J. Morowitz's *Energy Flow in Biology*, Academic Press, 1968.

heat flow back up the ladder, but this takes more energy, or work, than we got out of it in the first place and eventually adds even more heat, or entropy, to the environment.

If we assume that the universe is a closed system, then it is argued that it must eventually run down to a state of maximum entropy, or heat death, in accordance with the second law. The panorama of the evolution of life on this or any other planet is then only a temporary ripple of order. When the last star in every galaxy has burned out, and when the last chemical that stores free energy has been degraded to heat, the temperature throughout the universe will be constant. No activity of any type will be possible.

Although the total energy in the universe will remain constant in accordance with the first law, all energy will have been degraded to a common level of minimum free energy and maximum entropy. None of it will be available to perform any type of work to drive any chemical or physical processes. There would be no way of distinguishing one point in space and time from another. Time, space, and life would have no existence—no meaning.

This "heat death" hypothesis has intrigued philosophers and theologians. Some have described this as the meaning of "hell." Others have advanced it as a proof of the existence of God, by reasoning that if the universe is running down, then it must have been originally "wound up" in some nonequilibrium state. The theme has also been expressed again and again throughout literature. For example, Hotspur in Shakespeare's *Henry IV*, Part I, says

time, that takes survey of all the world must have a stop.

For the person who thrives on gloom and doom, worrying about the heat death of the universe is hard to beat. It diverts our attention from mundane problems to gloom on a cosmic time scale of billions of years. Is this gloomy projection a valid and legitimate application of the second law? There are two implicit assumptions in it.

First, it is assumed that the universe is an isolated system, one in which there is no matter or energy flow between the system and its surroundings. This further implies that the universe is a finite bounded body isolated from its surroundings. We do not know at this time whether the universe is finite or infinite, and we certainly can attach no meaning to the "surroundings" of the universe. If they existed, why should the second law necessarily apply? If the universe is finite, what are the properties of the boundary? Does energy or entropy leak in or out of this "boundary"? Is the "boundary" fixed or changing? Needless to say, we cannot answer these and other such fundamental questions.

Second, the laws of thermodynamics have been developed by observing an extremely small volume of space relative to the universe. We know so little about our tiny portion of the universe and have observed it for such a minute period of time relative to cosmic time, that extrapolating this meager knowledge to the entire universe seems highly speculative and perhaps somewhat arrogant. Such ideas may be useful and interesting, but they must be based on faith rather than on the science of thermodynamics.

In recent years cosmologists interested in the origin of the universe have questioned the validity of applying the second law on a universal scale. They argue that the concept of entropy throughout the entire universe is not defined clearly enough under all circumstances to make such broad generalizations. For example, how do we define the entropy of a gravitational field?

Furthermore, the three current contenders, or models, for explaining the origin of the universe appear to be at odds with one or both laws of thermodynamics, if they apply on a universal scale, as discussed below.

1. *The "big bang,"* or *expanding universe, model* postulates that the universe originated when all the matter in the universe collected in a superdense state and exploded. The universe has been expanding since that event some 12 billion years ago (assuming a constant rate of expansion). This model is consistent with the first and second laws after the "big bang," but it is silent about what happened or existed before the explosion.

2. *The pulsating universe model* postulates a series of alternating expansions and contractions occurring on about an 80-billion-year cycle—don't worry, the next reversal is not due for another 30 billion years. It answers the question, "What came before the big bang?" but it still does not explain how the whole cycle got started. This model avoids the need for creation and thus is consistent with the first law, but it alters the second law. During periods of expansion the second law as we know it would hold, but during contraction the reverse of the second law would apparently hold.

3. *The continuous creation,* or *steady state, model* postulates a forever expanding universe with matter in the form of hydrogen being continually created in the universe. The rate of hydrogen production has been estimated at the equivalent of 1 atom per year in a volume equivalent to that in the Empire State building. Even this "tiny" event would represent the "creation" of at least 10^{32} tons of hydrogen *per second* throughout the entire universe. This model denies the existence of a single moment of creation and therefore denies the first law as we know it.

For the moment, all we can say is that the second law applies in localized closed systems—such as the planet Earth.

It is interesting to note that two of the great principles of science—the *uncertainty principle*, developed from quantum mechanics at the microscopic level, and the *Second Law of Thermodynamics*, based on the macroscopic world—have been the source of much philosophical and theological discussion during the twentieth century. Many useful generalizations have come from these speculations, and no doubt the future will provide other broad applications. At the same time, it must be recognized that there have been numerous abuses and unwarranted extrapolations made by many nonscientists (and scientists) who have not understood some of the basic principles and limitations of both quantum mechanics and thermodynamics. In such cases these areas of science provide neither proof nor disproof, as many have attempted to claim, but only silence, because of their limited applicability.

9-3 Entropy as Time's Arrow

The concept and intuitive feeling of time is certainly one of the most important ideas of life, and men have constantly speculated on its meaning and significance.

One useful concept of time was presented by Sir Arthur Eddington in 1925. He suggested that the statistical concept of entropy underlies our concept of time. We feel intuitively that time is flowing forward like a river that cannot be reversed.

How do we arrive at this intuitive feeling? The answer may lie in the fact that the forward flow of time is determined by our observation that spontaneous processes go in the direction of increasing disorder, or entropy, when both the system and surroundings are considered. It is only by noting the direction in which disorder increases that we have any concept of the passage of time in the macroscopic world. Only if events are accompanied by a change in entropy can we determine the direction of time. In other words *entropy apparently determines the direction of time flow. Entropy is,* in Eddington's words, *the "arrow of time," or "time's arrow."*

This idea can be demonstrated by the running of a motion picture. If the film depicts events in which no objects are moved so that there is no directly observable entropy change, then it is impossible to tell whether the film is being run forward or backward. For example, suppose the film depicts a room barren except for a pile of blocks. No difference would be noticed whether the film were run forward or backward. If the blocks were scattered over the room by small children, this increase in entropy would be considered a measurement of the forward direction or flow of time. Suppose this segment of the film were run backward, with children walking backwards, blocks rising off the floor and the entire pile being spontaneously reassembled into a state of order. We would say that this represents reversal of time that would not occur in the real world.

Thus, entropy appears to be a useful indicator for the direction of time in the macroscopic world. Although this is a very important and far reaching idea, there are some inherent limitations and there are other ways of looking at time. Again we may run into the problem of extrapolating the second law too far. We have only observed phenomena for a blink of the eye relative to cosmic time. How do we know that time will not be reversed at some future date? Furthermore, the terms "direction of time" and "time's arrow" are vague terms that are used in a very loose and metaphorical sense. One may ask, "Whence and whither does time flow?" Many would answer "from the past to the future" but this merely begs the question. Time thus takes on the characteristics of some mystical or metaphysical river whose source lies in the infinite past and its destination in the infinite future.

Furthermore, according to relativity theory, every motion in space is relative and the present moment can be regarded as stationary with future events moving toward the past with equal and opposite velocity. This was eloquently stated by William James.

. . . the specious present, the intuited duration, stands permanent, like the rainbow on a waterfall, with its own quality unchanged by the events that stream through it.

In spite of some limitations and philosophical difficulties, the statistical concept of entropy as "time's arrow" is a very useful idea for describing the direction of time flow in the macroscopic world. Time is a mysterious and elusive idea that invites thought and speculation.[3] Albert Einstein reminded us that "the most beautiful thing we can experience is the mysterious. It is the source of all art and science."

---

[3] The interested reader is urged to explore a magnificent interdisciplinary study of the concept of time, *The Voices of Time*, edited by J. T. Fraser (George Braziller, Inc., 1966). A group of scholars discuss time from the standpoints of religion, philosophy, literature, music, history, psychology, biology, physics (relativity and quantum mechanics), thermodynamics, and cosmology.

9-4 The Origin and Evolution of Life

Where did we come from and how did we get where we are? How did life originate on earth, and how did it evolve to its present system of diverse and complex species—a state that has involved the evolution of some 200,000 different kinds of plants and over a million types of animals—all living in an interlocking network of chemical cycles and energy flow. We don't know the answers to these questions but considerable evidence has been gathered to support some tentative hypotheses. We shall explore the relationships of thermodynamics and kinetics to these important questions but first we must outline some of the current ideas about the origin and evolution of life.

The Characteristic Properties of Life

How did it all begin? Let us start by working back from what we now know about life. Although living organisms are composed of lifeless molecules, they possess some properties not found in collections of non-living matter. What are some of these major characteristics of life? *First*, we now recognize that living organisms have *a complex and highly organized internal structure*. For example, one of the smallest and simplest forms of life, such as the bacterium *E. coli*, probably contains at least 5000 different organic compounds, including 3000 different kinds of protein and perhaps 1000 different types of nucleic acids. A more complex organism, such as a human, may contain over 5 million different proteins. There are probably around 100 billion, or 10^{11}, different kinds of protein molecules and about 10 billion, or 10^{10}, different types of nucleic acids in the estimated 1.2 million species of living organisms. Since chemists have succeeded in synthesizing only about 10^6, or one million, different organic molecules, we clearly have a long way to go in understanding life at the molecular level.

One of the major achievements of modern biochemistry has been the discovery that the immense variety of organic molecules in living organisms are made up of small building block molecules that are linked together in long chains to form the two major classes of compounds upon which all life depends—*the proteins and the nucleic acids*. Thus, a *second* characteristic of life is that the *structural identity of each species of organism is found in its distinctive set of proteins and nucleic acids*.

Third, each part of a living organism, from individual chemical compounds to larger aggregates such as a cell nucleus or membrane, *appears to have a specific function*. Nucleic acids such as DNA and RNA—which compose the genes—are the key to self-replication in living cells. Proteins form the enzymes that control the rates of the complex chemical reactions occurring in the cell—including the production of chemical building blocks to produce more nucleic acids and proteins. The organic phosphates, such as ATP and ADP, are used for energy transfer in the cells. Nucleic acids and proteins thus form an interlocking system capable of replication and growth and change (mutation), so that genetic information is passed on as life evolves by natural selection.

Fourth, living organisms *extract and transform matter and energy from their environment* in order to build complex molecules and use them in an intricate, self-regulating system of chemical reactions required to maintain life. Living organisms are open thermodynamic systems—exchanging both matter and energy with their surroundings to maintain a steady state. A living cell exists essentially at constant temperature and pressure throughout the cell. Because of this they cannot use heat as a source of energy, since heat cannot do any work at constant pressure unless it passes from a zone of higher temperature to one at a lower temperature. Living cells must then obtain energy in some other form from their environment—for example, radiant energy from the sun or chemical energy in the nutrient molecules it receives. This chemical energy is then converted to other forms of energy and used to maintain the complex cellular processes necessary for life. A simple bacterial cell synthesizes simultaneously thousands and thousands of complex proteins and nucleic acid molecules in precise ratios and at specific rates. These are used in a complex and interconnected sequence of specific enzyme catalyzed reactions whose reaction rates are regulated by feedback to maintain the steady state of the living cell. The study of the kinetics and energetics of these life processes is known as *bioenergetics*, and it will be studied in more detail in the next chapter.

Fifth, perhaps the most remarkable characteristic of living organisms is their *capacity for self-replication*—the ability to reproduce themselves almost precisely over and over again for hundreds and thousands of generations. As we saw in the last chapter (Section 8-4), this genetic information is contained and preserved in the nucleotide sequence of a very small amount of the nucleic acid, DNA, weighing perhaps around one trillionth of a gram. We also saw that the one-dimensional information contained in the base pair sequence of the DNA is translated into three-dimensional protein structure by a three-letter code sequence.

Summary of the Characteristics of Life

1. Organisms have a complex and highly organized internal structure built up by linking together a small group of building block molecules.
2. Each species has a distinctive set of proteins and nucleic acids.
3. Each part of a living organism has a specific function.
4. Organisms can extract and transform matter and energy from their environment by a self-regulated system of enzyme-catalyzed reactions.
5. They have the capacity for precise self-replication.
6. They can evolve many diverse species by the mechanism of natural selection.

Finally, living systems *can evolve many diverse species by the mechanism of natural selection*. They not only reproduce, but they can mutate, or change, their patterns, and then reproduce these mutations. Natural selection can cause one kind of genetic information—for example, that producing people with brown eyes—to become more and more common in the gene pool of a population. *Evolution* consists of the changes in the gene pool with time. Contrary to the popular but erroneous interpretation of Darwin's theory of evolution by natural selection, the term "survival of the fittest" does not mean that species slug it out for food and resources in the battleground of life. Instead the key to natural selection is the *differential reproduction* of genetic types. A species that can develop resistance to changing environmental conditions and thus reproduce at a faster rate than another species has greater survival potential. It will survive as a species over some other species by producing a larger number of organisms in a given time period.

Chemical and Biological Evolution

In the past it has been traditional to view the origin of life as a result of design and to endow organisms with a mysterious life force. This doctrine, called vitalism, has been rejected by modern science. The modern view considers life on Earth as having been formed over billions of years through a natural, random sequence of chemical or molecular events governed by the First and Second Laws of Thermodynamics and by other physical laws.

A study of the origin and evolution of life involves studying both the *chemical evolution* of the molecules and chemical reaction systems necessary for life in the first cells and the *biological evolution* of the early cells to diverse species by natural selection.

Phases of the Origin and Evolution of Life
(about 4.8 billion years)

Phase 1 Chemical Evolution (1.7 billion years)

Part 1. Evolution of the nonliving molecules necessary for life—primarily the nucleic acids and proteins.

Part 2. Evolution of the systems of chemical reactions needed for maintaining and reproducing life (the early cells).

Phase 2 Biological Evolution (about 3.1 billion years)

Evolution of diverse species by the mechanism of natural selection.

We shall be concerned primarily with the first phase—*chemical evolution.*[4] In effect, this involves two questions:

1. How did we get the nucleic acids and proteins?
2. How did they form a system of interacting and self-duplicating molecules in a primitive first cell or protocell?

If we assume that these very large and complex molecules must form by the simultaneous collision and linking of the millions of atoms that comprise them, then life would, for all practical purposes, be impossible. Consider the probability of formation of a very small protein molecule—hemoglobin—a molecule consisting of only 539 amino acids. The number of different combinations of these amino acids—taking into account that some are the same—is about 4×10^{619}.

The important point is that out of the unbelievably large number of possible combinations of amino acids and other molecules essential for life, only certain particular combinations have occurred and evolved into more complex organisms. Early investigators considered the formation of even one protein molecule so improbable that for all practical purposes it was impossible. Two important points were overlooked. First was the concept of time. An outline of the time scale of chemical and biological evolution is given in Figure 9-2. It can be seen that the earth is believed to have been formed about 4.8 billion years ago (4.8×10^9). The period of chemical evolution is believed to have lasted about 1.7 billion years, or nearly one-third of the earth's history. Considering the billions and billions of years of time available, a vast array of random molecular events could have occurred, with the first living cells forming about 3.1 billion years ago. These early cells were not capable of synthesizing their own nutrients through photosynthesis. The blue-green algae—believed to be the first oxygen-producing photosynthetic cells—probably arose between 1.5 and 2.5 billion years ago. Until their appearance there was probably very little oxygen in the atmosphere, as we shall discuss in the next section. The development of higher plants and animals occurred over the past one-half billion or 500 million years—with Homo sapiens appearing only about 2 million years ago. In terms of a relative time scale of the history of the Earth man appeared only during the last 30 seconds of a 24-hour day.

Second, we now know that these molecules do not form all at once by the random collision of millions of smaller atoms or molecules. Instead, the process is stepwise, much like forming a long freight train by linking the cars one at a time. As we saw in the last chapter (Section 8-4), nucleic acids are long-chain molecules made by linking together subunits called nucleotides. Each nucleotide in turn consists of one of 5 nitrogen bases, one of 2 sugars, and one or more phosphate groups. Genetic information is encoded in the sequences of these nucleotides, and with only minor exceptions the same nucleotides are found in all known species. The differences between the genes of a man and, say, those of a dog lie only in the linear arrangements of these nucleotides.

---

[4] The reader interested in the details of the second phase—biological evolution by natural selection—should consult introductory biology texts.

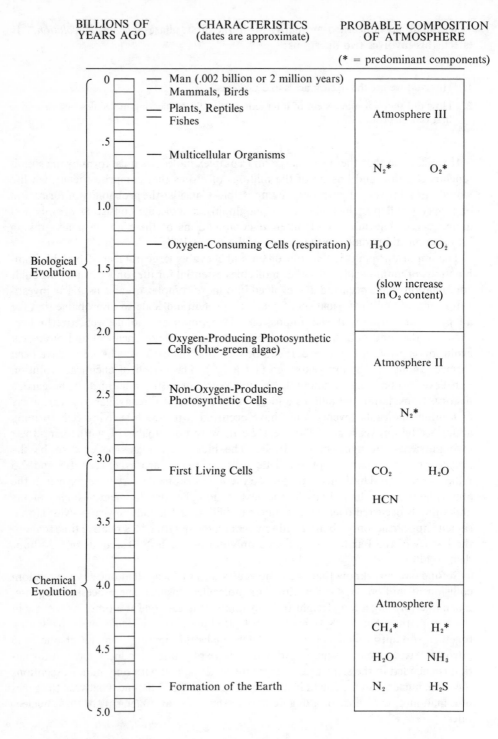

Figure 9-2 Approximate Time Scale of Chemical and Biological Evolution.

Proteins are also long-chain molecules made up by linking subunits called amino acids. There are 20 different kinds of amino acids in proteins, and again the same ones are found in all species, from viruses to man. The properties of protein molecules are determined by the linear arrangement of the different types of amino acids.

Thus, the problem of the origin of life is essentially reduced to asking how we could form amino acids, nitrogen bases, certain sugars (ribose and deoxyribose), and organic phosphate molecules (ATP and ADP). These molecules are made up of atoms of carbon, hydrogen, nitrogen, oxygen, and phosphorus. If these elements existed and combined in small molecules in the primitive atmosphere or surface of the earth, they could conceivably form these basic building blocks of life.

Chemical Evolution of Our Atmosphere and the Evolution of Life

The evolution of these molecules appears to be tied to the chemical evolution of the earth's atmosphere. Our present atmosphere consists primarily of nitrogen, N_2, and oxygen, O_2, with traces of CO_2, H_2O, and other gases. Did we always have this atmosphere and is it capable of producing the necessary molecules for life?

In 1936, the Russian biochemist, Oparin, suggested that the primitive earth's atmosphere may have been different from our present atmosphere. He suggested that oxygen was not present in the early atmosphere and that the atmosphere probably consisted primarily of gases such as methane (CH_4), hydrogen (H_2), ammonia (NH_3), water (H_2O), nitrogen (N_2), and hydrogen sulfide (H_2S).

This suggestion constituted a major breakthrough in the formulation of possible mechanisms for the origin of life. During the past 20–25 years an array of laboratory experiments—designed to simulate such early atmospheric conditions—have supported Oparin's hypothesis. These included investigations by Melvin Calvin, Harold Urey, and Stanley Miller in the fifties and the experiments in the sixties of John Oro, Melvin Calvin, Sidney Fox, Phillip Abelson, Cyril Ponnamperuma, and others.[5] They mixed these primitive gases (and others) in a container and exposed them to various conditions of heat, electrical discharge (thunderstorms), radioactivity, or ultraviolet radiation, believed to have existed on the primordial earth. They found that under such conditions amino acids, sugars, nitrogen bases, and some of the other chemicals necessary for life are formed rather easily.

A general picture of chemical evolution is thus emerging. It should be emphasized that it is by no means complete. Many details are missing or hotly debated, and the experiments continue in laboratories throughout the world.

Out of these experiments has come the proposal that our present atmosphere may have evolved in three phases. The original atmosphere would be made up primarily of methane (CH_4) and hydrogen (H_2) with lesser amounts of H_2O, N_2, NH_3, and

---

[5] The reader who is interested in a summary of the work on the chemical origin of life might consult the following outstanding references: M. Calvin, *Chemical Evolution*, Oxford University Press, 1969; D. H. Kenyon and G. Steinman, *Biochemical Predestination*, McGraw-Hill Book Co., 1969; Chapter 34 in *Biochemistry* by A. L. Lehninger, Worth Publishers, Inc., 1970; and the article "Chemical Origins of Cells" by Sidney W. Fox, K. Harada and others in *Chemical and Engineering News*, *48*, No. 32, (1970), 80.

$H_2 S$. However, the ultraviolet radiation from the sun would penetrate this atmosphere and could be absorbed by the water molecule. Experiments and thermodynamic and kinetic analysis have shown that ultraviolet (UV) light can cause H_2O to decompose or "photodissociate" into hydrogen and oxygen.

Photodissociation of Water

$$H_2O(g) \xrightarrow[\text{light}]{\text{UV}} H_2(g) + [O]$$

This hydrogen gas, along with that already present, is so light that it would not be held by the earth's gravitational field and would eventually escape into space. The oxygen, however, would remain and it could combine with methane to produce CO_2 and water and with ammonia to produce gaseous nitrogen (N_2) and water.

Atmosphere I to Atmosphere II

$$CH_4(g) + 2O_2(g) \longrightarrow CO_2(g) + 2H_2O(g)$$

$$4NH_3(g) + 3O_2(g) \longrightarrow 2N_2(g) + 6H_2O(g)$$

Study Question 9-4

a. Use the data in Chapter 3 (Table 3-1) to calculate the standard free energy changes at 300°K for the two reactions just formulated for the conversion of Atmosphere I to Atmosphere II. Comment on the thermodynamic feasibility of the reactions under these conditions.

b. Conditions in the early atmosphere were undoubtedly different from standard conditions. At that time the sun (which also goes through a life cycle) probably was much less luminous, and atmospheric temperatures could have been close to 0°C or lower. Calculate ΔG_T° for these reactions at this temperature and comment on thermodynamic feasibility.

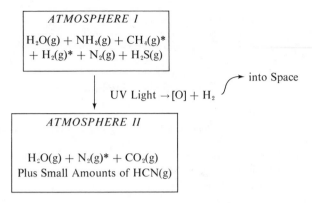

ATMOSPHERE I

$H_2O(g) + NH_3(g) + CH_4(g)*$
$+ H_2(g)* + N_2(g) + H_2S(g)$

into Space

UV Light → [O] + H₂

ATMOSPHERE II

$H_2O(g) + N_2(g)* + CO_2(g)$
Plus Small Amounts of HCN(g)

Figure 9-3   Atmosphere I to Atmosphere II. Species Marked with an Asterisk Were
Probably the Major Components.

c. On the other hand, temperatures on the earth's surface could have been higher
because of volcanic activity and radioactivity. Assuming the temperature on the
surface was 90°C, calculate ΔG_T° and comment on thermodynamic feasibility.

d. See if you can use some combination of gases in the primitive atmosphere (CH_4,
NH_3) to form hydrogen cyanide gas (HCN). Calculate ΔG° at 300°K, 273°K,
and 363°K, and comment on the thermodynamic feasibility of your proposed
reaction. For HCN(g), $\Delta H_{f298°K}^\circ = +31,000$ cal/mole and $S_{298°K}^\circ = 48$ cal/°K mole.

Over millions of years the primitive atmosphere (Atmosphere I) would have been
converted to Atmosphere II, as the ammonia and methane were consumed and
replaced by N_2 and CO_2 with the regeneration of water, as shown in Figure 9-3.
In addition, small amounts of hydrogen cyanide (HCN) were probably formed.

During this period of time there would have been a changing mixture of the gases N_2,
H_2O, CH_4, NH_3, CO_2, HCN, and H_2S. The experiments mentioned earlier have
shown that when these small gaseous molecules are exposed to various sources of
energy (particularly UV light), they can combine to form small organic molecules
such as sugars, amino acids, and nitrogen bases which would have dissolved in the sea
that then covered most of the Earth. In turn these small units can be linked and built
up into the more complicated molecules—the nucleic acids, proteins, and others—
necessary for life. From this "primeval soup" or broth of small organic molecules
and long-chain polymers the first cells could have arisen. This first part of chemical
evolution is summarized in Figure 9-4. Thus, the basic molecules necessary for life
can appear as the natural result of chemical evolution and they could appear on
any planet in the universe with the appropriate conditions.

Evolution of Cells

How were these chemical building blocks organized to form the first cells? A number
of different molecules would have to be brought together and a number of possibilities have been suggested. One involves the gradual buildup of complex molecules in

Chemical Evolution

Part 1: Formation of the Building Blocks of Life

Step 1: *Formation of the Elements*—Primordial nebulae formed from cosmic hydrogen. Nuclear fusion reactions probably resulted in the formation of elements.

Step 2: *Formation of Small Molecules in the Atmosphere*

Atmosphere I	Atmosphere II

$$
\begin{array}{c}
\text{H} \\
\text{O} \\
\text{C} \\
\text{N}
\end{array}
\Bigg\} \longrightarrow
\quad
\begin{array}{l}
CH_4(g) \\
H_2(g) \\
\\
NH_3(g) \\
\\
H_2O(g) \\
N_2(g) \\
H_2S(g)
\end{array}
\xrightarrow[\text{light}]{\text{UV}}
\quad
\begin{array}{l}
CO_2(g) \\
\\
N_2(g) \\
\\
H_2O(g) \\
\\
HCN(g)\text{—small amounts}
\end{array}
$$

Step 3: *Formation of Small Organic Molecules*

$$CH_4, H_2O \longrightarrow \text{sugars}$$

$$CH_4, H_2O, NH_3, \text{ or } HCN \longrightarrow \text{amino acids}$$

$$CH_4, H_2O, NH_3, HCN \longrightarrow \text{nitrogen bases}$$

Step 4: *Formation of Large Organic Molecules (Polymers)*

$$\text{sugars} + \text{sugars} \longrightarrow \text{polysaccharides}$$

$$\text{amino acids} + \text{amino acids} \longrightarrow \text{proteins}$$

$$\text{nitrogen bases} + \text{sugars} + \text{phosphate} \longrightarrow \text{ATP, ADP, and nucleotides}$$

$$\text{nucleotides} + \text{nucleotides} \longrightarrow \text{nucleic acids (DNA, RNA)}$$

Figure 9-4   A Proposed Outline of the First Part of Chemical Evolution. (The steps are not necessarily sequential. Steps 3 and 4 may have occurred during the transition from Atmosphere I to Atmosphere II in Step 2.)

puddles of "primeval soup or broth" scattered about on the surface of the earth. Under the sun's rays water might be slowly evaporated—thus concentrating the molecules and bringing them together.

On the other hand, Haldane and others have proposed that the necessary chemicals could have been brought together in the world's oceans by forming little bubbles, or spheres, containing a number of organic molecules insoluble in water—a process much like the formation of droplets when oil is added to water.

Recently Sidney Fox has shown that if a mixture of amino acids is heated, the individual acids can link together to form protein-like molecules—which he calls "proteinoids." When these proteinoids are dissolved in hot water and the solution is cooled, they cling together in little spheres about the size of small bacteria and are called "proteinoid microspheres." By the addition of chemicals to the solution, these proteinoids can be made to swell or shrink and to produce buds which can break off. They can also divide or link together in chains. Fox proposed that the first cells—or protocells—could have evolved by chemical reaction systems developing in such microspheres. From these early cells could have evolved the more complex cells that exist today.

There were probably many false starts. Possibly the early cells may have arisen from nonliving matter several times at different places or times and perhaps from different building blocks. However, evidence indicates that all present organisms evolved from only one line of cells that survived. It is estimated that as high as 99 percent of all species of organisms that existed are now extinct—with no modern descendants.

The early cells were undoubtedly quite different from modern ones. The nature and origin of these early cells is hotly debated and the question, which came first—nucleic acids or proteins, continues to intrigue scientists around the world. Oparin in the Soviet Union has suggested that the first cells functioned without nucleic acids and the genetic system formed later, while the American geneticist Mueller and his later colleagues suggest that a nucleic acid or "naked" gene was the first form of life and that they provided the information for the evolution of proteins at a later stage.[6] This latter hypothesis is supported by modern knowledge of the structure and self-replication of viruses and the fact that nucleic acids contain all the information necessary for their replication.

As the early cells evolved to more complex ones, they probably developed many different mechanisms for obtaining food or nutrients. Some probably obtained food by parasitism, or feeding on dead tissue—as do present-day bacteria. Others may have provided food by chemosynthesis—the chemical synthesis of necessary molecules from compounds in the surrounding medium. With the growth and multiplication of these early cells the seas covering much of the primitive Earth probably became depleted of organic compounds, and a new photosynthetic nutritional process involving simpler carbon compounds (like CO_2) probably evolved. These first photosynthetic cells probably appeared about 1.5 to 2.5 billion years ago.

[6] An excellent summary of this debate can be found in A. L. Lehninger, *Biochemistry*, Worth Publishers, Inc., 1970, pp. 782–789.

Evolution of the Present Atmosphere—The Oxygen Revolution

Our present atmosphere is different from Atmosphere II. It contains O_2 in place of large quantities of CO_2. It is composed primarily of $N_2(g)$, $O_2(g)$, and some $H_2O(g)$ with only small amounts (0.03%) of CO_2.

How was the O_2 formed and how was the CO_2 content decreased? We might expect that after the NH_3 and CH_4 in Atmosphere I were consumed, the photodissociation of H_2O by UV light into H_2 and O_2 would continue. This could account for the increase in O_2 but not for the decrease in CO_2. Furthermore, the photodissociation of H_2O would not continue. As oxygen accumulates, some of it is converted into ozone, O_3, in the upper atmosphere. Ozone strongly absorbs ultraviolet radiation, so that once ozone is formed, UV light is not available to continue the photodissociation of H_2O. Furthermore, at this stage, life might have been in danger of extinction, since UV radiation was probably one of the main sources for life in the early cells.

Another branch in the evolution of nutrition in cells probably took place—the development of photosynthesis. It uses visible, not UV, light to form fuel for cells and at the same time consumes CO_2 and produces oxygen.

As more and more photosynthesizing cells developed, Atmosphere II would have changed very slowly (probably over a period of 1 to 1.4 million years) to our present

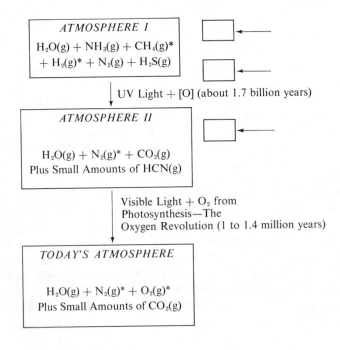

Figure 9-5 Possible Evolution of Our Present Atmosphere. Asterisks Indicate Major Components.

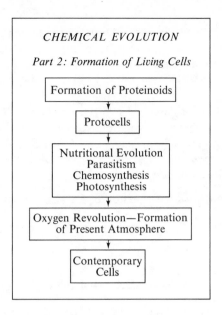

CHEMICAL EVOLUTION

Part 2: Formation of Living Cells

Formation of Proteinoids

Protocells

Nutritional Evolution
Parasitism
Chemosynthesis
Photosynthesis

Oxygen Revolution—Formation
of Present Atmosphere

Contemporary
Cells

Figure 9-6   A Proposed Outline of the Second Part of Chemical Evolution.

atmosphere, containing 20% O_2, as shown in Figure 9-5. Because of the increase in oxygen, this is sometimes called the "oxygen revolution."

This second phase of chemical evolution to form the basic units of life—the cells—is summarized in Figure 9-6.

From here the chemical stage is set for the biological evolution of diverse animal and plant species by natural selection.

9-5 Thermodynamics, Kinetics, and Chemical Evolution[7]

The formation of nonliving molecules in the first part of chemical evolution can be understood in terms of thermodynamics and kinetics. Once the simple molecules were formed, an extremely large number of reactions took place over the billions and billions of years in this early part of evolution. Some of these reactions were exergonic and rapid, while others were endergonic, or required long reaction times. Forcing such reactions requires energy, but fortunately at least three kinds of energy were probably available in the early atmosphere—radiation, ultraviolet light, and lightning. Another means for driving thermodynamically unfavorable reactions is by free energy coupling. The kinetic feasibility of these reactions was enhanced by two factors: (1) the long length of time available, so that even very slow reactions

---

[7] Much of this discussion is based on the excellent work of Harold F. Blum, *Time's Arrow and Evolution*, Princeton University Press, second Edition, 1968, and his article, "On the Origin and Evolution of Living Machines," *American Scientist*, *51*, 474–501 (1961).

could form reasonable concentrations of products over billions of years and (2) the relatively high temperatures, energy, and light sources that were available to provide the necessary free energy of activation.

Study Question 9-5   The formation of hydrogen cyanide, HCN(g), is now emerging as an important link in chemical evolution. Recent research indicates that it may be important for the synthesis of nitrogen bases (purines and pyrimidines), amino acids, and possibly compounds called porphyrins, which could have ultimately produced the chlorophyll so essential for photosynthesis.

a. Two proposed reactions for the formation of HCN in the early atmosphere are

$$CH_4(g) + NH_3(g) \rightleftharpoons HCN(g) + 3H_2(g)$$

and

$$CO(g) + NH_3(g) \rightleftharpoons HCN(g) + H_2O(g).$$

Calculate ΔG_T° for these reactions at 0°C, 100°C, and 1227°C and comment on their thermodynamic feasibility. For HCN(g), $\Delta H_{f,298°K}^\circ = +31,000$ cal/mole and $S_{298°K}^\circ = 48$ cal/°K mole.

b. It is also proposed that HCN could have formed in the lower atmosphere at low temperatures, where H_2O would freeze out by the reaction

$$3H_2(g) + N_2(g) + 2CO(g) \rightleftharpoons 2HCN(g) + 2H_2O(s).$$

Calculate ΔG_T° at 0°C and comment on thermodynamic feasibility.

Figure 9-7 is a free-energy diagram showing how various reaction paths could have led to an array of molecules over a long period of time.

The series of reactions running from $A \rightarrow B \rightarrow C \rightarrow D$ is the most probable pathway from a thermodynamic standpoint, since it goes to the lowest free-energy state. The pathway $A \rightarrow B \rightarrow E \rightarrow F$ is also thermodynamically favored because of the decrease in free energy. However, it is not favored from a kinetic standpoint, because of the high free energy of activation needed to go from B to E. Thus, C would be formed faster than E, and E might never accumulate in appreciable amounts. Suppose, however, that at some time in the evolutionary process a molecule is formed that can act as a catalyst to lower the free-energy-of-activation barrier between B and E, as indicated by the dotted line. In this case, E would form faster than C, and a new evolutionary pathway eventually leading to large quantities of F could occur. With the appropriate conditions another entirely different pathway could open up to form molecules represented by H. For example, quanta of light or electrical energy from discharge could be absorbed by E molecules so that they would have enough energy to surmount the energy barrier and form H.

Over billions of years the many such thermodynamic and kinetic reaction pathways would result in the formation of an enormous variety of chemical molecules. As specific catalyst molecules formed and free energy coupling occurred, certain reaction pathways could be "selected" or favored in terms of energetics and kinetics, and

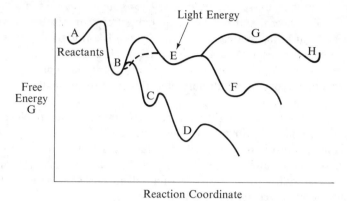

Figure 9-7   Free-Energy Reaction Pathway Profile in a Complex Reaction Mixture.
(After H. F. Blum, "On the Origin and Evolution of Living Machines,"
American Scientist, **51**, p. 479 (1961).)

large concentrations of some molecules could build up. These theoretical predictions
plus much experimental work have shown that it is highly probable that the small
and large organic molecules necessary for life did form and the stage was set for
the biological evolution of life by natural selection.

Biological Evolution by Natural Selection

Only thermodynamic and kinetic factors are necessary to establish the probability
of chemical evolution. Evolution by natural selection, however, must involve a third,
or spatial, component to account for the self-replication of molecules necessary for
life. The three components necessary for replication—spatial, thermodynamic, and
kinetic—can be associated with three groups of compounds. The *kinetic* function is
associated with proteins, since enzymes are proteins that promote kinetic feasibility
by lowering the activation energy. The *spatial* component is associated with the nucleic
acids, such as DNA and RNA, which form the master spatial template for the replication of molecules. The *thermodynamic* component is associated with energy-rich
compounds, such as ADP and ATP, which allow free energy transfer through transfer
of phosphate groups.

At this point a new type of "selection" can take over. In the chemical evolution of
nonliving molecules, the reaction pathway is a function of thermodynamics and kinetics, and a change in environmental conditions could cause the rapid formation of a
more favorable pathway.

In a living replicating system the response to the environment is different. A particular pattern can be preserved for very long times, being repeated over and over again
through many generations. The response to the environment is not so direct and

immediate, and the system can preserve its "molecular integrity" in spite of environmental stresses. However, stresses can become severe enough to cause some alterations, or mutations, in the pattern, and this new adapted pattern can also be preserved.

Natural selection of a new pattern is a slow process. It amounts to "picking out" altered, or mutated, patterns, or genes. A mutated gene is isolated from direct interaction with the environment, and if its replication, or differential reproduction, rate is high relative to other types of mutations, then this new form of life thrives and is successful.

In evolution by natural selection, the equilibrium thermodynamics of closed systems is not strictly applicable, since living or open systems exist in a steady state instead of thermodynamic equilibrium. The analysis of natural selection involves the use of nonequilibrium thermodynamics coupled with a feedback mechanism in a homeostatic system. Our understanding of natural selection will increase as these ideas of nonequilibrium thermodynamics are applied in the future.

In this chapter, we have seen how thermodynamics and kinetics can be applied to our ideas of time and to the origin and evolution of life on this planet. This discussion has necessarily been somewhat sketchy, for this important application of thermodynamics and kinetics to living systems is relatively new and many important developments lie ahead. The theme of this chapter—entropy, time, and evolution—is beautifully expressed in the Book of Ecclesiastes.

> For every thing there is a season, and a time for every purpose under the heaven;
> A time to be born, and a time to die; a time to plant and a time to pluck up that which is planted;
> A time to kill, and a time to heal; a time to break down, and a time to build up;
> A time to weep and a time to laugh; a time to mourn and a time to dance;
> A time to seek and a time to lose; a time to keep and a time to cast away;
> A time to mend and a time to sew; a time to keep silence, and a time to speak;
> A time to love, and a time to hate; a time of war, and a time of peace.

Ecclesiastes 3

9-6 Review Questions

9-1. What is negentropy?

9-2. What is meant by the statement, "Life feeds on negentropy?"

9-3. What is the largest source of entropy increase in our solar system? Explain.

9-4. Distinguish between open, closed, and isolated systems. Give examples of open and closed systems.

9-5. Distinguish between dynamic, or thermodynamic, equilibrium and the steady state.

9-6. How could you tell whether a system is in thermodynamic equilibrium or in a steady state?

9-7. Criticize the following statement: Life maintains itself by seeking a state of minimum free energy.

9-8. What is steady-state thermodynamics?

9-9. Does life violate the second law? Explain.

9-10. What is meant by the so-called "heat death," or "entropy death," of a universe?

9-11. Give several reasons why this could be an unwarranted extension of the second law.

9-12. Most texts state the second law as "the entropy of the universe is increasing." Criticize this statement.

9-13. Describe briefly the three major hypotheses for the origin of the universe and comment on whether each theory is in accordance with the First and Second Laws of Thermodynamics.

9-14. Explain what is meant by the statement, "entropy is time's arrow." Give an illustration of this concept.

9-15. Distinguish between "chemical evolution" and "evolution by natural selection."

9-16. Criticize the statement, "evolution by natural selection occurs by survival of the fittest."

9-17. Outline the four steps of the first part of chemical evolution.

9-18. Explain how we got from the proposed initial atmosphere to atmosphere II and then to our present atmosphere.

9-19. Outline a proposed sequence for the second part of chemical evolution—the formation of living cells.

9-20. What two factors can be used to account for chemical evolution?

9-21. What additional factor is needed to describe evolution by natural selection?

9-22. Use a free-energy profile diagram to illustrate how certain reaction pathways could have been favored, or "selected," during the period of chemical evolution.

9-23. Why was it apparently important for hydrogen cyanide to exist in the early atmosphere?

9-24. The coupling of amino acids to form long-chain proteins is an endergonic process. How could this unfavorable process have occurred in evolution? Is it not also feasible kinetically? How might this problem have been overcome?

9-25. Explain natural selection using a crude cybernetic or steady-state thermodynamic model.

Chapter 10

Bioenergetics— Thermodynamics and Kinetics of Living Systems

10-1 Sunlight—The Energy Source for all Life

One characteristic of living organisms is their ability to extract and transform matter and energy from their environment. How is energy transformed and stored by living organisms? How do plant cells transform simple molecules like CO_2 and H_2O into complex organic nutrients like glucose? How do plant and animal cells use these nutrients to maintain life? Answers to such questions are obtained through a study of bioenergetics—the study of the energetics and kinetics of life from the cellular level to the biosphere.

We must begin with the energy source that sustains all life on Earth—radiant energy from the sun. The life-supporting functions of this solar energy consist in warming the earth and in initiating the process of photosynthesis in plants, which provide nutrients to sustain all animal life.

What is the source of the sun's energy? The sun consists mostly of hydrogen maintained in its interior at temperatures which may be as high as 10 million °C. Under such high temperatures and pressures, four hydrogen nuclei can be "fused" together to form a helium nucleus. This process of *nuclear fusion*, which releases enormous quantities of energy, is the ultimate source of energy for life on this planet. The sun is thus a gigantic thermonuclear, or hydrogen, bomb undergoing continuous explosion that liberates about 10^{26} calories of energy every second. If we could completely harness this energy, each person on earth each second would have for his own personal use over 70,000 times the annual power consumption of the United States. The sun loses about 4.2 million tons of mass every second in order to produce this enormous amount of energy. Assuming an automobile weighs about one ton, or 2000 pounds, this means that every second the sun loses the mass equivalent to that found in 4.2 million automobiles. We should not become alarmed, however, about the sun running out of fuel. It has probably been in existence for at least 6 billion years, and there is enough hydrogen left to keep it going for at least 8 billion years more.

The sun radiates energy in all directions. In terms of the solar system, the earth represents a very small target, and only about two billionths of the sun's energy reaches the earth's outer atmosphere. Even this tiny fraction is equivalent in energy to about 100 million Hiroshima-size atomic bombs per day. But only a fraction of this is used to sustain life. Nearly half of the incoming solar radiation is reflected, scattered, or absorbed by atmospheric gases, clouds, and dust. More is absorbed or reflected by the land and oceans. As a result of these and other losses, a ridiculously small portion of the sunlight reaching us—about one to at most three per cent— is actually used to support life by photosynthesis in plants. Even so, plants use one hundred times more energy than that used by all manmade machines. We have much to learn from plants in terms of the efficiency of energy conversion, and it should be clear why the study of bioenergetics is so important.

10-2 Spaceship Earth—The Biosphere

We live on a spaceship containing all the supplies we will ever have to keep us alive. Adlai Stevenson eloquently summed up our predicament and our challenge a number of years ago when he said, "We travel together, passengers on a fragile spaceship, dependent on its vulnerable reserves of air, water, and soil; all committed for our safety to its security and peace; preserved from annihilation only by the care, the work, and I will say, the love we give our fragile craft."

What is our basic survival equipment? All life exists in the *biosphere* or *ecosphere*— a thin film of air, water and soil having a thickness only about one-thousandth the diameter of this planet. This life support system consists of three parts as summarized in Figure 10-1: above us—the *atmosphere*—an air envelope no more than seven miles high; surrounding us—the *hydrosphere*—containing the water in our rivers, lakes, oceans and that stored underground; below us—the *lithosphere*—a thin crust of soil, minerals and rocks extending only a few thousand feet into the earth's interior.

All life depends on the physical part of the biosphere for supplying the necessary water, minerals, oxygen, nitrogen, and carbon dioxide, which provide the chemical elements, or building blocks, that can be put together to form and maintain plant and animal life. If we liken spaceship Earth to a balloon, then all life is found and sustained on the thin layer of soil, water, and air extending only about nine miles up. This "thin and fragile envelope of life" can be compared to the skin of a balloon. This skin, or life-support system, on our ship is now being stressed by man at an alarming and rapidly increasing rate.

For all practical purposes the total amount of matter on our spaceship is fixed— with no matter entering or leaving the ship.[1] But energy from the sun enters and leaves the ship all the time.

---

[1] The amounts of matter represented in the meteors and other materials from space and the spaceships and satellites that we have put into space are negligible.

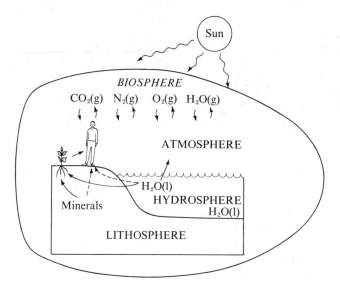

Figure 10-1   Spaceship Earth and the Major Components of Our Life-Support
System—the Biosphere.

Since no new matter is lost or added, certain chemicals necessary for life must be continuously cycled and recycled throughout the biosphere. Vital chemicals, or nutrients, such as carbon, nitrogen, oxygen, and water, are *cycled* through the biosphere, with the sun's energy being used to drive and sustain these chemical cycles—usually called *biogeochemical cycles* (*bio* for living; *geo* for water, rocks, and soil; and *chemical* for the processes involved). Note, however, that while certain chemicals are cycled, energy is not—it simply *flows* in one direction only through the biosphere, as shown in Figure 10-2.

Thus, all life on spaceship Earth, or on any closed system, depends on two important factors—chemical, or nutrient, cycling and energy flow.

Life on Spaceship Earth Depends Upon

1. Chemical cycling (C, N, O, H_2O and other cycles)
2. One way energy flow through the biosphere

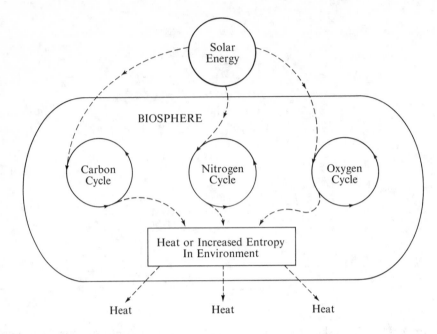

Figure 10-2   Life on Spaceship Earth Depends on the Cycling of Critical Chemicals and the One-Way Flow of Energy through the Biosphere.

The biosphere can be viewed as a series of subsystems, called *ecosystems*. *Ecology*, a term coined less than one hundred years ago from two Greek words that mean "the study of the home," is the study of the interrelationships of living things with each other and with their environment. A cell, a multicellular organism, a population, biological community, city, ocean, or nation, can each be considered as an *ecosystem* if the total environment is added in each case. The biosphere is then the sum of all the various ecosystems, or subsystems, on our planet.

This home for all living things was created by living things through the long process of evolution. As we saw in the last chapter, when the Earth was formed there was no oxygen in the atmosphere, and there were no natural systems for purifying waste. Green plants that evolved are responsible for our total oxygen production while plants, animals, and microorganisms build up the soil and renew the water.

The ecosystems that make up our biosphere are intricately interconnected—thus maintaining a certain degree of stability—known as a steady state or the "balance of nature." They form a giant and elaborate set of homeostatic systems that operate together in an even larger homeostatic system to maintain an overall balance, or steady state. When the system becomes unbalanced, it readjusts through negative feedback mechanisms to maintain stability. This dynamic balance is maintained by the one-way flow of solar energy through the system and its interaction with the major cycles of nutrients. If this balance is upset, we have a spaceship in trouble, and the ability of the biosphere to support life can become degraded or destroyed.

The principles of chemical cycling and energy flow can help us understand the major problems that can arise on our spaceship as well as how to deal with them or prevent them from occurring. One class of problems would involve disrupting one or more of the essential chemical cycles. Life depends not only on the existence of these cycles on a global and local scale but on the maintenance of a particular *rate of cycling*.[2] Trouble can therefore occur either by breaking the cycle or by altering the rate of cycling through adding too many nutrients (nutrient overload) or by removing too many nutrients (nutrient leaks) at some point in the cycle. These problems can occur on a localized scale (particularly in industrialized society) or eventually on a global scale so that the entire life support system, or at least the quality of life for everyone, is threatened. Hopefully, those of us in a "throwaway society" are finally beginning to realize that on a finite spaceship you can't throw anything away—there is no "away." An understanding of the chemical recycling processes in nature can enable us to preserve these cycles and guide us in recovering and recycling some of the valuable chemicals that we have been discarding or tying up in nonuseful forms.

A second class of problems involves disrupting the flow of energy through the spaceship. If the outer skin of the ship—in our case the atmosphere—were changed significantly—for example, by adding excess CO_2—more solar energy could enter than could be radiated back into space by the skin of the ship, and the atmosphere inside the ship would heat up. A $2°C$ rise could cause catastrophic changes in global weather patterns. On the other hand, decreasing the solar input, for example by adding dust as particulate matter, would cool the atmosphere inside the ship. It is sobering to realize that a reduction of only about 2 percent of the available solar energy can, in theory, lower the mean temperature of the atmosphere $2°C$ and produce an ice age.

Excessive heat (or entropy) buildup inside the ship can also occur through man's activities. Since all of man's activities end up by adding more heat to the atmosphere, then adding more passengers who use more and more energy is ultimately a death

---

Summary of Critical Problems on Our Spaceship

1. Disruption of essential chemical cycles on a local or global scale.
   a. Breaking the cycle.
   b. Changing the rate of cycling by nutrient overloads or nutrient leaks.
2. Disruption of energy flow on a local or global scale.
   a. Decreasing or increasing solar energy input by changing the properties of the atmosphere.
   b. Heat or entropy buildup in the life support system due to too many passengers using too much energy—we can't ignore or repeal the Second Law of Thermodynamics.

---

[2] For example, oxygen produced by photosynthesis is recycled about every 2,000 years while carbon dioxide produced by respiration in cells is recycled about every 200 years.

trap. On any spaceship this heat or entropy buildup eventually becomes the ultimate problem because of the Second Law of Thermodynamics. As we have seen, the maintenance of life always requires the creation of order in the system at the expense of a greater increase in disorder, or entropy (usually in the form of heat), in the environment. Life at any level is an ordering process, but life in an industrialized society requires a massive ordering of materials (cities, factories, cars, tractors, fertilizers, pesticides, and so on) with a drastic increase in disorder, or heat, in the local (for example, around cities) or global environment—as we are now discovering.

10-3 Energy Flow and the Carbon and Oxygen Cycles

The most important principle of ecology is that *everything is connected to everything*. Everyone and everything on this planet are interconnected in a fragile and complex web of life. Intruding into or disrupting an ecosystem in one place *always* has some complex, usually unpredictable, and frequently undesirable effect somewhere else in the system or its environment. This boomerang effect was stated eloquently by the English poet Francis Thompson:

> Thou canst not stir a flower without troubling a star.

The key to a deeper appreciation of this important principle lies in understanding how the component parts of our ecosystem are related to and dependent upon one another through nutrient cycling and energy flow. Although there are a number of biogeochemical cycles, we shall look at only two—the *carbon and oxygen cycles*.[3] There are about 20 elements,[4] or nutrients, essential for life. Nine of these—carbon, hydrogen, oxygen, nitrogen, phosphorus, calcium, magnesium, sulfur, and potassium —are required in relatively large amounts and are called *macronutrients*. The remaining eleven elements (iron, copper, iodine and others) are needed in relatively minute quantities—primarily as parts of enzymes—and they are called *micronutrients*. An additional 20 or more elements are culturally (technologically and industrially) required. The macronutrients, carbon and oxygen, can serve, however, to illustrate the principles involved in energy flow and nutrient cycling in an ecosystem.

The major components of an ecosystem are the nonliving *nutrients*, or *chemicals*, *energy*, and the living portion consisting of *primary producers* (plants), *consumers* (animals), and *decomposers*, as summarized in Table 10-1. The ecosystem is a vast network of interconnections: between plants that produce oxygen and animals and plants that use it; between the plants consumed by animals and the animals that consume other animals; and between the air, water, and chemicals that serve as the fabric connecting all living things.

---

[3] For more details on these and other chemical cycles see the September 1970 issue of the *Scientific American*—a superb issue devoted entirely to a discussion of the biosphere.

[4] The number keeps going up as scientists learn more about life.

Table 10-1 The Major Components of an Ecosystem.

Major Components of an Ecosystem

Nutrients: The nonliving matter, water, oxygen, carbon dioxide, organic compounds, and other chemicals used by plants and animals. The critical nutrients are cycled through the biosphere in the biogeochemical cycles.

Plants, or Primary Producers: Ranging in size from tiny phytoplankton in water to giant trees, they provide food for themselves and other organisms by converting carbon dioxide and water to sugars by photosynthesis.

Animals, or Consumers: Herbivores, such as deer, cows, rabbits, mice, grasshoppers, and sheep, are the primary consumers that feed on plants. In turn, small carnivorous animals, such as frogs, lizards, snakes, cats, and wolves, are secondary consumers that feed on the herbivores. Finally, top carnivores, such as lions, hawks, and fleas can feed on the smaller animals. Omnivorous animals, such as man, are both herbivores and carnivores, feeding on both plants and other animals.

Decomposers: Tiny organisms, such as bacteria and fungi, that complete the cycle of chemicals by breaking down the dead animals and plants and returning their nutrients to the ecosystem for reuse.

Energy: The solar energy that drives the entire system. It flows through the system, and at each level a small part of it is used to support life. Most of it is degraded to less useful forms and returned to the environment as heat, or entropy, in accordance with the Second Law of Thermodynamics.

The interdependence of all of these components is based on the cycling of essential chemicals and the flow of energy through an ecosystem, as shown in Figure 10-3. We can use this diagram to understand the cycling of carbon.

Carbon is a basic element in all organic compounds. The source of carbon for plants and animals is the carbon dioxide that makes up about 0.03% of our atmosphere and the CO_2 dissolved in the waters that cover two-thirds of the earth. Green plants (primary producers) utilize solar energy, CO_2, and water to produce simple carbon sugars (for example, glucose) and O_2 in a complex series of reactions known as *photosynthesis*. These simple sugars are then synthesized by plants into more complex carbon compounds, such as fats and sugars. Some of the carbon in these compounds is converted back to CO_2 and H_2O in the *respiration* reactions taking place in the plant cells. But much of it remains in the plant bodies until they die or are eaten by plant-eating animals—the *herbivores*. Again, part of the carbon in the plant

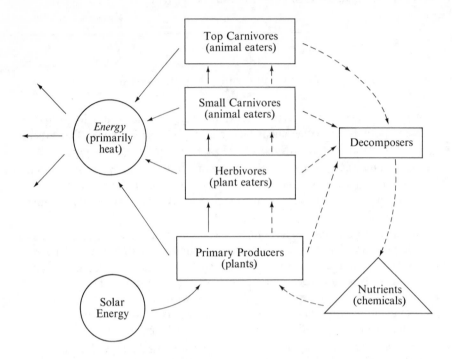

Figure 10-3   Interdependence of Plant and Animal Species in Chemical Cycling
(dotted lines) and Energy Flow (solid lines) through an Ecosystem.

food eaten by herbivores is converted back into CO_2 through respiration in the animal cell, but most of it remains until the animals excrete part of it as wastes, die, or are eaten by meat-eating animals—the *carnivores*. The same processes of respiration, wastes, and decay occur in the carnivores.

Usually the wastes of animals and the dead bodies of animals and plants are broken down into simpler substances by bacteria and fungi—the *decomposers*. Most of the carbon in them is then released as CO_2, back into the air or water, thus completing the cycle. Each year some 33 billion tons of carbon flow through this cycle on the face of the Earth.

Over millions of years, part of the carbon from decayed plants and animals has been incorporated into the earth's crust as fossil fuels—coal, gas, and oil—or as carbonate rock formations, such as limestone and coral reefs. Since the industrial revolution we have been burning the carbon in fossil fuels and returning it to the atmosphere as CO_2 at an increasing rate. As a result of this man-made disruption of the carbon cycle, the CO_2 content of the atmosphere has increased at 0.2 percent each year since 1958. At this rate, by the year 2000 the CO_2 content would increase 25% over 1958 levels. This might increase the surface temperature of the earth 0.5°C through the "greenhouse effect"—only a 2°C rise could lead to long-term warming of the planet. These estimates are uncertain, however, because we know so very little about our atmosphere. Indeed, the atmosphere may be cooling because of increased amounts of particulate matter from natural (for example, windblown dust, volcanoes) and man-made sources. Ironically the largest man-made source also comes from the burning of fossil fuels with the production of sulfur dioxide and its subsequent conversion to sulfate particles. Again we don't really know the long-term effects, but we do know that man now has the capacity to bring about changes in the atmospheric balance on a global and local (for example, around cities) scale. A recent authoritative study[5] of these and other major environmental problems stated:

Although we conclude that the probability of direct climate change in this century resulting from CO_2 is small, we stress that the long-term potential consequences of CO_2 effects on the climate or of societal reaction to such threats are so serious that much more must be learned about future trends of climate change.

Actually, the *oxygen cycle* is interdependent with the carbon cycle. When CO_2 and H_2O are combined by photosynthesis in plants, O_2 is produced, and it is then converted back to H_2O and CO_2 by respiration in plant and animal cells. A greatly simplified version of the carbon and oxygen cycles is shown in Figure 10-4. By stripping away some of the details, we can see more clearly how carbon and oxygen are cycled while energy from the sun flows through the biosphere and is degraded to less useful forms in the process.

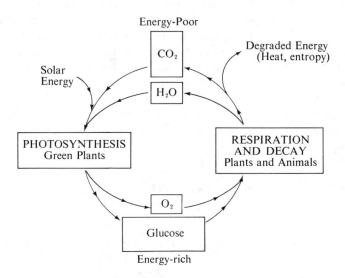

Figure 10-4 Simplified Version of the Carbon and Oxygen Cycles in the Biosphere.

[5] Report of the Study of Critical Environmental Problems, *Man's Impact on the Global Environment*, The MIT Press, 1970 (paperback). An excellent and readable summary of environmental problems by teams of prominent scientists. An additional source is *Global Effects of Environmental Pollution*, S. F. Singer (ed.), Springer-Verlag, New York, Inc., 1970.

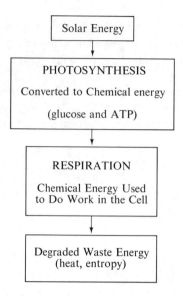

Figure 10-5   The One-Way Flow of Energy through the Biosphere.

But the important thing about these cycles is that throughout the entire process the First and Second Laws of Thermodynamics hold. The first law—the law of conservation of energy—means that total input and output of energy are exactly balanced. The energy is not destroyed but as it flows through the system, it is degraded from a form of energy capable of driving reactions and performing work into heat energy—consisting of the random, disorganized motion of molecules not available for performing work, as shown in Figure 10-5. Thus, the second law states that in any real process useful potential energy is lost.

Now that we have discussed energy flow and chemical cycling in the biosphere, let us examine these bioenergetic processes at the molecular level in the smallest unit of life—the cell.

10-4 Energy Conversion in the Cell—ATP and ADP

The cell is a sort of "chemical factory" that requires fuel (normally glucose) and energy for its functions. Free energy is available to perform work in the cell when glucose is burned to form carbon dioxide and water.

$$C_6H_{12}O_6(s) + 6O_2(g) \rightleftharpoons 6CO_2(g) + 6H_2O(l) \quad \Delta G^\circ_{300^\circ K} = -675,900 \text{ cal}$$

In the cell, this large amount of energy is not released at one time. It would be too much for the cell to handle, and too much of it would be wasted and lost as heat. The temperature of the cell would become so high that the enzymes that control the kinetics of cellular reactions would be inactivated—the cell processes would be disrupted and the cell destroyed.

Thus, the living cell needs (1) a means for capturing and storing free energy from its fuel molecules, (2) a means for releasing this energy in the appropriate amount when and where it is needed, and (3) a means for bringing about various nonspontaneous reactions that are essential to the cellular processes. The cell accomplishes these tasks through the capture and storage of energy in a special group of molecules known as the *high-energy phosphates*.[6] This free energy is released in small amounts and used to drive nonspontaneous reactions through a sequence of reactions that involve free energy coupling.

Glucose is converted to CO_2 and H_2O by a reaction mechanism involving at least thirty steps, each using a specific enzyme.

During this process, "small packets" of free energy are released for use. But it is important to remember that the first law is obeyed and the total energy released is exactly the same as the initial input, whether the energy is released all at one time or in "small packets." What is gained in the cellular processes is not more energy but a more efficient use of the energy available.

How is energy stored and transferred in the cell? A number of compounds serve this function, but the most important of these are *adenosine triphosphate*, or *ATP*, and *adenosine diphosphate*, or *ADP*. It may be recalled (Section 3-2) that ATP consists of three parts—a nitrogen ring compound known as adenine, the five-carbon sugar, ribose, and three phosphate groups, as shown in Figure 10-6.

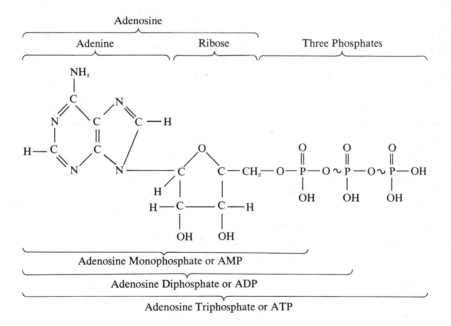

Figure 10-6 Structure of Adenosine Phosphates, ATP, ADP, and AMP.

[6] As we shall see shortly, this is an unfortunate name—a better choice would be high free-energy phosphates.

A simplified mechanical analogue for ATP can be written as follows:

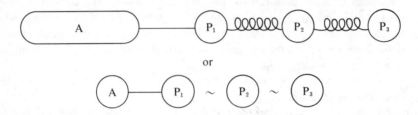

or

where the wavy lines ~ are used to indicate that these are high free-energy compounds. The symbol P (or sometimes P_i) is used to indicate a phosphate group. If one or both of the two end phosphate bonds are removed, for example, by reaction with water (hydrolysis), a large release of free energy occurs. Removing the third phosphate by reaction with water involves a much smaller release of free energy, as shown below.

$$\boxed{\begin{array}{l}
\text{(A)—(P_1)~(P_2)~(P_3)} + H_2O \rightleftharpoons \text{(A)—(P_1)~(P_2)} + H_3PO_4 \qquad \Delta G_1{}^\circ = -7300 \text{ cal} \\[2ex]
\text{(A)—(P_1)~(P_2)} + H_2O \rightleftharpoons \text{(A)—(P_1)} + H_3PO_4 \qquad\qquad\quad \Delta G_2{}^\circ = -7300 \text{ cal} \\[2ex]
\text{(A)—(P_1)} + H_2O \rightleftharpoons \text{(A)} + H_3PO_4 \qquad\qquad\qquad\quad \Delta G_3{}^\circ = -3400 \text{ cal}
\end{array}}$$

This reaction is reversible, and it can go to the right to release free energy (ATP + $H_2O \rightarrow$ ADP + P) or absorb and store free energy in the ATP by going to the left (ADP + P $\rightarrow$ ATP + H_2O). When a reaction occurs in the cell to release about 7300 calories or more, this energy can be used to add a phosphate group to ADP to form a molecule of ATP—thus forming a compound (ATP) with a high free energy of hydrolysis. This addition of a phosphate is called *phosphorylation*.

Energy Storage—Phosphorylation

$$\text{ADP} + \text{(P)} + \text{energy} \rightleftharpoons \text{ATP} + H_2O \quad \Delta G^\circ = +7300 \text{ cal}$$

At another time, when energy is needed for a cellular activity, one or two of the phosphate groups in ATP can be hydrolyzed with a release of about 7300 calories for each mole of ATP that is hydrolyzed.

Energy Release—Hydrolysis

$$ATP + H_2O \rightleftharpoons ADP + H_3PO_4 + energy \quad \Delta G^\circ = -7300 \text{ cal}$$

and

$$ADP + H_2O \rightleftharpoons AMP + H_3PO_4 + energy \quad \Delta G^\circ = -7300 \text{ cal}$$

This ΔG° is the standard free-energy change for the reaction under standard conditions that are not prevalent in the actual cell. The actual free energy of hydrolysis under true cellular conditions may be as high as $-12,500$ calories/mole.

Study Question 10-1 The hydrolysis of ATP has a $\Delta G^\circ_{300^\circ K}$ of approximately -7300 calories per mole. The standard enthalpy change, $\Delta H^\circ_{298^\circ K}$, is approximately -5000 calories per mole. Calculate the standard change in entropy ($\Delta S^\circ_{298^\circ K}$) for this reaction and comment on whether the reaction represents an increase or decrease in disorder. Would it be more or less spontaneous at a higher temperature? Explain what would happen at temperatures above 37°C.

The ATP-ADP system can be regarded as an energy bank account, and ATP is often called the universal energy currency for living things. Many different reactions can deposit energy in the form of ATP, and later other reactions can draw on the free energy available from this molecule by converting it back to ADP. This energy can then be transferred to other compounds or used to provide the free energy to drive nonspontaneous reactions through free energy coupling, to synthesize more complex compounds, or to perform the work needed for muscular contraction, nerve conduction, transport of material across cell membranes, or any type of cellular process requiring energy or work.

Since the hydrolysis of ATP is a reversible reaction, there is a continuous cycle for the formation and breakdown of ATP molecules, as summarized in Figure 10-7.

As was mentioned earlier, the terms " high-energy " and " energy-rich " phosphate bonds are somewhat misleading. They incorrectly imply that high energy is in the phosphate bond and that when the bond is broken, this energy is released. Actually,

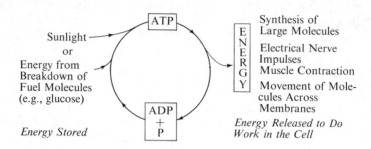

Figure 10-7   The ADP-ATP Cycle.

relatively large energies are required to break chemical bonds. The "high energy content" is thus not meant to describe the bond energy but merely to indicate that when the ATP undergoes reaction, it has a fairly large and favorable ΔG of reaction —that is, there is a relatively high difference in free energy between the reactants and products in the hydrolysis reaction of ATP. These so-called high-energy compounds are not wonder molecules supercharged with mysterious energy. It is simply that they are molecules whose molecular structures give relatively high free energy changes upon hydrolysis, and the term "high free-energy molecules" would be a more accurate description.

In summary, ATP is the primary means for storing chemical energy in the cell. By the releasing of this energy in small amounts, waste is reduced, and the cellular processes are more efficient.

Although ATP is the key energy source in the living cell, other phosphate compounds are also used for energy transfer. One of these is creatine phosphate (Cr ∼ P), shown in Figure 10-8.

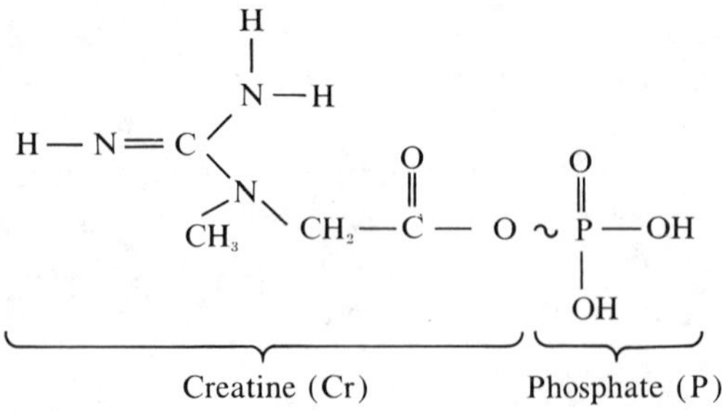

$$H-N=C \underset{\overset{|}{CH_3}}{\overset{\overset{\displaystyle H}{\overset{|}{N-H}}}{\bigg\langle}} \quad CH_2-\overset{\overset{\displaystyle O}{\|}}{C}-O \sim \underset{\overset{|}{OH}}{\overset{\overset{\displaystyle O}{\|}}{P}}-OH$$

Creatine (Cr)   Phosphate (P)

Figure 10-8   Structure of Creatine Phosphate (Cr ∼ P).

A phosphate on ATP can be transferred to creatine to form creatine phosphate.

$$ATP + Cr \rightleftharpoons ADP + Cr \sim P$$

or

$$\text{(A)}–\text{(P}_1)\sim\text{(P}_2)\sim\text{(P}_3)+ Cr \rightleftharpoons \text{(A)}–\text{(P}_1)\sim\text{(P}_2)+ Cr \sim\text{(P}_3)$$

When all of the ADP available in the cell has been converted to ATP and the energy "checking account" is full, then ATP begins to transfer phosphate to creatine phosphate, which serves as an auxiliary, or "savings," account for energy. This transfer also provides ADP, which can be converted to ATP to refill the "checking" account. When large amounts of energy have been used in the cell and ATP levels are low, phosphate in the creatine phosphate "savings" account can be transferred to ADP to produce more ATP.

10-5 Energy–Transfer Reactions

We have seen that ATP and other phosphates can store and provide the energy needed for carrying out cellular processes. But what are the processes by which the cell transfers energy? Two of the major energy transfer processes are *biological oxidation* and *oxidative phosphorylation*, which is a coupling of biological oxidation to the ATP-ADP transfer reactions.

Biological Oxidation

The cell gets most of its energy by the oxidation of glucose and other foodstuffs. Oxidation can occur in a number of ways. One involves direct reaction with oxygen,

---

No. 1: Oxidation by Direct Reaction with
      Oxygen

$$X + O_2(g) \longrightarrow XO_2$$

---

where a hypothetical substance X is oxidized to XO_2. But oxidation does not have to involve oxygen. It may also occur by *hydrogen removal* and *electron removal*. A hydrogen removal reaction involves the transfer of hydrogen from one substance (X) to another (Y). In the process, X is oxidized and Y is reduced.

No. 2: Oxidation by Hydrogen Transfer

$$XH_2 + Y \longrightarrow X + YH_2$$

A third oxidation process is electron removal. When an electron is removed from a doubly charged ion of iron, Fe^{2+}, it is converted to a triply charged ion of iron, Fe^{3+},[7] and oxidation has occurred.

No. 3: Oxidation by Electron Transfer

$$Fe^{2+} - e^- \longrightarrow Fe^{3+}$$

Actually, a closer analysis would reveal that all three processes involve a loss of one or more electrons,[8] and a more general definition of oxidation and reduction is

Oxidation

Loss of one or more electrons

Reduction

Gain of one or more electrons

[7] This results from the fact that the positive charge due to the protons in the nucleus of both Fe^{2+} and Fe^{3+} is the same but Fe^{3+} contains one less electron (or unit of negative charge).

[8] For our purposes, the details of electron removal in the other two processes are not important. The interested reader should consult any introductory chemistry textbook.

All three types of oxidation occur in the cell, but most cellular oxidations occur by either hydrogen transfer or electron transfer, with emphasis on hydrogen transfer mechanisms. Substances like Y are called hydrogen carriers.

$$XH_2 + Y \underset{\text{hydrogen carrier}}{\rightleftharpoons} X + YH_2$$

In the cell a sequence of reactions can occur involving the passing or transfer of hydrogen from one hydrogen carrier to another. In each step a little energy is released, and this can be tapped off and stored by the formation of ATP molecules. At the end of the chain of reactions, the last hydrogen carrier, for example, Z, is oxidized completely to get rid of the hydrogen.

Oxidation by Hydrogen Transfer Mechanism

Step 1: $XH_2 + Y \rightleftharpoons X + YH_2 + \text{energy}$

Step 2: $YH_2 + B \rightleftharpoons Y + BH_2 + \text{energy}$

Step 3: $BH_2 + Z \rightleftharpoons B + ZH_2 + \text{energy}$

Step 4: $ZH_2 + (O) \longrightarrow Z + H_2O + \text{energy}$

_____

Net: $XH_2 + (O) \longrightarrow X + H_2O$

Figure 10-9 shows the release of energy at each step as the hydrogen is passed down a chain of carriers, until at the very end the last carrier transfers its energy to oxygen to form water. The energy from the various steps can be tapped off and put into an energy bank account in the form of ATP for later use. Each step is catalyzed by a specific enzyme.

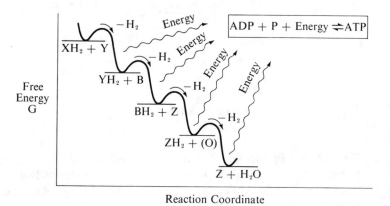

Figure 10-9   Oxidation and Energy Transfer and Storage by a Hydrogen Transfer Mechanism.

A number of important cellular hydrogen carriers have been identified—including NAD, FAD, the quinones (Q) and the cytochromes.[9] Thus, in the cell a typical hydrogen transfer sequence is $XH_2 \rightarrow NADH_2 \rightarrow FADH_2 \rightarrow QH \rightarrow$ four different cytochromes $\rightarrow H_2O + X$.

Oxidative Phosphorylation

The purpose of sending these hydrogens down the line is to release energy in small amounts so that it can be tapped off to produce ATP. Now we can couple this biological oxidation process by hydrogen transfer with the phosphorylation process involving the conversion of ADP to ATP. The combination of these two processes is called *oxidative phosphorylation*. The coupling mechanism is somewhat like the gear coupling that links the engine to the wheels of an automobile. Oxidation by hydrogen transfer causes the engine to turn. This in turn spins the wheels to convert ADP to ATP, as

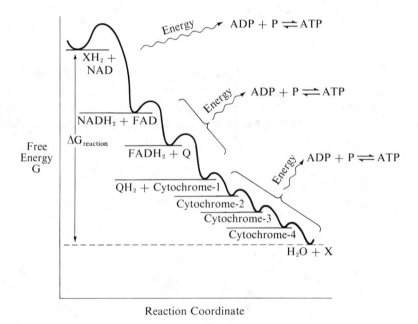

Figure 10-10   Free Energy Profile for Mechanism of Oxidative Phosphorylation in Living Cells. Oxidation Occurs by Stepwise Transfer of Hydrogen from One Carrier Molecule to Another. The Energy Released Is Stored in 3 Moles of ATP.

[9] The full chemical names and structures of these different hydrogen carriers are not important for our discussions here.

summarized in Figure 10-10. Here we see a good example of the application of thermodynamics and kinetics to a complex reaction sequence in the living cell. The net free energy change, ΔG_{net}, determines overall thermodynamic spontaneity, while the activation energy for the individual step determines the rate of each reaction and thus the kinetic feasibility. Actually, each step is catalyzed by a specific enzyme that lowers the activation energy so that each step can proceed at the rate needed for cellular activity.

Let us now take a broad look at the bioenergetics[10] of the processes of respiration and photosynthesis that make up the carbon and oxygen cycles discussed earlier in this chapter (Section 10-3).

10-6 Cellular Aerobic Respiration

Living cells fall into one of two classes. *Autotrophic* cells obtain energy from sunlight and, through photosynthesis, combine water and CO_2 to form food and release O_2. *Heterotrophic* cells then use O_2 to burn food and release CO_2 through the respiration process—thus completing the cycling of carbon and oxygen. Respiration can occur with oxygen (aerobic respiration) or without oxygen (anaerobic respiration). We shall be concerned only with *aerobic respiration*. As shown below, it should be thermodynamically spontaneous at standard conditions and it if were 100% efficient, it would give a maximum yield of 675,900 calories of free energy per mole of glucose.

Aerobic Respiration (net equation)

$$C_6H_{12}O_6(s) + 6O_2(g) \rightleftharpoons 6CO_2(g) + 6H_2O(l)$$

$$\Delta H^\circ_{298^\circ K} = -667,000 \text{ cal} \quad \text{(yes)}$$

$$\Delta S^\circ_{298^\circ K} = 63 \text{ cal/}^\circ K \quad \text{(yes)}$$

$$\Delta G^\circ_{300^\circ K} = -675,900 \text{ cal} \quad \text{(yes)}$$

Study Question 10-2   Draw a free energy profile diagram for the net reaction for aerobic respiration.

[10] The reader who is interested in more information on bioenergetics should consult the excellent and readable book *Bioenergetics*, 2nd ed., by A. L. Lehninger, W. A. Benjamin, Inc., 1971.

How is this energy released and used in the cell? There are at least seventy steps in the respiration mechanism with each step being catalyzed by a specific enzyme. The overall process consists of two major phases: *fuel breakdown* and *oxidative phosphorylation*, which we have seen consists of the coupling of hydrogen transfer with energy transfer to form ATP molecules.

The fuel breakdown phase consists of two major stages: *glycolysis*, in which the 6-carbon sugar glucose is broken down into two smaller 3-carbon sugars with the net production of 8 molecules of ATP for each molecule of glucose, and the *Krebs*, or *citric acid*, *cycle* in which the fragments are broken down further into CO_2 in a cyclic process with the production of 30 molecules of ATP per molecule of glucose, as summarized in Figure 10-11.

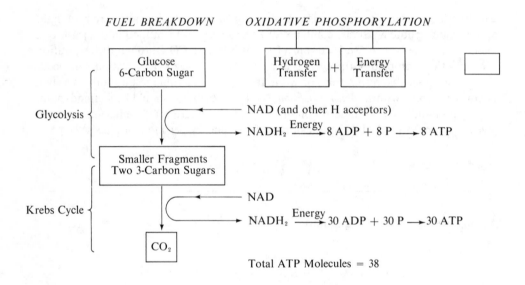

Figure 10-11   The Main Phases of Cellular Respiration.

What is the energy yield in terms of ATP molecules produced in the entire sequence? The breakdown of one molecule of glucose by the aerobic respiration process yields a net gain[11] of 38 moles of ATP. The standard free energy change for the formation of one mole of ATP from ADP is about 7300 calories. Thus, the total energy trapped during the oxidation of one mole, or 6×10^{23} molecules, of glucose is 38×7300, or approximately 277,400 calories.

---

[11] Actually 40 moles of ATP are produced per mole of glucose, but the energy from 2 moles of ATP is needed to start the glycolysis phase, so that the net gain is $40-2$, or 38 moles of ATP.

Net Reaction—Cellular Respiration

$$C_6H_{12}O_6(s) + 6O_2(g) + 38\ ADP + 38\ P \longrightarrow 6\ CO_2(g) + 6\ H_2O(l) + 38\ ATP$$

$$\Delta G^\circ_{maximum}$$
$$\Delta G^\circ_{stored\ in\ ATP} = -277{,}400\ cal/mole = -675{,}900\ cal/mole$$

We saw earlier that the maximum energy yield, assuming 100% efficiency, was $-686{,}000$ calories, as compared with the actual yield of approximately $-277{,}400$ calories. The efficiency of aerobic respiration can be calculated as follows:

$$\%\ efficiency = \frac{actual\ yield}{maximum\ yield} \times 100$$

$$\%\ efficiency \cong \frac{-277{,}400}{-686{,}000} \times 100 \cong 41\%\ at\ 1\ M\ concentrations$$

Approximately 41% of the energy content of the original glucose molecule ends up as useful energy available to the cell for work, with about 59% lost as heat, or entropy, in the process. Actually, the efficiency in the cell is higher. These calculations are based on standard concentrations of ADP, ATP, and P at 1 molar. The actual concentrations existing in the cell probably yield a ΔG value close to $+12{,}500$ cal/mole for the formation of ATP. A more realistic value for the energy stored might be $38 \times 12{,}500$, or 475,000, calories with an efficiency as high as 70%.

$$\%\ efficiency \cong \frac{-475{,}000}{-675{,}900} \times 100 \cong 70\%\ at\ nonstandard\ concentrations\ in\ the\ cell$$

In a gasoline engine, 15–25% of the energy liberated during the combustion of the gasoline is converted into work to power the car, and the very best steam engines seldom exceed an efficiency of 45%. One means of cutting down (but still not eliminating) air pollution from automobiles would be to find a process for powering a car with an efficiency approaching that in a living cell.[12]

Our familiar free energy diagram can be used to provide us with a profile of the kinetics and thermodynamics of the respiration process, as shown in Figure 10-12. The overall rate of the reaction is determined by the first step to form the activated glucose, which is accomplished by using ATP to add energy in the form of a phosphate to the glucose molecule. This first step in the glycolysis phase is a good example of

---

[12] The "reheating," or afterburner, in jet engines is designed to increase efficiency by reusing waste heat.

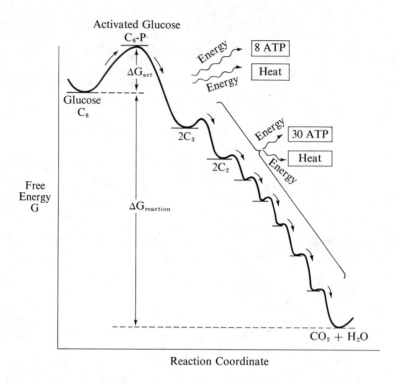

Figure 10-12   Free Energy Profile for Simplified Mechanism of Aerobic Respiration.

free energy coupling being used to bring about an endergonic, or thermodynamically unfavorable, reaction, as shown below.

Reaction 1: glucose + P $\rightleftharpoons$ glucose-6-phosphate (activated) + H_2O

$$\Delta G^\circ \cong +3000 \text{ cal (unfavorable)}$$

Reaction 2: ATP + H_2O $\rightleftharpoons$ ADP + P $\Delta G^\circ \cong -7300 \text{ cal (favorable)}$

Net: (Free
Energy   glucose + ATP $\rightleftharpoons$ glucose-6-phosphate (activated) + ADP
Coupling)

$$\Delta G^\circ_{net} = -7300 + 3000$$

$$= -4300 \text{ cal (favorable)}$$

After this activation step, specific enzymes lower the activation energy of succeeding steps so that these steps proceed relatively rapidly. As the glucose molecule proceeds down the energy hill, small amounts of energy are released at each step and stored in the form of ATP, which is used by the cell when needed to perform work. The overall process is exergonic. Now we can turn our attention to photosynthesis, the endergonic, or energy-requiring, half of the carbon and oxygen cycles.

10-7 Photosynthesis

The problems faced by a plant are essentially the same as those of an animal—the need to trap energy to synthesize larger molecules needed for life. Plant cells, like animal cells, break down glucose to CO_2 and water and trap the released energy as ATP by the process of respiration. With a supply of glucose and nitrogen, plants can live for a long time in the dark—but they prefer sunlight.

Animals get their glucose by eating plants (herbivores) or other animals (carnivores). Most plants, however, are autotrophic and have the ability to absorb radiant energy from the sun and convert CO_2 and H_2O to glucose in the process of *photosynthesis*. The absorbed light energy is converted to chemical energy and stored in ATP and the hydrogen carrier, NADPH. This chemical energy is then used to reduce carbon dioxide to form a basic sugar unit, (CHOH) or (CH_2O), plus oxygen, which is released to the atmosphere. The basic sugar unit is built up by a series of steps to form a sugar such as glucose. The *net* equation for photosynthesis is

---

Net Reaction for Photosynthesis

$$6CO_2(g) + 6 H_2O(l) \xrightarrow{\text{light}} C_6H_{12}O_6(s) + 6 O_2(g)$$

$$\Delta H^\circ_{298^\circ K} = +667,000 \text{ cal} \quad (\text{no})$$

$$\Delta S^\circ_{298^\circ K} = -63 \text{ cal/}^\circ K \quad (\text{no})$$

$$\Delta G^\circ_{300^\circ K} = +675,900 \text{ cal} \quad (\text{no})$$

From a thermodynamic standpoint, the formation of glucose from CO_2 and H_2O is an endergonic, or "uphill," process that requires an energy input in the form of light quanta. The net free energy profile for the process is given in Figure 10-13.

The total amount of carbon converted by photosynthesis to large fuel molecules is probably at least 100 billion tons per year. Simultaneously it renews the oxygen in the atmosphere and removes some of the CO_2 built up by respiration and man's fossil fuel burning activities. The cycling rate is such that, on the average, every molecule

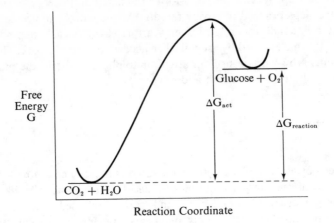

Figure 10-13   Free Energy Profiles for Net Photosynthesis Process.

of CO_2 in the atmosphere is incorporated into glucose by photosynthesis about once every 200 years, and every oxygen molecule approximately once every 2,000 years.

We usually think of the familiar green plants on land as carrying out photosynthesis, but about one third of all photosynthesis is carried out by the "grass of the sea"— the microscopic plants called phytoplankton found in our oceans, rivers, and lakes. The rate at which we in the U.S. are bulldozing, chopping, and paving over our land with asphalt and concrete—the equivalent to the state of Rhode Island is paved every six months—can eventually provide another potentially serious disruption of the carbon and oxygen cycles on local and global levels.

Man can now upset the delicate ecological balance in the world. For instance, the pollution of the world's oceans is one of our most serious problems, and it is projected to increase sevenfold before the year 2000. We are presently dumping over 500,000 different man-made chemicals into our oceans, rivers, and lakes; and only a handful have been tested to determine their effect on photosynthesis in phytoplankton.[13]

In spite of accounts in the popular press, recent calculations indicate that we are in no real danger of running out of oxygen from these intrusions into the carbon and oxygen cycles. The danger lies more in upsetting the food producing capacity of our land and oceans. With two-thirds of the world's population already hungry and mal-nourished or both and a projected doubling of world population in only 35 years, this would be catastrophic.

Since man could upset the process of photosynthesis, an understanding of its mechanism could provide some help in coping with this environmental problem.[14]

[13] For a startling scenario showing what could happen to the oceans and to man if he continues to disrupt the ocean ecosystem, see the article entitled "Eco-Catastrophe!" by Dr. Paul Ehrlich, Professor of Biology, Stanford University, in *Eco-Catastrophe*, Canfield Press (Harper & Row, Publishers), 1970.

[14] Scientific understanding of photosynthesis alone will not solve our environmental problems— they are also problems of man's political, economic, and social behavior. However, if man does decide to minimize ecological damage, then this knowledge will be crucial.

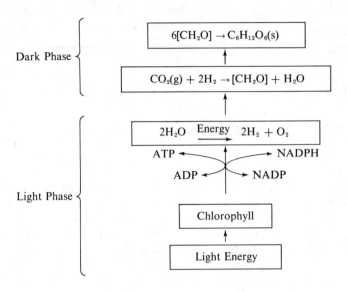

Figure 10-14   Simplified Summary of Photosynthesis Process.

Although the complete mechanism is still unknown, it is known that there are over one hundred steps in the process, with each step involving catalysis by a specific enzyme. The complex process of photosynthesis is divided into two stages, called the *light reactions* and the *dark reactions,* as summarized in Figure 10-14.

The *light phase* involves the absorption of radiant energy from the sun by chlorophyll (or other light trapping molecules). Through a sequence of reactions called cyclic and noncyclic *photophosphorylation,* this electron energy is converted to chemical energy and stored in ATP and the hydrogen carrier NADPH.

In the second stage of *dark reactions* the chemical energy from the ATP and NADPH, formed in the light stage, are used to convert carbon dioxide to glucose (or other organic compounds) plus oxygen. These reactions can proceed in the absence of light. The overall balance sheet shows that three molecules of ATP and two molecules of NADPH are required for each molecule of CO_2 reduced. Thus, the net equation in terms of the ATP and NADPH can be written as

---

Net Equation for Photosynthesis

$$6\ CO_2(g) + 6\ H_2O(l) + 18\ ATP + 12\ NADPH \xrightleftharpoons{light} C_6H_{12}O_6(s)$$

$$+\ 6\ O_2(g) + 18\ ADP + 18\ H_3PO_4 + 12\ NADP$$

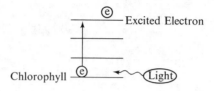

Figure 10-15   Effect of Light on Chlorophyll.

What happens in the *light phase* when light of the proper energy strikes a chloro-phyll molecule in a plant cell? The exact answer is not known, but a crude reaction mechanism has been developed. Chlorophyll is a complex molecule consisting of carbon and nitrogen rings joined together in a doughnut shape with magnesium (Mg) in the center. When light hits the molecule, an electron in the magnesium absorbs energy and is excited to a higher energy level, as shown in Figure 10-15. The energies, or wavelengths, absorbed correspond to red, violet, orange, yellow, and blue light—leaving green as the color, or energy, not absorbed. Thus, leaves (and supposedly some toothpastes) appear green, because chlorophyll has absorbed all but the green wavelengths of light—hence, the green is reflected to your eyes.

The events in the light phase consist of converting the energy in the "excited" electrons to chemical energy in the form of ATP and the hydrogen carrier NADPH as shown symbolically in Figure 10-16.

We have seen that light is trapped in the light phase of reactions and converted to chemical energy in the form of ATP and NADPH. These two chemicals then serve as the energy source for driving the sequence of reactions in the *dark phase*. This phase involves the combination of CO_2 with hydrogen from NADPH and energy from ATP to build up glucose in a complex cyclic reaction sequence known as the

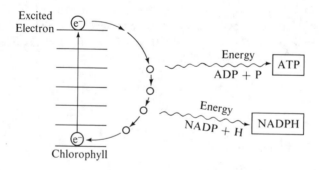

Figure 10-16   Symbolic Representation of the Light Phase Reactions of Photo-synthesis. (The actual mechanism, which is not completely known, is much more complicated and occurs in two parts known as cyclic and non-cyclic photophosphorylation.)

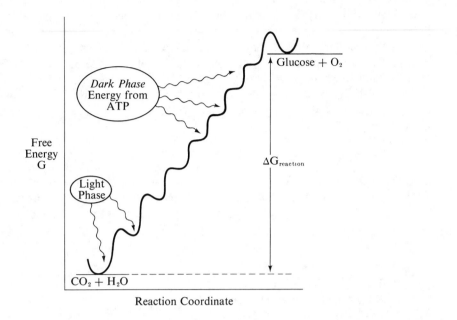

Figure 10-17   Crude Free Energy Profile for the Photosynthesis Mechanism.

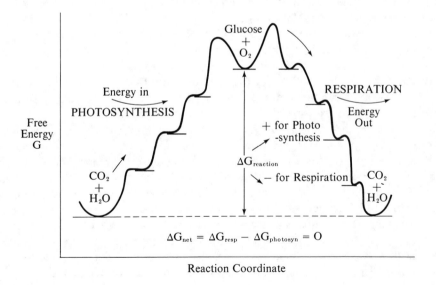

$$\Delta G_{net} = \Delta G_{resp} - \Delta G_{photosyn} = 0$$

Figure 10-18   Free Energy Diagram for the Photosynthesis-Respiration, or Carbon, Cycle.

carbon-fixation cycle, or *Calvin cycle*,[15] after Melvin Calvin and his co-workers at Berkeley who charted this path (for which Calvin received the Nobel prize in 1961).

The light and dark phases of photosynthesis interact like a complex machine with each part driving another. A crude representation of a free energy profile for the photosynthesis mechanism is given in Figure 10-17.

Figure 10-17 for photosynthesis and Figure 10-12 for respiration can be combined to show a free-energy profile for the photosynthesis-respiration, or carbon cycle, as shown in Figure 10-18.

Study Question 10-3 Did you thank a green plant today? Suppose all plant life on this planet were destroyed, leaving only animals. Assuming that no further evolution occurs, describe what would happen on this planet. (Give some details in terms of the carbon and oxygen cycles.)

10-8 Preserving Ecological Diversity—Have You Thanked a Green Plant Or an Alligator Today?

Why should you thank a green plant, an alligator, or a hippopotamus today and every day? Why should we be concerned about preserving all of the diverse species of plant and animal life on this planet? The answer goes far beyond a desire to preserve beauty and unspoiled, or natural, areas, although these are very important reasons. We are now realizing that our very survival depends on preserving diversity and complexity in the earth's ecosystem, the biosphere. This vast network of chemical and biological interactions generates and preserves our soil and air and maintains the purity of our water, and green plants supply our oxygen and our food. We have just seen that interrupting the processes of energy flow and chemical cycling or changing the rate at which they occur can threaten the thermodynamic and kinetic stability of the life processes at the localized (organism) or global (ecosystem) level.

But there is another reason for preserving plants and animals. They represent the biological capital provided by evolution. We are beginning to understand that *species diversity normally means ecosystem stability*. The more species present, the greater the possibility for adapting to changing conditions. We now recognize the importance of *the conservation of genetic information*—the idea that we ought to preserve every single species of life somewhere in its natural habitat because we presently have no way of knowing what animal or plant is now essential or may prove valuable or essential in the future. A living species is an irreplaceable resource; once extinct, it can never be recalled. Modern man has exterminated some 200 species of plants and animals in the last 150 years, and today over 500 others are on the danger list.

[15] In 1965 research indicated that most but not all plants follow the Calvin cycle. A second mechanism—the Hatch-Slack cycle—has been proposed for these plants. For our purposes it is not necessary to look at the details of either of these cycles.

Ecological Backlash

What happens when we remove a species? Unfortunately, we like to think of species as either good or bad and consider our role as one of wiping out the villains. We recklessly spread broad spectrum poisons across the land and place bounties on certain animals. The backlash from this "cowboy" approach is now becoming apparent.

Consider the alligator, a key species that is being systematically exterminated primarily by poachers who sell the hides for making fancy pocketbooks and shoes. In only ten years, between 1950 and 1960, Louisiana lost 90% of its alligator population, and the alligator in the Florida Everglades is now threatened. Who cares? The alligator is a key factor in preserving the entire ecological balance of the Everglades—a balance upon which most of the increasingly urbanized state of Florida depends. The alligator is a hole digger. In dry spells his deep pools or "gator holes" provide water for enough of the birds and animals so that they can begin a massive breeding cycle to repopulate the glades after the drought. Alligators also make large nesting mounds that are popular sites for nests of herons, egrets, and other birds essential for the life cycle in the glades. As alligators move from their gator holes to nesting mounds they help keep the waterways open, and they preserve a balance of game fish by consuming large numbers of predator fish such as the gar.

In parts of Africa, it was decided that the hippopotamus and the crocodile were to be eliminated. Then the streams began clogging up, fish that provided essential protein for a protein-starved population began disappearing, and a debilitating disease, schistosomiasis, or snail fever, spread among the population. Investigation revealed that removal of the hippopotamus and crocodile was the major factor bringing about these disastrous changes. Like the alligator, the hippopotamus and crocodile produced deep holes that maintained life in dry spells. They also kept the streams from clogging by removing plant growth and stirring up the silt. Without this the streams filled up with silt and plant growth and became shallow and warm—ideal for the massive breeding and spread of the snail which serves as a host for a parasite flatworm that causes schistosomiasis—a kind of poor man's emphysema. This disease, which makes a person so weak he can hardly hold up his arm, has also occurred as man has expanded irrigation to feed more mouths. It thrives in shallow water streams and irrigation ditches and is now the world's most prevalent infectious disease. Ironically, the Aswan Dam in Egypt, an engineering marvel, is proving to be an ecological disaster. The streams are silting up, and the increased irrigation it allows has resulted in schistosomiasis for over one half of the population in some parts of Egypt. During the time it took to build the dam, the increase in Egyptian population surpassed the additional food it allows to be grown.

In Brazil, DDT was used to kill off mosquitoes. This eliminated a type of wasp that fed on caterpillars. The caterpillars, without a predator, multiplied and ate the thatch off the roofs of the houses.

In the western part of the United States, official government policy called for the extermination of the coyote. As a result the rodent population, which the coyote helped control, has multiplied astronomically. The rodents then attack the roots of prairie grass and destroy the range for sheep and cattle used to support man.

Two major species—insects and rodents—are found all over the world but are usually kept in check by natural predator-prey relationships. By drenching the world with broad spectrum herbicides and insecticides, 700 million pounds each year, we seem determined to turn the world over to the rats, flies, and cockroaches and numerous disease-carrying microorganisms that are always waiting in the wings. If we manage to eliminate predators such as hawks, owls, and coyotes, our entire continent would be knee-deep in mice within a year. Without sea gulls, pelicans, and beach birds we couldn't go near our beaches because of the stench of dead fish and human garbage. Without land birds, regardless of how much pesticide we use, the Earth would be covered with insects. The 700,000 known species of bugs with fantastic reproductive potential can adapt quickly to almost any poison or level of radioactivity. Without natural predators only two bugs could cover the Earth fifty feet deep with insects in only four months.

Or suppose that one of the 500 man-made chemicals we add to the environment each year eventually poisons certain types of soil bacteria crucial to the nitrogen cycle. Plants cannot build proteins directly from nitrogen gas in the atmosphere. Certain types of bacteria and blue-green algae convert the N_2 to nitrogen in the form of ammonia and nitrates, which serve as plant nutrients. When plants and animals die, other denitrifying bacteria complete this nitrogen cycle upon which all life depends, by converting the protein and other nitrogen-containing substances back into N_2. So far we have been incredibly lucky—we hope.

Man Simplifies the Ecosystem

What is man doing to this essential ecological diversity and complexity? We attempt to rearrange nature to support immediate needs. We *simplify the ecosystem* so that we can control it. Paradoxically, this threatens our long-range survival by making the ecosystem more vulnerable to the very stresses we are imposing.

Without realizing it, we have linked much of our productive technological society to features that are incompatible with preserving the stability of our life-support system. We bulldoze fields and forests of thousands of interrelated plant and animal species, and cover them with asphalt for a shopping center, highway, or other forms of the almost totally artificial environment we call a city. Farming, once diversified, now consists primarily of monocultures—single crops of wheat, rice, or corn covering vast areas as far as the eye can see. Because of a lack of diversity, single crops must be supported by massive and ever increasing doses of fertilizers, much of which washes into our lakes and streams and threatens to overload and disrupt the nitrogen and oxygen cycles. Because a monoculture crop can be wiped out rather easily by only a single species of insect, we attempt to protect it with more and stronger pesticides in a continually escalating form of chemical warfare rather than in using or developing biological pest controls. As the insects quickly breed resistance that we couldn't match even in several hundred million years of additional evolution, some of these same poisons are magnified up the food chain to accumulate in the highest concentrations in our bodies.

This is not to suggest that all of these practices are bad and that we shouldn't and won't keep on simplifying parts of our ecosystem. We must do some simplification to survive, but our problem is twofold. First, we don't know which parts are really essential, and we don't know which of the man-made chemicals will be harmful in the long run. We lack the knowledge explosion that would help us deal with the pollution explosion. Professor Paul Ehrlich has likened the biosphere to a massive and intricate computer built by linking and cross-linking a vast array of transistors and other electrical components—a mysterious and amazing system that we do not really understand. Man is rapidly simplifying the complex computer network upon which life depends by randomly pulling out transistors and overloading and disconnecting various parts and circuits. Slowly we are beginning to realize that we are playing a massive game of "ecological Russian roulette," hoping that the computer whose workings we don't comprehend will not break down when we simplify its essential complexity to suit our purposes. Ecologists don't know how or when our life support system will break down, but they know that if we continue on our present course, sooner or later it will be disrupted.

As we newcomers on this planet eliminate threads from the magnificent tapestry of life built up over millions of years, we must remind ourselves that the biosphere is not only more complex than we think, but more complex than we can ever imagine. The essence of survival and freedom requires humility and cooperation rather than arrogance and domination.

> What has gone wrong, probably, is that we have failed to see ourselves
> as part of a large and indivisible whole. For too long we have based
> our lives on a primitive feeling that man's "God-given" role was to
> have "dominion over the fish of the sea and over the fowl of the air
> and over every living thing that moveth upon the earth." We have
> failed to understand that the earth does not belong to us, but we to
> the earth.

> *Rolf Edberg*

10-9   Review Questions

10-1.   What is the source of solar energy? Will the sun "run down"? Explain.

10-2.   What is the biosphere? What are its three parts? Draw a diagram showing how they are related.

10-3.   Why can earth be considered as a closed spaceship?

10-4.   Upon what two things does life on spaceship Earth depend?

10-5.   Distinguish between micronutrients and macronutrients.

10-6.   What is a biogeochemical cycle? Name several of these cycles.

10-7.   Diagram and explain the carbon cycle and the oxygen cycle.

10-8.   Distinguish between carnivores and herbivores.

10-9.   What is ATP? ADP? AMP? How do they differ?

10-10.   What is free energy coupling? Give an example.

10-11. How does ATP transfer energy to other molecules? What is the function of creatine phosphate in the cell?

10-12. Draw and explain the ATP-ADP cycle.

10-13. Explain carefully why the term "high-energy phosphate bond" is a misleading name.

10-14. Show three methods for oxidation in biological systems. What is oxidation? What is reduction?

10-15. Using reactions and a free energy profile diagram, show how oxidation can occur by a hydrogen transfer mechanism.

10-16. Distinguish between aerobic and anaerobic respiration.

10-17. Describe briefly the two phases of the process of aerobic respiration.

10-18. Draw and label net and detailed free energy profile diagrams for the following:
a. aerobic respiration,
b. photosynthesis,
c. the aerobic respiration-photosynthesis cycle, or carbon cycle.

10-19. Discuss the overall efficiency of the respiration process.

10-20. Distinguish between the light and dark reactions in the photosynthesis process.

10-21. What are the major components of an ecosystem? Draw a diagram showing the relationship between these components.

10-22. Explain what is meant by chemical cycling and energy flow through an ecosystem.

10-23. What is the so-called "ecological boomerang" principle?

10-24. What is man's primary effect on the ecosystem? This effect is perhaps both good and bad. Explain.

10-25. Explain specifically why we should preserve species such as the alligator and the crocodile. Try to determine why we should preserve a balance of snails, fleas, cats, robins, tigers, and even rats.

10-26. Distinguish between producers, consumers, and decomposers.

10-27. Explain why you should "thank a green plant every day."

10-28. Distinguish between herbivores and carnivores. Give two examples of each.

10-29. What is ecological diversity? Why should it be preserved?

10-30. What is meant by the term "conservation of genetic information?"

10-31. Give several reasons why we should be concerned about preserving any and all living species of plants and animals on this planet. Which reason is probably the most important? Why?

10-32. List the major problems that can occur on any spaceship. Relate these problems to the Second Law of Thermodynamics. What are the implications of these problems for life on our spaceship? How do they relate to your everyday life?

Chapter 11

Thermodynamics and Spaceship Earth

11-1 Around the Bend on a J-Curve

The two most significant facts of our existence as a global society are:

---

1. We all live on a spaceship of finite size and resources.
2. Our life-support system is threatened, because we have gone around the bend on a J-curve of exponentially increasing population, resource consumption, and pollution.

We seemingly live in the midst of a crisis of crises—one exponential curve, or J-curve (Figure 4-2, page 80), piled on top of another—all interacting synergistically, so that for many the entire fabric of our society and our life-support system seem to be coming apart at the seams. Something has gone wrong during the past few decades. Increased control over nature is not producing safety and peace of mind. As we grow richer materially, the environment grows poorer. As we acquire more leisure, we find less available space and beauty to enjoy. Technological innovations create problems which require even bigger and more expensive countertechnologies to correct their impact on the environment.

Regardless of what we do or where we go, we can no longer escape the symptoms of our ecological crisis—our senses are bombarded with constant reminders. You can feel it in the tremor of anxiety that runs through our society—in the increasing unrest, frustration, and alienation, in the deterioration of public services, and in the progressive loss of privacy and freedom. You can smell it in our choking air and rotting rivers, lakes, and oceans.

Those predicting the end of man or the world are premature, but spaceship Earth is in trouble—many are starving now, the casualties will surely increase, and maintaining or increasing the quality of life will be a formidable task.

Cowboy to Spaceship Rules

What does it mean to live around the bend on a J-curve on a spaceship? We have only recently become aware of the problem; its profound implications are still matters of some conjecture. What we are beginning to perceive is that our situation will require a radically new set of survival rules. It is as if we had been brought up to learn and perfect the rules for baseball only to find that our survival now depends on finding and learning new rules for a different and much faster game. As the economist Kenneth Boulding put it, we have been living by "cowboy, or frontier, rules," and we must now discover and convert to "spaceship rules."

We have envisioned the Earth as a place of unlimited frontiers and resources, where ever-increasing consumption and production inevitably lead to a better life and where success is measured by our Gross National Product. Many are now recognizing that to hold rigidly such "frontier rules" will assure a steady decrease in the quality of life. We don't know yet what the new rules should be; much less how to change some of our present political, economic, technological, and ethical rules or systems to new ones.

More Implications of the J-Curve

The implications of living around the bend on a J-curve boggle the mind. Many ideas and practices that seemed reasonable when we were on the earlier part of the curve before the bend now seem invalid—a sudden catastrophic change has occurred. As Alvin Toffler[1] puts it, we experience "future shock." Consider the simple example of doubling a page of this book over and over and noting the increase in thickness. Doubling it once, twice, three, or even ten times results in a change in thickness that we can easily comprehend—we are on the lower part of the J-curve. But if we could physically double this piece of paper only 38 times, its approximate thickness would reach from New York to Los Angeles—some 3000 miles. Suddenly something almost incomprehensible happened as we went around the bend on the J-curve. Once around the bend and heading almost straight up, the shock becomes even greater. If we could

[1] Alvin Toffler, *Future Shock*, Random House, Inc., 1970.

double the paper only about 45 times, it would reach from the earth to the moon—some 240,000 miles, and doubling it only about 53 times would give a thickness reaching from here to the sun—some 93 million miles. Going around the bend on a J-curve is truly a mind-stretching and shattering experience.

The Population Bomb

In terms of the overpopulation problem on our spacecraft, it means that the birth rate is about two times the death rate so that we are now adding about 200,000 additional passengers each day—1.4 million each week—70 million each year. This is an annual increase equivalent to the entire population of the planet 4000 years ago. It means adding the equivalent of the entire U.S. population every three years. In the past 24 hours India alone increased by 40,000 and the U.S. added 6000 people and 12,000 automobiles.

All of these new passengers must be fed, clothed, and housed; and each will use a portion of our finite resources and add to the global pollution. This is occurring when two-thirds of our present passengers are either hungry or malnourished, and three out of four don't have adequate housing and are without a safe and adequate water supply—when there are more hungry and weakened people on this planet now than there were human beings in 1850.

This difference between birth rate and death rate determines our present predicament, but an even more important long range factor is the momentum built up by going around the bend on a J-curve. The most critical factor in our future population growth is the *age distribution* of the population. Probably the most alarming fact that one can state is that approximately 37 percent of the people already on this planet are under 15 years of age, and in the developing countries the average is 45 percent. As a result, even if birth rates decline sharply, there will still be a large increase in population for many, many years unless death rates climb.

For example, the Bureau of Census has recently projected that even if every couple in the United States decided from now on to replace themselves by having only two children[2] instead of the present average of around three, and if we stop all immigration into this country (some 400,000 per year), the population of the U.S. would still not level off until the year 2037 with about 275 million Americans. If we don't achieve the two-child average until 1980, then our population would level off at about 300 million in 2045, and if this replacement average is not reached until the year 2000, population growth would stop in 2065 with a population of over 350 million, almost twice our present size. Like passengers on a supersonic jet we no longer have the option of getting off at a stop a hundred miles down the road. Once around the bend on a J-curve, a rational approach to limiting population growth by lowering the birth rate requires long-range planning and deliberate action on the scale of 70 to 100 years.

Add to the J-curve of increasing agony in the underdeveloped countries, the fact that we in the developed countries are threatening the entire life-support system by

[2] Actually replacement requires an average of 2.11 children.

rounding the bend on two other J-curves, of increasing use of energy and finite resources and the resulting pollution. During four days in April, 1970, millions of people watched and prayed as the lives of three astronauts were saved on Apollo 13, a spaceship in trouble. During those same four days back on spaceship Earth, some 116,000 human beings died of starvation or malnutrition; half were children under five.

The Time Factor

The most significant aspect of this problem is the *time factor*. Several generations ago, when we were on the lower part of the curve, the world seemed enormous, and man's problems, although serious, seemed manageable. There were hundreds of years available to absorb a new idea or technological invention. Because of the relatively small number of people, their impact on the environment was not serious; "frontier" rules were both necessary and useful. But now that the bend on the J-curve has been rounded, it is as though the globe has shrunk; our problems seem endless, and the time factor for dealing with them is apparently measured in only a few years.[3]

The S-Curve

What will happen? In a closed system with finite resources, the growth of population and resource consumption will not continue indefinitely, and the "J" curve *will* be leveled off to an "S" curve (see Figure 4-4, page 82) as the system is brought into equilibrium. Either the birth rate must go down drastically—not slowly as it has been for the past 50 years—or the death rate must rise dramatically.

The figures in Table 11-1 give some indication of just how much the death rate would have to increase in order to achieve population equilibrium, if we don't use the birth rate approach. They show that a population crash involving the death of one-half billion to several billion would have to occur to bring population and resources into equilibrium.

The only real question is not whether the J-curve will be leveled to an S-curve, but when and how much pain and misery we will have to endure in the process. If we attempt to feed the world and provide health care without simultaneously and drastically decreasing the birth rate, we only defer death for an even greater number at a later date.

In our plastic, technological cocoon we have become insulated from nature,[4] ignorant of the meaning of large numbers and growth rates, and desensitized to the agonies and needs of our fellow passengers.

---

[3] For a listing of our planetary problems in order of their importance, with estimates of time to disaster for each problem, and a summary of what we must do, see the superb article entitled "What We Must Do" by J. R. Platt, *Science*, **116**, 1115 (1969).

[4] For example, many affluent Americans going from one heated or air conditioned environment to another have the illusion that air conditioning removes air pollutants. It doesn't even remove all of the particulate matter, much less the deadly gases.

Table 11-1   Implications of Going around the Bend on a J-curve.

Some Past Disasters	Number Killed	Present World Population Growth Replaces This Loss in Approximately
Pakistan tidal wave, 1970	100,000	12 hours
All Americans killed in all our wars	600,000	3 days
Great flood-Hwang-Lo River, 1887	900,000	$4\frac{1}{2}$ days
Famine in India, 1769–1770	3,000,000	2 weeks
China famine, 1877–1878	9,500,000	7 weeks
Present global famine	15,000,000 per year	$2\frac{1}{2}$ months
All wars in the past 500 years (some 250 wars)	35,000,000	6 months
Bubonic plague (Black Death) 1347–1351	75,000,000	13 months

This summarizes our problem. We are finally beginning to realize the nature and complexity of the problem. This awareness is our hope and our challenge. But to act wisely we must become ecologically and thermodynamically informed.

What is the use of a house if you haven't a decent planet to put it on?

Thoreau

11-2   Algaeburgers, Anyone?—Thermodynamics, Food Chains, and Overpopulation

Let us look more closely at what happens to energy as it flows through an ecosystem. In the last chapter we listed the major components of an ecosystem (Table 10-1, page 259) and showed how they were related to chemical cycling and energy flow (Figure 10-3, page 260).

A very useful but simplified model pictures people, particularly those in affluent countries, as living at the end of food chains. For example, grass ⟶ grasshopper ⟶ frog ⟶ trout ⟶ man. Energy is transferred from organism to organism in the various levels. A typical food chain might have three to five levels, called *trophic levels*. Plants, or producers, are eaten by herbivores, or primary

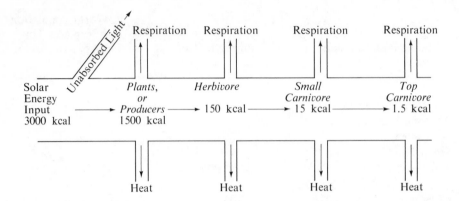

Figure 11-1 Simplified Version of the Energy Flow through a Food Chain (all values are approximate).

consumers; small carnivores, or secondary consumers, eat the herbivores; and then top carnivores, or tertiary consumers, eat the small animals.

How much of the initial input of solar energy into plants is actually available to keep species at the top of the chain alive? Figure 11-1 shows the flow of energy through a typical food chain. The energy input at one end must equal the total energy output in accordance with the First Law of Thermodynamics. However, we can see that in accordance with the second law, most of the energy input is degraded to useless heat

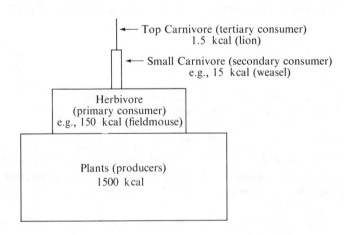

Figure 11-2 Simplified Food, or Energy, Pyramid Showing Energy Flow through an Ecosystem (all values are approximate).

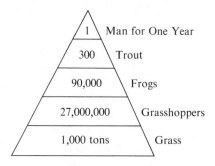

Figure 11-3   Pyramid of Numbers in a Food Chain.

energy that ends up in the environment. There is a considerable loss of energy at each level, with relatively little actually available to support life. Typically, about 80–90% of the energy is simply wasted and lost as heat to the environment at each step. In other words, 10–20% of the energy[5] is stored in the living tissue available for transfer to the species at the next level.

This energy loss at each step can also be expressed in an *energy*, or *food*, *pyramid*, as shown in Figure 11-2.

Another way of looking at this is in terms of the numbers of organisms at each level that are required to support a certain number at the next higher level. This also usually leads to a pyramid—*a pyramid of numbers*, as shown in Figure 11-3. Bruce Peterson has shown the implications of this particular example. Three hundred trout are required to support one man for a year. The trout in turn must consume 90,000 frogs, that must consume 27 million grasshoppers, that live off of 1000 tons of grass. Because of the 80–90 percent loss of energy at each level of the pyramid, a small number of large organisms at the top can normally only be supported by a large number of smaller organisms at the lower levels.

Food Webs

Omnivorous animals, such as man, may eat at all three consumer levels. We are primary consumers when we eat lettuce, secondary consumers when we eat lamb, and tertiary consumers when we eat herring. While it is convenient to talk of linear food chains, as shown in Figure 11-2, such isolated food chains do not actually exist. In nature, many different food chains cross-link and intertwine to form a complex system called a *food web*. A greatly simplified version of a food web is shown in Figure 11-4.

[5] Actually the figure varies somewhat with different species and portions of the chain. For example, up to 15% of the energy might be transferred from herbivore to carnivore and slightly higher figures might occur in transfer of energy from one carnivore to another. Figures 11-1 and 11-2 are designed only to illustrate the basic principle of energy loss in a food chain.

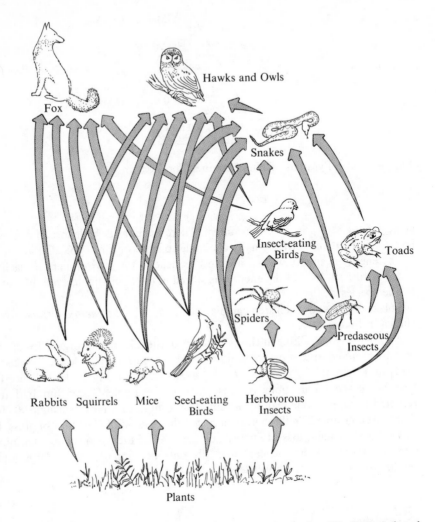

Figure 11-4   Diagram of a Hypothetical Food Web (greatly simplified). Reprinted
from *Elements of Biological Science* by William T. Keeton by permission of W. W. Norton & Company, Inc. Copyright © 1969, 1967
by W. W. Norton & Company, Inc.

Food Chains and the Second Law

The Second Law of Thermodynamics explains the phenomenon that the useful energy
available decreases drastically as one goes up a food chain toward man and other top
carnivores.

   This loss of energy as it proceeds through the food chain has far-reaching implications for the overpopulation problem. The shorter the food chain between producer

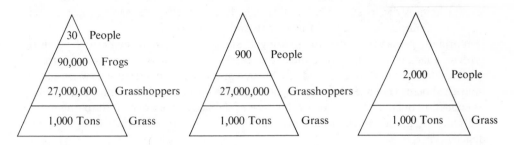

Figure 11-5   By Eating Further Down the Food Chain a Larger Population Can
Be Supported.

and ultimate consumer, the less the energy loss and the greater the relative energy value of the food. For example, going back to Figure 11-3, we can see that more people can be supported by shortening the chain, so that 30 men could be supported for a year if they ate frogs instead of trout and 900 people could be supported for a year by consuming about 27 million grasshoppers. Finally, if we lived as low as possible on the food chain by eating grass, a population of 2000 people could be supported for a year, as summarized in Figure 11-5.

A starving population will be better off, at least in terms of total intake of calories, by eating further down the food chain. It then becomes obvious why most people in the world today must live on a diet of grain rather than meat. Meat will not support a very large population. For example, it takes about 18,000 pounds of alfalfa to support one 2000-pound steer, which might support a 150-pound person for a short time. The second law makes it clear why a pound of steak costs so much more than a pound of corn or wheat. Each year in the U.S. it takes about 600 million tons of feed to produce only 84 million tons of meat, eggs, and milk. As the human population gets larger and larger, most people will have to move to a grain diet based on wheat or rice. The North American diet is about 25 percent livestock, the European diet 17 percent, and the Asian diet about 3 percent livestock. As the population gets even larger, we may have to move even further down the food chain and perhaps eat algae—algae soup, algaeburgers, algae pie, or some form of algae at every meal. We who eat very inefficiently and expensively at the top levels of the food chain forget that most people in the world have very little, if any, meat. We should also realize that our type of diet represents a massive waste and drains off valuable nutrients that could be used to cut down starvation; not to mention the massive pollution that modern agriculture directly and indirectly imposes on the environment.

Protein Hunger and Exploitation

Unfortunately, eating only plants (such as algae) in order to get enough calories in a starving world will not work. Most starvation is not caused by a lack of calories but is the result of malnutrition and diseases that result directly or indirectly from malnutrition. Unfortunately, plants that contain the most calories generally contain the least protein. A diet based on a plant protein from a single species does not contain all of the amino acids essential for growth and good health.

World hunger is primarily protein hunger. Since the growth of the brain depends on protein, a lack of protein before the age of three to five usually means some degree of mental retardation. A recent study in Mexico showed that children who were undernourished before the age of five averaged 13 points lower in I.Q. No amount of feeding or education in later life can repair this type of damage. From one-third to one-half of the children in the world today are vulnerable to brain damage through protein deficiency.

We who eat at higher levels on the food chain are systematically draining off the protein from our fellow passengers who must attempt to survive on a grain diet. We in the U.S. drink half the world's milk and eat three-fourths of its meat. We take fish from protein-starved South America and feed it to our chickens and cats. Each year dogs in the U.S. consume an amount of protein that would feed 4 million human beings; and this figure is going up—some 10,000 dogs are now born each day in the U.S.

Professor Georg A. Borgstrom,[6] a distinguished food scientist, has graphically demonstrated the massive protein exploitation of the underdeveloped countries by the developed countries—an exploitation undertaken under the guise of "humane"

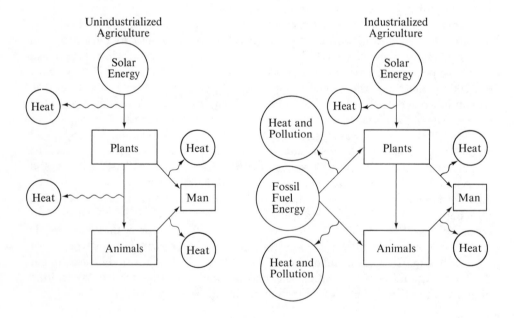

Figure 11-6   Modern Agriculture Is Based on Extensive Use of Fossil Fuels.

[6] See G. A. Borgstrom, "The Dual Challenge of Health and Hunger—A Global Crisis," Population Reference Bureau, Selection No. 31, (January 1970).

foreign aid. In 1968, the developed countries shipped approximately 3 million tons of *low grade* grain protein to the underdeveloped countries to help them fight protein starvation. During this same year the developed nations took out of the under-developed nations 4 million tons of *high grade* grain protein. Our "humanity" resulted in a net loss of 1 million tons and an exchange of low grade for high grade protein.

On the other hand, it is argued that the most effective aid we can give to developing countries is by purchasing raw materials from them. This enables them to purchase some of the capital equipment they need for their own economic development. As usual, the truth may lie somewhere between gross exploitation and true aid. More likely the problem is that we are not really paying a just price for the value of these resources. The per capita GNP in the U.S. has been rising steadily and is now more than $4500, but our per capita cost of raw materials is less than $100—a price no higher than that at the beginning of the century.

The increased yield in modern agriculture is not based on making more efficient use of solar energy. Instead it is based on supplementing solar energy with massive inputs of fossil fuel energy to grow more food, as shown in Figure 11-6.

If fossil fuels and nuclear energy were removed, industrialized countries like the U.S. would very quickly have an almost total collapse of our specialized, monoculture agriculture. We would have to ask farmers in underdeveloped nations to show us how to survive using unindustrialized agriculture. Of course, our population or at least our level of food consumption would drop sharply. In the U.S. we have only about one year's supply of food on hand.

Indifference is the essence of inhumanity.

George Bernard Shaw

11-3 Technological Optimists Keep Forgetting about Thermodynamics

Our circuits are already overloaded with a multitude of "world-shaking" problems, and we want desperately to believe that science or superman or some unknown factor X will magically and simply solve these complex problems. We are frequently lulled into a false sense of security by glowing reports from technological optimists who fail to understand ecology, thermodynamics, and the mathematical implications of a J-curve.

We read in the newspaper about some breakthrough, not realizing the thermodynamic and economic differences between discovering something in the laboratory and carrying it out on a worldwide scale. The first law—the conservation of energy—tells us that we can't get something for nothing. A man-made process on a worldwide scale always requires a massive input of energy, money, and time. The second law tells us that whatever we do to order our system—such as growing more food; mining materials; making cars, tractors, planes, and plastics; building houses, factories and cities; producing more fertilizer and pesticides—will always result in a net increase in disorder, or entropy, in our environment. We fail to take this "thermodynamic factor" into account. In addition, man's modern agricultural and technological processes almost always simplify the ecosystem and thus threaten its stability. Finally, because of the J-curve, our impact is no longer local, but global.

The Land Myth

Some suggest that we can grow the food we need by cultivating more land. In advanced countries, such as the U.S., the reverse is happening as prime farm land is being taken over for urbanization. In the U.S. we cover two acres per minute with factories, houses, and stores—over a million acres each year. Highways are now equivalent to paving the entire state of Indiana. We now have one mile of road for every square mile of countryside. California produces about 25% of the nation's table food, 43% of our fresh vegetables, and 42% of our nut and fruit crops. Every day California loses 300 acres of agricultural land, and by conservative estimates, at least one-half of California's prime cropland will go to housing and industry by the year 2000; some experts estimate 80%. In some places new land can be brought under cultivation, but because it is poorer and has less topsoil, it requires more mechanization, irrigation, fertilization and pesticides—all taking their toll on the environment. Since 1900 the average depth of topsoil in the U.S. has decreased from nine inches to six inches. In this Age of Garbage the projected amount of solid refuse in the U.S. between 1965 and 2000 would cover the state of Delaware ten feet deep.

The Water Myth

Water is the real key to life and to growing food. One half of the water used in the U.S. is for agriculture. Our spaceship has a large fixed load of water, but most of it is not *usable* water. Over 97% of our water is in the salty oceans. Robert and Leona Rienow[7] have provided a striking example of just how little water on this planet is really usable. Imagine a 12-gallon jug as representing the total water supply on the planet. The entire supply of fresh water would then be about 5 cups. However, all but two-thirds of a cup of this are tied up in glaciers. When you remove the additional fresh water sources deep below the earth's crust, along with that contained in the top soil and in the atmosphere, and the amount we have polluted, then the useful water that we have amounts to only about *six drops* in the bottom of the 12-gallon jug.

In the past 50 years the per capita use of water in this country has increased twenty-five fold, and our water demands are expected to double by 1980. We expect, or think, we shall be using 650 billion gallons of water per day by that time. The Senate Water Resources committee indicates that by 1980 we shall have allocated every last drop of our water from natural sources. You can see why a *usable* water famine is predicted in the U.S. by 1980.

But can't we desalt the ocean water? Very optimistic estimates by Westinghouse

---

[7] R. Rienow and L. T. Rienow, *Man Against His Environment*, Ballantine Books, Inc., 1970. Highly recommended reading.

indicate that desalination could supply 1/30 of U.S. water needs by 1984. And we forget the second law which tells us that building and maintaining a vast network of desalination plants would cause a significant entropy increase in our environment— not to mention the resulting mountains of salt that might blow into the sky and affect our weather or be dumped back into the ocean, thus threatening marine life near the coasts. A typical desalting plant might process one million gallons per day. To meet our projected needs by the year 2010, we would need 270,000 of these plants along our coastline. Even if the capacity of each plant is somehow increased by a factor of ten, we would still need 27,000 plants. This would take care of all of our shoreline. And to be useful all of this fresh water produced would have to be pumped uphill for hundreds of miles at fantastic expense. We can't get something for nothing.

The Ocean Myth

Some have the naive idea that the ocean will continue to absorb all the waste we pump into it and at the same time produce all the food we need. There are several reasons why this is another tranquilizing myth. First, it is too expensive in money and energy (don't forget the first law), even if we had the time (don't forget the J-curve). Second, most of the sea is a biological desert. An area about as large as the state of California produces about half of the world's fish supply. The major portions of the sea that are rich in plant and animal life are near the coasts. Ninety percent of all salt water fish are taken in relatively shallow coastal waters. The spawning ground for most of the ocean's life is the estuarine zone where fresh water and salt water mix. The estuaries are probably the richest and most productive life habitats on this planet—habitats we are busily dredging, poisoning, polluting, covering with beach cities, and with nuclear power and desalination plants. Pollution of the world's oceans is one of our most serious crises, and it is projected to increase 700 percent by the year 2000. Making very optimistic projections, probably the best we can hope to do is double or, conceivably, triple the protein from the ocean, which would still provide only about 20% of the world's protein.

Third, getting food from the ocean primarily through fish is extremely inefficient from a thermodynamic standpoint, since edible fish are at the top levels of the food chain, (Figure 11-2). It is like trying to feed the world on land by harvesting lions. As a result, the ocean can provide only about 3% of the calories for the population of 6 to 7 billion expected by the year 2000. Finally, even if massive harvesting of the sea were feasible, this large-scale tampering with the ocean ecosystem—which we are already engaged in by dumping in over 500,000 man-made compounds—could result in an ecological catastrophe that could destroy or drastically decrease life in the ocean. We must use the ocean, but we must use it wisely.

The Synthetic Food Myth

If the land and the sea will help us but won't save us, then why can't we feed the world with synthetic food such as the astronauts ate on their trips? Again, this ignores the First and Second Laws of Thermodynamics. Reference to synthetic materials is somewhat misleading. We have never made any new materials, but rather discovered ways of rearranging existing substances in new combinations. To do this on the massive scale needed (the J-curve) would be more expensive in terms of money and energy than trying to grow food on the land (the first law). Furthermore, the resulting air and water pollution and the increase in heat and entropy in the environment from building and maintaining the vast number of chemical plants needed would seriously disrupt the biosphere (the second law).

The Education Myth

Can we educate people in the world so that they will recognize and deal with the problem? Optimistically assuming that we could provide the money and teachers, we still have to reckon with the J-curve. We are losing the educational battle. Over forty percent of the people in the world today are illiterate, and the percentage is increasing— 25 million illiterates are added each year—250 million in the last 10 years. Add to this the fact that up to one-half of the children born today—who manage to make it to school age—will probably have suffered brain damage from protein deficiency. These are the people we must reason with in the future. Any progress in education will probably be made by short-circuiting the present educational system, which is agonizingly slow to change. This might involve use of transistorized and battery-operated television sets with a global satellite broadcasting system.

The Evolution Myth

Many ask, "Can't we evolve new lungs or other physiological changes so that we can adapt to a polluted environment?" In addition to blindly accepting the idea that we should go on polluting our atmosphere at an increasing rate, this view rests on a naive view of biological evolution. As we have seen earlier (Chapter 9), the time needed for major evolutionary change through a majority of the population's gene pool is measured in hundreds of millions of years; not in the small number of years imposed by the J-curve. This view also does not recognize that for such changes to occur, billions and billions of people would have to die until the proper mutations occurred in a small number of reproductive individuals. The idea that our present problems should be left to the course of biological evolution probably imposes the biggest death sentence of all.

The key lies in social evolution which can take place quickly; not biological evolution. We can and must change our way of thinking about our spaceship and our fellow passengers. We must learn to live in harmony with nature rather than attempting to conquer it.

The Space Myth

I have saved the most naive and absurd myth for last. It is the last stronghold of the "cowboy," or "frontiersman," who, after making a garbage dump out of this spaceship, thinks we can solve our population problem by shipping people off to other planets. With all of our money and technology, we have only managed to make about 12 percent of the surface of this planet habitable; trying to live on most of the so-called wide open spaces that people see from airplanes would result in mass starvation. Let's face it, most countries can't feed their people because they have a poor piece of real estate. But let's wave the biggest technological magic wand of all and do something we can't even do on Earth. Let's make all of those unlivable and unappealing planets out there completely habitable. Dr. Garrett Hardin, a prominent biologist, using extremely optimistic assumptions (including an economy fare ticket), estimated that if we used every dollar earned for one year (our total GNP), we would have enough money to send to nearby habitable planets just *six days'* increase of the population on this planet (the J-curve again). It would take over 60 times our present GNP each year just to stay even.

We would need to ship away about 70 million passengers a year, or 9000 persons per hour, and of course the exhaust from the millions of spaceships leaving each year would completely wreck our environment, not to mention the additional entropy and pollution from the industrial plants needed to manufacture these spaceships. Undoubtedly, the people making this trip would have to give up any freedom to have as many children as they want during the trip; thus the really "big breeders" would remain on Earth.

The value of our space program relative to the problems here on Earth is a hotly debated issue, but the expense involved can probably be justified solely on the fact that it has demonstrated to us all that: (1) we do indeed live on a spaceship and (2) we are down to our last planet, and we had better treat it right. The importance of having these fundamental ideas imbedded in the depths of our consciousness cannot be over-estimated as a key to dealing with our ecological crisis.

Growing More Food is Not Enough

Most experts agree that the only effective way of increasing our food supply is by increasing the yield per acre, as is being done with some of the new high-yield varieties of rice and wheat in what is called the "green revolution." Unfortunately, these new varieties also require much greater quantities of water, fertilizer, and pesticides. Although growing more food is extremely important in that it buys a little time, this will *not* solve the population problem. It merely treats the symptom, not the disease.

As long as the population continues to increase at its present rate, any major gain in food production is only a temporary lull before the food supply is eventually over-whelmed with a tidal wave of new people to feed. To give everyone in the world today the bare minimum diet would require instant doubling of today's world food supply, and still *everyone* would go to bed hungry and malnourished. If everyone in the world today tried to eat at the U.S. level, two-thirds of the people in the world would probably die. Agricultural optimists have for years talked about feeding everyone in the world with this miracle food or that, but the truth is they haven't even fed half of those here now; not to mention the additional billions expected in only 30 years. By the year 2000 we will need to triple the world's food supply, an increase in only 30 years greater than has occurred in the 10,000 years since agriculture began (the J-curve again).

Given that we could wave a magic wand and somehow produce enough food, our problem would still not be solved. Having food is one thing, but distributing it evenly to the people throughout the world is another. This requires a sophisticated system of worldwide fertilizer plants, food storage and processing centers, road networks, planes, trucks, railroads, docks, airports, and many other expensive systems. Needless to say, these systems do not exist in most of the world, particularly in the countries that have the largest rates of population growth.

Even more serious, if we could somehow grow and distribute food to the six to seven billion passengers projected by the year 2000, we have to take into account the thermodynamic cost imposed by the second law. Modern agriculture depends heavily on mechanization, irrigation, fertilizers, chemical control of weeds and pests (herbicides and pesticides), animal feedlot practices (in which 10,000–50,000 animals or up to 250,000 chickens or other poultry are raised in a confined area), and ecosystem simplification. As a result, modern agriculture may represent the most severe and comprehensive disruption of the biosphere.

For example, the use of inorganic nitrogen fertilizer increased 700 percent from 1 million tons in 1950 to 7 million tons in 1969; 50 million tons is the projected need by the year 2000. Nitrate pollution and eutrophication of our lakes from fertilizer runoff is already one of our most serious water pollution problems. Half of our water pollution and two-thirds of our solid wastes in the U.S. come from agriculture. Fertilizer plants either directly or indirectly add additional pollution to the air and water. The production of 1 million tons of fertilizer requires 1 million tons of steel and about 5 million tons of fuel, mostly fossil fuels.

In the past animal wastes were returned naturally to our soil to aid in the soil-rebuilding process. Modern feedlot or animal factory practices eliminate this natural recycling of nitrogen and produce mountains[8] of animal wastes, much of which over-load our water with nitrates. Animal wastes in the U.S. each year are equivalent to the waste from a human population of 2 *billion* people. Each year our animal wastes would fill a train reaching from here to the moon and halfway back.

Crop and animal yields can be increased significantly, but many agricultural optimists have forgotten or do not understand the environmental penalty extracted by the second law. As agricultural expert Lester R. Brown[9] said: "The central question is no longer 'Can we produce enough food?' but 'What are the environmental consequences of attempting to do so?'"

It is the top of the ninth inning. Man, always a threat at the plate, has been hitting nature hard. It is important to remember, however, that NATURE ALWAYS BATS LAST.

Paul Ehrlich

[8] Residents in a California town finally sued a farmer because the mountain of animal wastes on his farm interfered with TV reception.

[9] Lester R. Brown, "Human Food Production as a Process in the Biosphere," *Scientific American*, **223**, 161 (1970).

11-4 Overpopullution!—Thermodynamics and Our Life-Support System

We now realize that there are two types of overpopulation. The *first type*, which we have in the underdeveloped countries, is overpopulation relative to the food supply. This type already means death for 10 to 20 million human beings each year.

A *second type* of overpopulation is even more serious, because it threatens to destroy or disrupt our life-support system. This type occurs in the developed countries, particularly the United States. It is overpopulation relative to consumption of energy and renewable and nonrenewable resources and to the increase in pollution and entropy that automatically results. We might label this second type of overpopulation as *overpopullution*—too many people consuming too many finite resources and thus polluting the environment. The Western nations plus Japan and Russia account for only about one-quarter of the world's population but use about 90% of its natural resources—clear evidence that the developed nations are the greatest threat to the environment. The problem is that pollution is a multiplicative function, not an additive function.

Overpopullution

Pollution = population × per capita consumption

In other words, if population is doubled and consumption is doubled over a time period, then pollution might increase by a factor of four, or even more depending on the weighting of the two factors. It also means that in order to reduce pollution to any significant degree you *must reduce population and consumption simultaneously*. Cutting the population in half and then doubling the consumption, or halving consumption and doubling population, would lead to no improvement in environmental quality. Furthermore, since we have seen that it takes 70 to 100 years to reduce population by using the birth rate approach, dealing with the population factor cannot be delayed.

The interaction of all of these variables can be seen in Figure 11-7, showing a computer simulation by Professor Jay W. Forrester at MIT. This model shows what might happen even if we increase our capital investment to save the environment by 20 percent. It demonstrates why the overpopullution problem is not just a matter of money and technology, like our space program. *There is no purely technological solution to this problem.* Instead, the birth rate must come down drastically by deliberate human action or eventually the product of resource consumption and increasing population will cause pollution to rise sharply with a predicted dieback

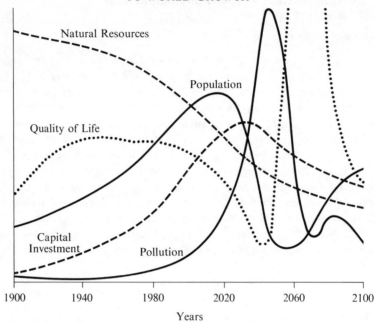

COMPUTER SIMULATION OF POLLUTION LIMIT
TO WORLD GROWTH

Figure 11-7 Computer Projection of Overpopullution Based on Work at MIT by Professor Jay W. Forrester. Model Assumes a 20-Percent Increase in Capital Investment to Combat Environmental Deterioration with the Eventual Result That About 6 to 7 *Billion* People Die over About a 40-year Period between 2010 and 2050.

of about *6 to 7 billion people* (almost twice the present population of the Earth) over approximately a 40-year period between 2010 and 2050. *In other words, the population-pollution, or overpopullution, problem is ultimately an entropy problem.*

The Most Overpopulluted Country in the World

In terms of the second type of overpopulation, based on entropy, the most over-populluted country in the world is the United States. With only 6 percent of the world's population this nation annually uses about 50 percent[10] of the world's non-renewable resources and 30 to 40 percent of all the raw materials produced. By the year 2000, we will need for our anticipated growth all of the raw materials produced by the non-communist world.

We seem destined to add at least 75 million more Americans by the year 2000—

[10] Estimates run from 40 to 75%, depending on how one defines nonrenewable resources.

equivalent in terms of people and resources to creating a new city for 250,000 people every 40 days without stop for the next 30 years. This is occurring when we can't even maintain a quality life in our present cities, much less build new ones. By 1980 the number of potential mothers in the U.S. will have doubled over the 1960 figures because of the baby boom in the '40s.

Many people point their fingers at the poor as the culprits. It is the rich and middle-class Americans, not the poor, who are threatening to kill us off. They are the mega-consumers and megapolluters, who occupy more space, consume more of each natural resource, disturb the ecology more and pollute directly and indirectly the land, air, and water with ever increasing amounts of thermal, chemical, and radioactive wastes. Professor Wayne H. Davis[11] and others have estimated that the average middle-class American has at least a 50-fold greater impact on the environment than a typical peasant in a village in India. Assuming that the average poor person in America is three to five times better off than an Indian peasant, this means that in terms of pollution the typical middle-class American family with three children is equivalent to a poor American family with somewhere between 30 and 50 children.

The interaction between people and consumption to produce pollution can also be seen in the Department of Health, Education and Welfare projections on air pollution from the automobile, as shown in Figure 11-8. The number of Americans is increasing,

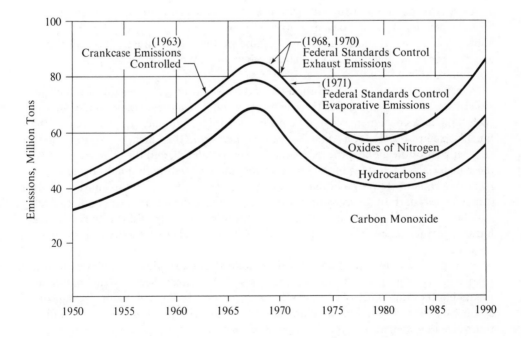

Figure 11-8   Without Further Restrictions, Auto Pollution Will Again Increase. (Source: Department of Health, Education and Welfare.)

[11] W. H. Davis, "Overpopulated America," *The New Republic*, January 10, 1970.

but the number of cars is increasing twice as rapidly. As a result, restrictions we have imposed now, if strictly enforced, will lead to a temporary decrease in air pollution until about 1980; then air pollution will rise again as the multiplication of the two factors overtakes the decrease.

One great American myth is that growth always means progress. This is a useful and necessary "frontier" rule, when one is on the lower part of the J-curve. Once around the bend, however, continued growth cannot mean progress; it can only lead to a steady degradation of life. As Stewart E. Udall put it, "At this moment in history we need to realize that bigger is not better; slower may be faster; less may well mean more."

And I brought you into a plentiful country, to eat its fruit, and its goodness, but when ye entered, ye defiled my land, and made mine heritage an abomination.

Jeremiah 2:7

11-5 The Law of Conservation of Pollution

To exist is to pollute and to exist at our level of affluence is to be a super-polluter. Strictly speaking, talk of "cleaning up our environment" and "pollution-free" cars and industries is a scientific absurdity because of the law of conservation of matter and the First and Second Laws of Thermodynamics. We can decrease pollution but not eliminate it.

In terms of the law of conservation of matter, we never really get rid of pollution. Americans are discovering that there is no such thing as a throwaway society and no such thing as a consumer. We don't consume anything. We only borrow some of the Earth's fixed resources for a while. We extract them from the Earth, transport them to another part of the globe, process them into a new material, and then discard or reuse these materials. Each stage of the process adds some form of pollution to the environment. We can only move pollution from one part of the environment to another where it may be less harmful or transform it from one chemical substance to another which may be less harmful. Remember that we are speaking not only of direct harm to man but of harm to the environment and to all of the other species upon which our lives depend.

We can collect dust and soot from the smoke stacks of industrial plants, but this solid must then go into either our water or soil. We can collect garbage and solid wastes but they must either be burned (air pollution) or be dumped in our rivers, lakes, and oceans (water pollution) or dumped on the land (soil pollution and water pollution when it washes away). We can expand sewage treatment plants to deal with water pollution, but if we use present primary treatment methods this can ironically threaten to "kill off" our rivers and lakes by adding even another load of nitrates and phosphates.

Phosphorus, which is vital for all life, is probably the first chemical that will become unavailable in useful forms, unless we can learn to extract it from the ocean bottom. Yet we throw it away in detergents, which end up in our rivers and lakes and hasten their oxygen death. This waste is particularly unnecessary, since these detergents are not really necessary to get our clothes clean.[12] The overall result is becoming clear. Lake Erie, now a giant cesspool, serves as our early warning system. All lakes eventually die, but it is estimated that man has advanced the death of Lake Erie by probably 150,000 years.

We can reduce air pollution from automobiles by going to electric engines, but this increases air and water pollution by the drastic increase in power plants needed to generate the electricity to recharge the car batteries every night. We can shift to nuclear fission power plants not dependent on fossil fuels, but this increases the thermal pollution of water and the possibility of releasing radioactivity into the environment.

We are slowly coming to the realization that *we can never do one thing*: that *there is no "away" on a spaceship*, and that ultimately *there is no technological solution to pollution*, although technology can help if we use it wisely. Our problem is to choose wisely between the lesser of several "thermodynamic evils." The only way to have a completely pollution-free world is to eliminate every living thing. But somewhere between the end of the spectrum and our present situation lies a reasonable balance that can preserve ecological stability in the biosphere.

11-6 Are We Running Out of Resources?

While it is obvious that we will eventually run out of natural resources such as fossil fuels and mineral resources such as iron, lead, uranium, and others, we should realize that our spaceship has the same amount of each element it had millions of years ago, and this amount will not change in the future. Our problem is not one of using resources up but of converting them to less useful or available forms. For example, once a fossil fuel is burned to carbon dioxide or monoxide and water, the solar energy that was stored over millions of years in the form of chemical energy is no longer available. The first law tells us that to take CO_2 and H_2O and to try to convert them back into fossil fuels will take at least the same amount of energy we got out of them. Actually it will take more energy because our energy-producing processes are very inefficient. Furthermore, the second law tells us that all attempts to do this have an unfavorable impact on the environment. In other words these resources, while potentially reusable, are in practice essentially nonrenewable.

What will happen? As we deplete these natural resources, several things may occur. First, the cost of obtaining these resources will rise and as a result some resources may be recycled because the cost of recycling will then become economically feasible. Second, if technologically feasible we will shift to other sources of energy and materials. For example, one can envision a gradual shift over the next 150 years from fossil fuels

---

[12] Withdrawal from "detergent addiction" and clean clothes can be achieved as follows: First strip the detergents from your clothes by washing in hot water to which has been added about 4 tablespoons of washing soda. From then on wash a normal load of clothes with one cup of pure soap (flakes or powder) to which has been added 2 to 4 tablespoons of washing soda, depending on the hardness of your water. Nitrilotriacetic acid, or NTA, which is being used in place of phosphates by some detergent manufacturers, may have equally undesirable ecological consequences. It has been banned by the government.

to nuclear fission to nuclear fusion to solar, geothermal, and tidal energy—provided, of course, that the technology can be developed and the environmental costs are not too great. The other alternative is a rapid drop in the amount of energy available.

In terms of materials, we might expect a shift from metals to composite materials such as structural plastics, reinforced with glass fiber and other resin materials. This shift, unfortunately, is in direct conflict with our present attempt to use up the Earth's fossil fuels. They are needed to make plastics and in the long run this is probably their more important use; a factor which we are presently ignoring.

How long will our usable supplies of certain resources last? Nature took about 500 million years to store immense amounts of energy in the form of oil and natural gas, 1 billion years to produce coal deposits, 2 billion years to produce iron and 4 billion years to produce lead deposits. We are depleting the fossil fuels in a span of only several hundred years. In the last 50 years man has used more minerals and fuels than he did in all of his previous history. Whereas early man consumed about 2000 calories per day, the average U.S. citizen now uses directly or indirectly about 200,000 calories daily—a thousand-fold increase. The developed nations use about 77% of the world's coal, 81% of its oil, 95% of its natural gas, and about 80% of its nuclear energy.

Total energy use in the world is expected to increase by a factor of five between 1970 and 2000 and by 800% in the U.S. According to the U.S. Bureau of Mines the use of coal, oil,[13] and natural gas will increase by a factor of two to four in the next 30 years, and the use of uranium for nuclear power will increase by a factor of 15. We are caught in an energy dilemma by our insatiable demand for energy, with power needs doubling every ten years. At this impossible rate, in about 200 years all available land in the United States would be taken up by power plants.

The projected usable[14] supply of some of our important fuels and minerals is given in Table 11-2.

The long-term energy supply, not considering environmental effects, may be favorable. Nuclear fission is being phased in and is expected to supply 40 to 50% of our energy by the year 2000. This is severely limited, however, by the supply of uranium, which might last for only about 30 years. Scientists hope to alleviate this problem by the development in the next 10–20 years of breeder reactors that regenerate additional nuclear fuel. The real problem with nuclear fission is its environmental impact in the form of thermal water pollution, and the potential danger from the massive amount of radioisotopes that will be produced and have to be stored safely for hundreds of years. The problem of thermal pollution should not be underestimated, and unfortunately the Atomic Energy Commission has not carried out the necessary research and evaluation in this area. By the year 2000, some projected nuclear plants will require cooling water in such large quantities each day that there are only five rivers in the U.S. with adequate flow to meet these needs. It is estimated that by the year 2000 power plants will need half of all the water runoff in the country for cooling and all of the runoff in summer months.

Most aquatic animal life cannot exist in water above 30–35°C. Raising the tempera-

---

[13] Anyone who thinks the great powers might pull out of Southeast Asia should be aware that one of the largest untapped sources of oil in the world is there—in South Korea, Taiwan, the Philippines, Malaysia, Burma, Cambodia, North and South Vietnam. Five billion dollars are to be spent on oil exploration in Southeast Asia in the next 12 years (*Time*, April 12, 1970).

[14] Some can be recovered but only at high economic and environmental costs.

Table 11-2 Estimated Recoverable Supplies of Certain Resources.

Resource	Years of Exploitable Reserves Left
Aluminum	570
Oil	300?
Coal	250
Iron	250
Natural gas	65
Phosphorus	60
Tin	35
Uranium	30
Copper	29
Zinc	23
Lead	19

* These estimates were obtained primarily from John McHale, *The Ecological Context*, George Braziller, Inc. 1970 (a magnificent book), and G. A. Mills, H. R. Johnson and H. Perry, "Fuels Management in an Environmental Age," Environmental Science and Technology, **5**, 30 (1971).

ture of water also reduces the solubility of gases, just like heating a soft drink. The amount of oxygen dissolved in water decreases about 17% between 20 and 30°C. At the same time the higher temperature increases the metabolic rates of organisms which in turn increases the need for oxygen and makes the organism more susceptible to some diseases and poisons. In addition, algal blooms from excess nitrate and phosphate are increased with increasing temperature. In other words, excessive thermal pollution severely disrupts an aquatic ecosystem.

One hope for the future lies in the development of controlled nuclear *fusion* as an energy source, as opposed to the more dangerous and environmentally harmful nuclear *fission*. Nuclear fusion would produce little or no radioactive wastes and operate at higher efficiency so that thermal pollution of air and water would be reduced, but of course not eliminated. Unfortunately, we are putting most of our money and research effort into developing nuclear fission, rather than fusion.

Another approach, which is also receiving relatively little support, is in improving the efficiency of existing processes so that less of the energy is wasted. Present overall efficiency of energy conversion is about 6–10 percent. The internal combustion engine is about 12-percent efficient and the automobile, with additional energy losses due to friction and other sources, is only about 5-percent efficient, so that 95 percent of its energy is lost as heat to the environment. A steam turbine engine, on the other hand, has an efficiency of about 40 percent and drastically reduces air pollution. Fuel cells potentially have an efficiency up to 80 percent.

The "cleaner" sources of energy which we have not been able to tap effectively are geothermal (tapping into the heat inside the earth), tidal, and direct use of solar energy. Solar energy is the cleanest because the primary entropy increase occurs in the

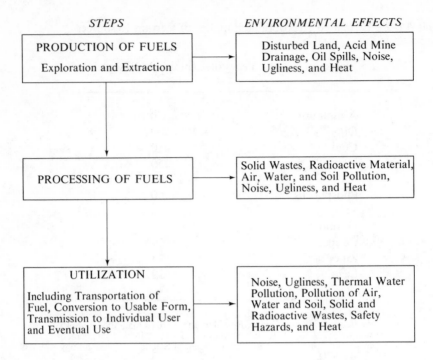

Figure 11-9   Pollution and the Use of Power.

sun—a gigantic nuclear fusion reactor some 93 million miles away. All of these possibilities are a long way off, if they are ever developed, and cannot be expected to help us significantly in the very critical short-range transition period.

In summary *the limitation of energy consumption in the next 30 to 100 years does not seem to lie in any critical shortage of resources but in the impact on the environment of using these resources*, as shown in Figure 11-9 and earlier in Figure 11-7. In other words the energy and resource problem is primarily an entropy problem.

Since all of our energy conversion processes result in adding heat to the environment, the ultimate pollutant is heat. When man's activities generate more heat than the atmosphere can radiate back into space, the atmosphere will begin to heat up. If our present energy use continues to increase about 7% per year, then the critical 3°C rise in temperature of our atmosphere could occur somewhere in the next 100 to 165 years.[15]

We are thus led to consider two fundamental questions. What level of population should we have on our spaceship and what patterns of energy and resource consumption should we develop so that all—not just some of the optimum number of passengers—can lead quality lives with freedom and dignity?

Professor J. A. Campbell[16] and others have made estimates of the world population

---

[15] For detailed discussions of heat limits see A. B. Cambel, "Impact of Energy Demands," *Physics Today*, December 1970, p. 41, and S. F. Singer, "Human Energy Production as a Process in the Biosphere," *Scientific American*, **223**, 174 (1970).

[16] See J. A. Campbell, *Chemical Systems*, W. H. Freeman & Co., 1970.

that could be supported at present U.S. levels of consumption by existing resources, as summarized in Table 11-3.

Table 11-3 Maximum World Population Potentially Capable of Being Supported at Present U.S. Levels of Consumption of Existing Resources.

Present World Population = 3.5 billion
Projected World Population by 2030 = 10 billion

Resource	Population (in billions)
Maintaining or increasing the overall quality of life	.5–2
Preserving the quality of our life-support system	1–5
*Heat buildup	10
Food	30
Oxygen	100
*Space	100
*Energy	100
*Water	100

* See J. A. Campbell, *Chemical Systems*, W. H. Freeman & Co., 1970.

These estimates assume that we adopt entirely new or much improved methods for using resources and that the rate of use does not increase over our *present* U.S. levels and that these methods do not wreck or disrupt our environment. They are probably somewhat deceiving because they represent maximum levels. For example, the projection for 30 billion people in terms of food is based on the estimated total photosynthetic capability of plant life on the planet. It assumes that (1) everyone in the world is on a vegetarian diet—no meat would be available for anyone because it is wasteful of food energy; (2) there is no other animal life or insect life to compete with man for the plants; and (3) that we would consume the entire plant, leaf, stalk, and so forth.

In terms of maintaining the quality of life and the quality of our life-support system, we have probably already far exceeded the maximum population figure.

11-7 The First and Second Thermodynamic Revolutions

The developed countries have engineered what might be termed the *first thermodynamic revolution*. It consists of a dramatic increase in material goods, broad political participation, and education for a high percentage of their citizens. It has been based, however, on improving the system at the expense of the surroundings, and the thermodynamic debt required by the second law is now coming due.

Two-thirds of the world's population have yet to participate in this first thermodynamic revolution. People talk glibly about underdeveloped countries becoming developed by following the American approach to industrialization. What would happen if the present level of American industrialization were extended throughout the world? Within a short time the planet would be uninhabitable. To provide automobiles at the U.S. level of one to every three persons would require 2300 million tons of steel when the world production is only 500 million tons. Our atmosphere would contain about 200 times more sulfur dioxide and 750 times more carbon monoxide and carbon dioxide. Our lakes, rivers, and oceans would be loaded with 175 times more chemical wastes, and thermal pollution would completely disrupt our aquatic ecosystems. Two-thirds of the world's forests would be eliminated, and each year 30 million acres of farm land would be converted to cities and highways.

The only hope of the underdeveloped countries and our only hope for preserving and increasing the quality of life lies in our ability to bring about a *second thermodynamic revolution*. It would be an ecological revolution that involves taking the Second Law of Thermodynamics seriously. It would mean that our agriculturists, engineers, doctors, and indeed all of us take into account the impact of any of our actions on the surroundings—not just for the present but also for future generations who will inherit this beautiful planet.

At the same time that the first thermodynamic revolution threatens the survival of our life-support system, the second thermodynamic revolution has provided us with the concept of cybernetics, a revolution in the use and handling of information. If used wisely, it can allow us to analyze complex systems and anticipate problems and solutions rather than responding only to crises.

The key for the underdeveloped countries is to skip over the first thermodynamic revolution and move directly into the second thermodynamic or cybernetic revolution. Trying to follow our pattern of steel production, heavy industry, and massive centralization would be a death trap for everyone. By concentrating their resources on using light metals, polymers, nuclear fusion, information handling, and by avoiding investment in military establishments, they could catapult over the industrialized nations in a relatively short time and perhaps show us the way to diversity and harmony on a finite planet.

A check list from Genesis 1: 28.
   √ Be fruitful, and multiply
     and replenish the earth,
   √ and subdue it!
   √ and have dominion over the fish of the sea,
   √ and over the fowl of the air,
   √ and over every living thing that moveth upon the earth.

Adapted from an ad appearing in Time *magazine by the Barbetta Miller Advertising, Inc., Bill Gentry, copywriter, and Ed Szep, art director.*

Chapter 11

11-8 Review Questions

11-1. What does it mean to say we have "gone around the bend on a J-curve"?

11-2. Distinguish between "cowboy, or frontier," rules and "spaceship" rules and discuss their implications for those of us on Earth.

11-3. Criticize this statement: We don't really need to worry about the population problem, because on the average the worldwide birth rate has been going down since about 1900.

11-4. Discuss the "time factor" and its implications on the lower part of a J-curve and on the upper part of a J-curve (around the bend) of population increase.

11-5. Distinguish between a J-curve and an S-curve for population growth. How will we get to an S-curve?

11-6. What is a food chain, or energy pyramid? Draw a typical one. How does it occur and what is its significance to man and the overpopulation problem?

11-7. Explain the reasoning behind the following statement: "Man may soon have to eat only algaeburgers." Explain why this would not really work.

11-8. What is a food web? Distinguish between a food web and a food chain.

11-9. Discuss the relationship between a food chain and the Second Law of Thermodynamics.

11-10. Why is protein malnutrition really more serious than a shortage of calories?

11-11. It was stated that the developed nations, such as the U.S., are involved in protein exploitation of the underdeveloped nations. Explain and give an example.

11-12. Explain in terms of the First and Second Laws of Thermodynamics and rounding the bend on a J-curve, why, in terms of the overpopulation problem,
   a. the land will not save us,
   b. the oceans will not save us,
   c. synthetic food will not save us,
   d. education will not save us,
   e. going into space will not save us,
   f. biological evolution will not save us.

11-13. Does this mean that we should not increase our use of these areas? Explain.

11-14. In terms of the Second Law of Thermodynamics explain why the poorer nations of the world must necessarily be primarily vegetarian.

11-15. Give several reasons why increasing food production (although we must do this) is not the solution to the overpopulation problem.

11-16. What are the two basic types of overpopulation? Where is each type primarily found and what are the primary consequences of each type?

11-17. Explain these statements: To exist is to pollute. To exist at the U.S. level is to be a super-polluter.

11-18. What is "overpopullution"?

11-19. What is the "Law of Conservation of Pollution"? Give several examples.

11-20. Criticize these statements: We will have a nonpolluting engine in 10 years. We will clean up our environment. Our product is pollution free.

11-21. Explain in terms of the laws of thermodynamics how the United States can be considered as the most overpolluted country in the world, even though we have a much smaller population than many other countries.

11-22. Give the estimated maximum populations that might be supported in terms of air (oxygen), heat buildup, food, water, space, and energy. What are the assumptions involved in these estimates? Why might the estimates be misleading?

11-23. What two factors probably give the smallest desirable population for earth? Compare these figures with our position now; our estimated position in terms of population by the year 2000.

11-24. Explain the statement, The population problem is ultimately an entropy problem.

11-25. Why will the U.S. population probably increase significantly for the next 50 to 75 years even though our birth rate might go down?

11-26. What is meant by zero population growth? Does that mean that a couple should have no children? Explain.

11-27. Distinguish between the first and second thermodynamic revolutions.

Chapter 12

Concluding Unscientific Postscript

Søren Kierkegaard, the famous existentialist philosopher, wrote a work entitled *Concluding Unscientific Postscript*. This title expresses the approach in this last chapter. The ideas presented here are not based on the science of thermodynamics. Instead, they are interesting and hopefully useful implications that can be drawn from the laws of thermodynamics through reasoning by analogy. They are not meant in any way to be a presentation of established scientific facts. This is an attempt to explore possible applications of the concept of entropy to problems at the interfaces between science and society.

12-1 The Case For Hope

The psychologist Rollo May says that people are losing their ability to care about anyone or anything because they feel overwhelmed. They have a feeling of powerlessness—their lives seem to be managed by impersonal and uncontrollable forces. They feel alienated, polarized, fragmented, and isolated from nature and from other human beings, as a result of a society emphasizing specialization—an atomization and encapsulation of the human spirit. More are finding the goals they sought to be without meaning and feel they don't know where they are going or where they should go.[1] Couple this with the fact that a large portion of our finest minds, energies, and resources are being absorbed by killing and devastation. Add to this our immense power to destroy ourselves by nuclear or ecological holocaust, and it is not surprising

---

[1] One is reminded of the airline passengers who heard the following announcement over the loudspeaker: "This is your captain. I have both good and bad news. The good news is that we are making rapid progress at 600 m.p.h. The bad news is that I don't know where we are or where we are headed. This is a recording."

that many wonder if there really is any hope. Have we booked passage on the *Titanic*? Should we all become swinging hedonists and have one big party while the ship inevitably goes down?

The most important message of this chapter and this book is that the answer to these questions is a resounding NO—probably the most significant NO in the history of mankind. Teilhard de Chardin once said, "It is too easy to find excuses for inaction by pleading the decadence of civilization or even the imminent end of the world." There are probably three human forces that can kill us: *blind technological optimism*, following those who say don't worry, science or some fairy godmother will save us; *gloom and doom pessimism*, giving up because of those who say there is no hope; and *our own greed, apathy, and refusal to face reality*, giving up concern and involvement through an easy but fatal fatalism or acting on the basis of a naive view of reality.[2]

What is the case for hope? It does not, of course, rest on provable facts or rational analysis. Paradoxically, hope rests on the very idea of hope. John W. Gardner, former Secretary of Health, Education and Welfare and now chairman of Common Cause, the people's lobby, has eloquently stated why hope is one of the greatest driving forces in life.

No doubt the world is, among other things, a vale of tears. It is full of absurdities that cannot be explained, evils that cannot be countenanced, injustices that cannot be excused. Our conscious processes—the part of us that is saturated with words and ideas—may arrive at exceedingly gloomy appraisals, but an older, more deeply rooted, biologically and spiritually stubborn part of us continues to say yes to hoping, yes to striving, yes to life. All effective action is fueled by hope.[3]

In a fast changing, exceedingly complex world we have apparently lost our confidence. Paradoxically, it was our overconfidence that the world was essentially infinite and that we have dominion over it, that led us to the present crisis. Hope lies in our ability to rekindle a more realistic confidence—a new confidence based on a revolution in human consciousness that truly accepts our limitations. It means accepting and acting on the basis that we live in a finite world. The purpose of the previous chapter was not to overwhelm you with gloom and doom, but to overwhelm you with the fact that we do indeed live on a spaceship—there is no escape from this fact. Because of this and the Second Law of Thermodynamics, there can be no technological advance without the social or environmental cost of resource consumption and waste accumulation, however delayed the presentation of the final bill may be.

My purpose has been *not* to show you what the future will be, but to show you where we appear to be heading, if present trends continue. George Orwell's famous book *1984* was designed to aid us in preventing a possible future. The history of mankind demonstrates the capacity of man to change radically his world view and to act quickly and often unexpectedly on the basis of a new consciousness. Biological evolution takes

---

[2] McLandburgh Wilson said, "The optimist sees the doughnut, but the pessimist sees the hole." Perhaps we should add that the naive realist sees the doughnut and the hole but the ecological realist tries to see the doughnut and the hole and their relationship to the rest of the world.

[3] John W. Gardner, *The Recovery of Confidence*, W. W. Norton & Company, Inc., New York, 1970.

Chapter 12

millions of years, but cultural change can occur rapidly, even more rapidly today because of speed and potential power of mass communication. Possible futures based on extrapolation of present trends probably actually occur only when man gives up and assumes they are inevitable.

Adopting a Spaceship Consciousness

Accepting and adopting a spaceship consciousness means not only that we can no longer have or do everything we want, but that we cannot afford the dangerous *notion* that we think we can. This dangerous dream must be replaced with an ecological dream —of man living in harmony with nature and his fellow man in a closed system but not a static system. A balanced ecosystem is based on exciting diversity and complexity in a maze of continuous change—but change within the limits imposed by the system. Achieving this steady-state dream instead of the false drive of simplification of nature and accelerating change and growth is not only possible but necessary. It means we must make choices; we can only have more of this if there is less of something else.[4] We can only increase the quality of life by limiting the number of passengers and their energy consuming activities. J. Irwin Miller,[5] chairman of the board of Cummins Engine Company, has expressed the reality which we must face.

The real price of the future is our willingness to grow up and become an adult people— to make choices rather than to avoid them. When we ask what our national priorities should be and how we should allocate our national resources, we are posing questions we have never in our history supposed we would have to ask. We never thought Americans would have to choose. For the first time we are beginning to realize that we can't have everything. The price of the future, then, will be found in the things we give up *in order to gain the things that are compatible with the limitations of living on a spaceship.*

Charles Reich,[6] in his controversial book, *The Greening of America*, indicates that our refusal to face reality is the true source of powerlessness and he also calls for a revolution in consciousness. He categorizes Americans as existing in three levels of consciousness. Consciousness I is the traditional outlook of the farmer, small business-man, and worker, based on a simple but decent life with reverence for morality, hard work, self-denial, patriotism, and limitations on big government, big business, and the rate of change. Reich indicates that this represents a dream world that no longer exists in our industrialized state based on big business and government. Consciousness II is the modern outlook of an organizational society based on planning, liberal reform, and continued growth and change through technology. Consciousness III, which Reich advocates in place of I and II, is a new humanism that rejects or is suspicious of

---

[4] It is not quite as bad as Ralph Lewin indicated when he said, "Everything I like is either illegal or immoral, pollutes the environment, or increases the population."

[5] J. Irwin Miller, "Changing Priorities: Hard Choices, New Price Tags," *Saturday Review*, January 23, 1971, p. 36. Emphasis added.

[6] Charles A. Reich, *The Greening of America*, Random House, Inc., 1970.

logic, rationality, analysis, thought, and organization and replaces it with being true to oneself, respect for the absolute worth of every human being, a sense of spontaneity, community, honesty, peace, sharing, and caring. He envisions this revolution occurring merely by allowing it to sweep spontaneously through the minds of our youth and citizens.

Unfortunately, Reich's vision, while appealing, is not large enough. It too founders on unreality. It does not take into account our finiteness or the second law which always requires continual organization and action for survival and a quality life, or the J-curve which does not provide us with the luxury of waiting around for something to happen to us. His view is also narrow in that it does not recognize that the true strength and hope for America and the world is found by preserving the essential diversity found in all three levels of consciousness along with many other qualities his oversimplified classification omits. The resistance to change found in Consciousness I, the organizational skills in Consciousness II, and the sensitivity to human dignity in Consciousness III are all essential ingredients in a steady-state world. They and other elements must be combined and used in a synergistic pluralism—a new consciousness based on a blend of new humanism with an ecological understanding of the need for diversity, stability, and organization in a steady-state system.

Awareness is Increasing

There are three levels of awareness of our ecological crisis. The *first level* is that of discovering the symptom—*pollution*. In only a few years' time the entire country has become acutely aware of pollution, in both its direct and in many of its more subtle forms. Polls now indicate that pollution is the number one issue for most citizens. While this is most encouraging it is also dangerous. As soon as we discover a problem, we want to fix blame and we want a quick solution. We are now engaged in an unhealthy and counter-productive phase of the environmental crisis—a pollution witch hunt. In our search for villains we have targeted industry, government, technology, doctors and health officials, the poor, Christianity[7]—anyone, of course, but ourselves. We have not accepted the fact that we are the enemy and that industry and government would not exist and produce things unless we demanded them or allowed ourselves to be duped into thinking we need them. Indeed, we must point out and curtail irresponsible acts of pollution by large or small organizations, but we must at the same time change our own ways. We have all been drilling holes in a leaky boat. Arguing over who is drilling the biggest hole only diverts us from working together to keep the boat from sinking.

Another danger in remaining at the pollution awareness level is that it leads people to view the crisis simplistically as a moon shot problem. Spend 30 billion dollars and go to the moon; spend 300 billion dollars (the bill is much higher) and clean up the

---

[7] For an interesting debate on whether the Judaeo-Christian ethic is the villain see Lynn White, Jr., "The Historical Roots of Our Ecological Crisis," *Science*, *155*, 1203 (1967); F. Elder, *Crisis in Eden—A Religious Study of Man and Environment*, Abingdon Press, 1970; L. W. Moncrief, "The Cultural Basis for Our Environmental Crisis," *Science*, *170*, 508 (1970); and R. T. Wright, "Responsibility for the Ecological Crisis," *Bioscience*, *20*, 851 (1970).

environment. Have technology fix us up, send me the bill at the end of the month, but don't ask me to change my way of living. I hope that you are convinced that using technology and spending enormous amounts of money will be absolutely necessary, but that this will not solve the problem. Hopefully, by dealing with the symptoms, this will buy us enough time to deal with the disease. Dealing only with pollution as a solution to our environmental ills is like trying to cure cancer with band-aids and aspirin.

As this becomes more and more apparent over the next 10 to 20 years, we may enter another very dangerous phase of the environmental crisis. Because there is no technological cure, people will want to blame technology and perhaps even destroy the machines in a neo-Luddite[8] revolution. If this should occur, it would seriously aggravate the crisis and probably lead to mass starvation in the U.S. We saw earlier (in Figure 11-6) that modern agriculture is almost totally dependent on technology, particularly fossil fuels. Technology will be absolutely essential in dealing with the ecological crisis. The problem is not technology, but our unwise use of technology.

Many have already moved to the *second level* of awareness—*the overpopullution level*. If you ask what causes pollution, the answer seems obvious—people. But it is not just people. It is also their level of consumption; people × per capita consumption = pollution. There are some[9] who are engaged in a misleading debate as to whether population or consumption is the culprit. This is like arguing over which is more important, your left foot or your right foot or what stops a speeding car, you or your brakes? Both of these variables are interconnected in a multiplicative function. Consumption is now going up faster than population, but all consumption is tied to people in a thousand different interconnecting links. In an affluent society the addition of a person amplifies consumption considerably, and these two variables together amplify pollution to an even greater extent, as shown earlier (in Figure 11-7). To make matters more difficult, we have seen that population control through decreasing the birth rate normally involves a time lag of 50 to 100 years. At the overpopullution level the answers seem obvious. We must simultaneously reduce the number of passengers and their levels and wasteful patterns of consumption.

But this will not happen in a rational and planned manner, unless a reasonable number of our leaders and citizens move to the *third and final level* of ecological awareness—*the spaceship Earth level*. At this level we recognize that population and consumption will not be reduced unless there is a major change in our political, economic, social, and ethical systems. We are faced with four unpleasant political and personal choices, each one more unpleasant than the succeeding one. (1) *Voluntary changeover* to spaceship rules by sacrificing some things and freedoms now in order to preserve some freedom, and probably survival; (2) *semi-voluntary changeover* to spaceship rules based on mutually agreed-upon incentives, taxes, sanctions, and laws that protect our environment and ourselves from infringement by others. This is, of course, the basis of our common law system—mutually agreed-upon limits that are enforced. At present our rights to clean air, drinkable water, space, beauty, and diversity are not protected or insured by mutually agreed-upon laws. They have been assumed as the common property of everyone, and as Garrett Hardin has so aptly

---

[8] The Luddites were people who destroyed their machines.
[9] See A. J. Coale, "Man and His Environment," *Science*, *170*, 132 (1970), and B. Commoner, M. Corr, and P. J. Stamler, "The Causes of Pollution," *Environment*, *13*, 2 (1971).

demonstrated in his article "The Tragedy of the Commons,"[10] whenever everyone is responsible for something, no one is responsible. Without mutually agreed-upon limitation and enforcement to prevent people from abusing the commons, it is destroyed; (3) *repression* and complete loss of freedom if we don't change or change fast enough and are forced to move into a martial law and rationing system as things get out of control; and (4) *the death rate solution*—letting nature take its course with a massive dieback, probably involving billions. The longer we wait, the more likely these last two choices become.

Fortunately, as a result of our space program almost everyone in our society has the concept of a spaceship etched into his consciousness. Some are now beginning to see what drastic changes this spaceship outlook requires. Others are also seeing that it prepares or forces us to really accept our finiteness and interdependence with one another and with nature. If this occurs, it can represent mankind's transition into adulthood.

The People are Stirring

There are grounds for cautious optimism that such a value revolution is underway in this country. People are stirring, questioning, listening, and organizing. In spite of all the noise, confusion, and the gap between rhetoric and reality (the R and R gap), more and more Americans are beginning to question our institutions and purposes. They are asking, "What is true wealth? What have we done wrong? What should be the true aims of our affluent nation?" It is particularly significant that some of our youth are educating their elders by making them take a fresh look at these crucial questions.

Some politicians are beginning to listen and act and to stop treating the average voter as a naive child who wants to be told that there are simple solutions, that we are not really in trouble, and that more and more progress based on a blind use of technology will lead to the promised land. There is a new awareness that the key to leadership lies in telling people the truth—that we can't have everything, that we are in deep trouble, that we must make some significant and difficult changes, that for everything we want to preserve we will have to give up something, that the heaping of crisis upon crisis need not be taken as a forecast of doom but the birth pangs of a new world where we finally face up to the issues of what man is and what his place is in the world.

As Patrick Moynihan, former counselor on urban affairs to the President of the U.S., said: ". . . the essence of tyranny is the denial of complexity. What we need are great complexifiers—men who will not only seem to understand what it is they are about but who will also dare to share that understanding with those for whom they act." The American public recognizes, as H. L. Mencken put it: "for every problem there is a solution—simple, neat and wrong." When more of our leaders recognize this, they will be able to mobilize our vast reservoir of human energy, creativity, and conscience in a massive *spaceship Earth program*.

Such a program will not just be another "Manhattan project" or "moon shot"

---

[10] Garrett Hardin, *Science*, *162*, 1243 (1968).

based on the assumption that with money and technology we can accomplish anything. The problem we face now is far more complex. It will require a massive and organized effort of our best minds from all fields to determine our priorities, to determine what the rules are for living on a spaceship, to develop social innovations along with technological innovations, to determine what "tire patches" or "band-aids" must be put on now to give us the time needed to make fundamental changes in our economic, political, and technological institutions. It will, of course, take enormous amounts of money, but mostly it will take ecological awareness, conscience, and organized action on the part of individual citizens.

Slowly, citizen groups are forming to protect the environment, to challenge the priorities and assumptions of elected leaders, to insist on equal rights for all Americans— male or female, black, red, brown, or white—to protect the consumer, to lobby for the common man as a counterbalance for the powerful lobbies of organized business and other interests, to use existing laws in a creative way to protect the poor, the disadvantaged, and indeed all of us. In the past few years we have seen the development of a powerful ecological lobby, the rise of public service lawyers, inspired and led by Ralph Nader, the formation and merging of consumer protection groups, and the formation of a people's lobby, Common Cause, led by John W. Gardner. While many sit on the sideline talking about how the system can't be changed, these groups *are* changing the system. Others are talking about forming a new ecological-consumer political party.

These are not mass movements as yet, but they are growing. Political analysts indicate that it probably requires dedicated political activity by only about 10% (perhaps less)[11] of the population to bring about significant change. In the 1970 elections a number of the candidates with the worst environmental records were defeated.

I am not suggesting that everything is going well; rosy and naive optimism is not in order. We have only begun to recognize our predicament—much less decide on courses of action. There will be much confusion and disagreement, and many will hop off of the ecology bandwagon once they realize the really fundamental changes it requires. But there is hope. There is a thin pencil-beam of light far down a complex and treacherous tunnel.

The ecological, or second thermodynamic, revolution will be the most all-encompassing revolution in the history of mankind. It involves questioning and altering almost all of our ethical, political, economic, sociological, psychological, and technological rules or systems. No one could ask for a more challenging and meaningful way in which to devote his life. It will not involve dramatic breakthroughs, but only day-to-day hard work, many setbacks, extremely bitter disputes, anguish, and the joy that comes from caring for the Earth and our fellow passengers.

The capacity for hope is the greatest fact in life. Hope is the beginning of plans. It gives men a destination, a sense of direction for getting there, and the energy to get started. It enlarges sensitivities. It gives proper value to feelings as well as to facts.
Human default or degradation can reach a point where even the most stirring visions lose their regenerating or radiating power. This point, some will say, has already been reached. Not true. It will be reached only when men are no longer capable of calling out to one

[11] Many estimate only 5% or less, but it probably takes an additional 5% to counteract the 5% who will organize against almost any movement.

another, when the words in their poetry break up before their eyes, when their faces are frozen toward their young, and when they fail to make pictures in the mind out of clouds racing across the sky. So long as men can do these things, they can be capable of indignation about the things they should be indignant about; they can be audible about the things they should be talking about, and they can shape their society in a way that does justice to their hopes.

Norman Cousins
Editor, *Saturday Review*

12-2 A Spaceship Earth Program

How can we convert from our present "frontier rules" to "spaceship rules"? No one really knows, but a tentative outline of a spaceship Earth program might involve the following:

A Spaceship Earth Program[12]

1. Identify the major causes of our ecological crisis.
2. Set up a worldwide ecological monitoring and early warning system, perhaps as elaborate as our missile defense system.
3. Begin to formulate the rules for living on a spaceship.
4. Develop and begin instituting careful plans for the long-range transition (perhaps 30-50 years) to spaceship rules. Each proposal should be evaluated for scientific, political, administrative, and economic feasibility, ethical acceptibility and effectiveness along the lines suggested by Bernard Berelson.[13]
5. Identify short-range problems and institute temporary "band-aids" to buy the necessary time.
6. Move from voluntary methods to incentives to legal restrictions coupled with a massive awareness and persuasion program. Monitoring of the success of each phase should be set up to determine when or if it will be necessary to move to the next level in order to prevent catastrophe.

---

[12] This author and a colleague, Professor George A. Fouke, and 35 students have been engaged in developing a survival manual for spaceship Earth which is in unpublished form. P. R. Ehrlich and R. L. Harriman have published a plan to save spaceship Earth entitled, *How To Be a Survivor*, Ballantine Books, Inc., 1971.

[13] B. Berelson, "Beyond Family Planning," *Science*, **163**, 533 (1969).

Much work has already been done on identifying causes. Unfortunately, many flood us with quick and overly simplistic cures to our environmental ills. Each has a particular "culprit" and insists that if we will only deal with that particular "root cause," happy days will be here again.

The major root causes that have been identified appear to be:

Major Causes of the Environmental Crisis

1. *Overpopulation*
   a. Relative to food—Underdeveloped countries
   b. Relative to resource consumption and pollution—Overpopullution in developed countries.
   Both types are aggravated by failure of medical and health authorities to be concerned with simultaneously decreasing death rate and birth rate.
2. *Population distribution*—the population "implosion," or urban crisis.
3. *Overconsumption and wasteful patterns of consumption*—throwaway society, planned obsolescence, producing unnecessary and harmful items, consuming more than one's fair share of resources, very little recycling of essential resources.
4. *Unwise use of technology*—failing to take into account the impact of our activities on the environment, ignoring the Second Law of Thermodynamics, asking only if we *can* do something not if we *should* do it, blind faith in technology.
5. *Crisis in management*—the failure of our political and economic systems, the growth mania or bigger-is-always-better syndrome, slow response of government, a counterpunch society lurching from crisis to crisis with no long-range planning, misplaced priorities, and refusal to establish spaceship priorities.
6. *Oversimplification of the ecosystem*—failure to observe ecosystem stability through maintaining ecological diversity, failure to recognize that everything is connected to everything.
7. *I-centered behavior*—tragedy of the commons, lack of responsibility for present and future passengers and the condition of the ship, now centered, acting only as a passenger rather than a crew member, gloom and doom and fatalistic cop-outs, the enemy is the other guy.

Hopefully, you now recognize that there is no simple or quick cure. The causes are all connected in a complex, synergistic manner that is not understood. All of these causes (and any others we may discover) must be dealt with *simultaneously*, using both a short-range and a long-range approach. Focusing almost exclusively, as we are now doing, on the symptom of pollution hopefully buys a little time. But doing only this or focusing on only one cause, insures an even more serious and perhaps fatal outbreak later.

Much research needs to be done on how the major causes are interrelated. Are they all multiplicative? Are some additive? How should each factor be weighted in terms of time required for change, present impact, and future impact? In Figure 12-1, I have indicated a very crude model, assuming that all of the factors are multiplicative and are weighted equally. Further research will undoubtedly reveal other connections and establish weighting factors. This model is presented only to remind us that all of these factors are interconnected.

What We Must Do

We don't yet know what we can and must do, but it is possible to outline some suggestions for debate and thought. The important thing is not to rely on using just one approach or dealing with only one variable. A multi-faceted approach at many levels is needed.

1. *Reduce the Number of Passengers*

   a. It is the responsibility of the United States, being the richest nation, leading consumer of the world's resources, and biggest polluter, to set the example.
   b. A massive TV and media persuasion and education program to reduce family size.
   c. Spaceship Earth curriculum including population dynamics and control introduced into education from kindergarten through college.
   d. Primary emphasis on all education, persuasion, or incentive programs on middle and upper classes, who have 10 to 20 times more impact on the environment than the poor.
   e. Establish a new Department of Population and Environment to determine optimum population and plan and monitor effectiveness of all programs.
   f. Extensive adult education program with population dynamics as part of job training.
   g. Crash research program to develop improved birth control methods that work for years and can be reversed.
   h. Free birth control devices and free legalized abortion for all citizens.
   i. Make adoption easier.[14]
   j. Expand family planning but do not depend solely on it because its primary aim is to help couples have the number of children they want when they want

---

[14] Actually this may not be desirable if abortion is legalized, since the number of children available for adoption would probably decrease sharply.

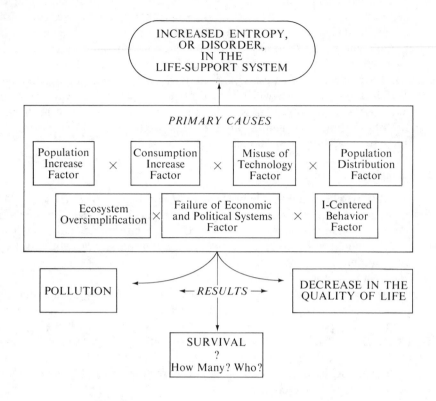

Figure 12-1 A Very Crude Model of Major Causes and Results of the Ecological Crisis on Spaceship Earth.

them. The problem is that too many people want too many children relative to the ship's capacity to provide them with food, water, and air.

k. A new medical ethic.[15] To save lives in the long run, each life prolonged or saved now ethically obligates medical and health personnel to prevent or persuade couples from having children. The ratio needed to restore balance is probably to prevent four births for each life saved.

l. Economic incentives: $500 bonus per year to each woman not having a child until age 30 and to women between 30 and 50 who have no more than two children; no income tax for poor who do not have a child during a given year; income tax deductions only for a maximum of two children; extra deduction for each year a couple does not have a child.

m. Guaranteed annual income to all poor families regardless of number of children, since most taxes and penalties penalize the children in poor families.

n. Storefront health, birth control, welfare and legal aid units in neighborhoods, supported by a vast fleet of mobile units bringing these services to the people.

[15] It can be argued that this is not a new medical ethic. It merely means that the Hippocratic oath requires the saving of lives on both a short-term and long-term basis.

o. If persuasion and incentives do not work, mutually agreed-upon nonvoluntary methods may have to be introduced. For example, child taxes and child licenses. The right to have as many children as one wants is in conflict with the right and necessity to have breathable air and drinkable water. After all, the number of husbands or wives we can have is limited by law.

p. Plan for a new age structure. U.S. society with a stabilized population (some 60-100 years from now) will have an altered age structure, with more old and fewer young people. Transition to this type of society should be planned carefully. In terms of maintaining the steady state required on a spaceship, this type of age structure should be most beneficial.

2. *Redistribute the Passengers*

a. Eco-catastrophes would probably first occur in urban areas. In the U.S., 70% of the people are living on 1-2% of the land, and urban areas are doubling in population at the same rate as India. One new city for 250,000 Americans must be created every 40 days just to keep up with present growth, much less overhaul our decaying cities.

b. Establish new cities of 100,000 to 250,000.

c. Use bonuses, income tax reductions, jobs, low-cost housing, and other incentives to encourage industries and people to move to new areas that can support more people.

d. Don't fall into the trap of thinking this is the only real population problem in the U.S. Redistributing passengers relieves localized population and pollution pressures, but the total amount of pollution for the entire ship is not altered significantly—it is merely spread out.

e. Most of our vast sparsely populated areas are uninhabitable for large numbers of people. Alaska, Oregon, Nevada, Utah, Idaho, N. Dakota, Wyoming, Montana, West Virginia, and Maine cannot support large numbers of people because of climate, topography, accessibility, and lack of resources. It may be easier and it is certainly cheaper to get citizens to have fewer children than to get them to move to other areas.

3. *Reduce the Level and Patterns of Consumption and Recycle Critical Materials*

a. Probably the most immediate problem is to turn down the power in order to prevent a catastrophic rise in pollution (Figure 11-7). When Apollo 13 was in trouble the first action was to cut back on the use of power.

b. Crash program to develop methods for evaluating short- and long-term impact and social costs of all products and technological projects on the environment.

c. Institute the necessary laws and enforcement to achieve the goal stated by President Nixon in his 1970 State of the Union address in which he said,

We can no longer afford to consider air and water common property, free to be abused by anyone without regard to the consequences. Instead we should begin now to treat them as scarce resources which we are no more free to contaminate than we are free to throw garbage in our neighbor's yard. This requires comprehensive new regulations. It also requires that to the extent possible the price of goods should be made to include the cost of producing and disposing of them without damage to the environment.

d. Redirect growth by instituting a pollution tax on all products and on all manufacturing, municipal, and agricultural activities. Pay for the true cost of each product, which includes environmental degradation when product is made, cost of making and distributing (our present price system), and cost of recycling or disposing. This would show everyone the real cost of the things we buy. Research on recycling and minimizing environmental impact would be stimulated, planned obsolescence would be sharply curtailed, growth of some industries and activities would be stimulated while other more harmful ones would be leveled off or decreased by the market mechanism. It would become profitable not to pollute—the reverse of our present system. Part of the extra costs to consumers would be returned by the decrease in bills for health, pollution, cleaning, painting and other hidden effects we are now paying for. For example, the national garbage bill in 1970 was 3 billion dollars; emphysema is our fastest-growing disease; cancer is primarily a disease of industrialized nations; pollution costs us 10-12 billion dollars each year. It is estimated that a 50% reduction in air pollution would reduce the nation's health costs by $2 billion annually.

e. Emphasis should be on reducing energy and resource consumption by the middle and upper classes.

f. A crash research program to develop more effective pollution control and methods for recycling. The ideal container is a beer bottle made out of pretzels. Recycling paper (60% was recycled in the U.S. during World War II) would reduce our massive solid waste problem by 60-80% and eliminate much devastation of forests (50% recycling would save 92 million acres of forest land each year) and decrease water and air pollution from paper-making industries. Recycling a stack of newspapers only 36 inches high saves one oxygen-producing tree.

g. Bonuses and tax deductions to industries developing new pollution control and recycling methods during the next 10-15 years.

h. Transfer SST and other "frontier" funds to developing a new automobile engine and to instituting mass transportation systems.

i. Alter our packaging methods. Packages inside of packages are undesirable. Emphasize reusable containers such as grocery and shopping tote bags as used in Europe, return to the metal or plastic lunch pail, milk and soft drinks in bottles, not cartons and cans.

j. Reducing our suicidal increase in electricity. Many Europeans live quality lives at half of our electrical output. The eventual cost in pollution, misery, and bad health of increasing our power consumption far outweigh the need for continuing our wasteful and extravagant use of power.

k. Ration energy consumption and entropy production to limit power production and pollution. For every high–entropy producing convenience (car, air conditioner and so on) you must give up another one. Plants, individuals, and particular areas or cities might be given only so many energy and entropy units.

l. Raise the cost of power and water significantly to discourage use. Both are now much too cheap relative to their true impact on the environment. In spite of advertisements, electricity or any source of energy is not clean, particularly if you trace it back to the power plant.

m. Emphasize the quality of products rather than quantity.

n.   Institute controls to promote the wise use of technology. Determine the short- and long-term potential environmental impact of all projects including the preservation of ecological diversity. Don't do some things even if we can do them.

4.   *Develop a Spaceship Governmental System and a Steady-State Economic System*

a.   Establish a Department of Population and Environment and a Department of Transition to guide and coordinate the changeover to spaceship rules.

b.   Institute careful long-range planning instead of a short-term crisis response to problems. Anticipate and prevent problems rather than counterpunching.

c.   Reorganize all branches of government at all levels to make them more responsive (but not overly responsive); a homeostatic system fluctuating within realistic limits (see Figure 8-4).

d.   A new bill of rights including environmental rights to clean air, water, and so forth.

e.   Pollution tax.

f.   Guaranteed annual income to prevent poor from bearing real brunt of pollution and other taxes.

g.   Crash program of research by top economists to design steady-state economic systems and plan an orderly transition to a new system. A stabilized population and economic system does not mean economic stagnation. It means redirected growth based on quality rather than quantity.

h.   A fundamental reordering of priorities deemphasizing the arms race and defense spending. World military expenses have escalated from 1 billion at the end of World War II to 200 billion in 1970. The U.S. now spends 8 million dollars per hour on defense and a thousand times less on environment. In cost, one prototype bomber is equivalent to 75 fully equipped, 100-bed hospitals; one atomic sub equals 10,000 high schools; one standard jet bomber equals school lunch for 1 million children.

i.   Ecology, not pollution, should become our number one election issue. Everyone is in favor of reducing pollution, but moving the issue to the over-popullution and spaceship levels requires people to take controversial stands. A new breed of ecologically aware and committed leaders must be put into office at all levels. Ecology, consumer, and other groups should join together rather than being defeated by divide and-conquer-tactics.

j.   Education and persuasion program to show that bigger is not always better. The next time someone says we must clear off this land, park, or wilderness for an airport or power plant because we are supposed to have 100,000 new passengers or users of electricity in the next 10 years, ask just one question. Suppose we don't build it. Then we wouldn't have all of those passengers and users crowding into our living space and adding more to our stress, noise and pollution, would we?

This is a crude, incomplete and undoubtedly controversial program. It has been presented to start the crucial debate and evaluation that we have avoided for too long.

What can you do personally? Become ecologically informed. Immerse yourself in spaceship, not cowboy, thinking. Live a simpler life based on reduced energy consumption and entropy production. Above all, become politically involved. Join forces with national ecologically-oriented organizations. Do the little things.[16] It is all of the little things, the individual acts of consumption, litter and so on, that got us into this mess. It is also the only way out. Avoid the extrapolation-to-infinity syndrome as an excuse for not doing anything—the idea that if we can't change the entire world, then we won't change any of it. The world is changed by changing the person next to you. Environment begins at home. Whatever you do, convince two other people to convince two other people to convince two others. . . to do the same thing. Start a reverse J-curve of awareness and action. Carrying out this doubling process only 28 times would convince everyone in the U.S., and you only need to convince about 10 percent for action. After 32 doublings everyone in the world would be convinced.

> When there is no dream, the people perish.
>
> *Proverbs*

12-3 Entropy and Evil

Frank L. Lambert,[17] a chemist, has proposed an interesting way of looking at evil in terms of thermodynamics. He defines two aspects of evil. One involves evil as an unwarranted disruption of an individual's dynamic pattern of life and thought—a randomization, or disordering, of our patterns. The second aspect of evil is the opposite. It is a crystallizing, or subhuman ordering, of life, like the Nazi tyranny of putting people in concentration camps. This type of evil represents excessive order.

From a thermodynamic standpoint, life can be viewed as an improbable arrangement maintained at a steady state—a plateau of high free energy maintained between the two opposing tendencies of order and disorder, as illustrated in Figure 12-2. The living organism maintains the steady state by dynamic homeostasis, through information feedback concerning its environment. If the environmental stress is too great, then death can occur either by excessive order ("crystallization") or by excessive disorder.

Physiologically man consciously or unconsciously attempts to maintain his biochemical system at nearly the same level in spite of environmental changes. Psychologically he tries to maintain a steady state on the mental level by holding to an acquired and accepted set of beliefs and actions[18].

From this standpoint, "evil" can be considered as anything that causes a violent disruption of the biochemical or psychological steady-state pattern of man. Mild changes can be helpful: they could lead to new steady states which would aid physiological and psychological growth or adaptability. However, intrusion leading to

---

[16] For suggestions of what you can do see P. R. Ehrlich, *The Population Bomb*, Ballantine Books, Inc., 1968; P. R. Ehrlich and R. L. Harriman, *How To Be a Survivor*, Ballantine Books, Inc., 1971; P. Swatek, *The User's Guide to the Protection of the Environment*, Ballantine Books, Inc., 1970; G. de Bell, (ed.), *The Environmental Handbook*, Ballantine Books, Inc., 1970 and G. Calliet, P. Setzen, and M. Love, *Everyman's Guide to Ecological Living*, The Macmillan Co., 1971. All are paperbacks.

[17] F. L. Lambert, "The Ontology of Evil," *Zygon, Journal of Religion and Science*, *3*, 116 (1968).

[18] C. J. Jung, "On Psychic Energy," in *Collected Works*, Vol. 8, Princeton University Press, 1960.

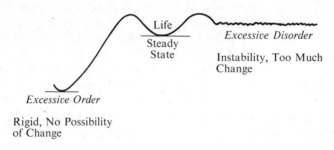

Life

Steady
State

Excessive Disorder

Instability, Too Much
Change

Excessive Order

Rigid, No Possibility
of Change

Figure 12-2   Life as a Steady State.

excessive order or disorder can be interpreted as "evil" relative to the life patterns of the individual.

Even if political or social evil can be accounted for as a disruption of the life pattern by outside events or other persons, there is still man's willing violation of his own ideas of the "good." Why does man apparently want to "sin"—to break the code even though he knows it is wrong? Lambert suggests that the tendency to break codes and to find new, unstable, and sometimes dangerous situations is part of our genetic inheritance. The tendency to find a new steady state by seeking new patterns may be accounted for by the genes we carry from ancestors whose adventurous nature may have enhanced their ability to survive and procreate.

Study Question 12-1   Comment on the validity and usefulness of Lambert's thermodynamic approach to evil. What are its limitations? Give three examples of how it could be applied to situations we normally consider as evil. How do you think Lambert would define "good"?

12-4 Entropy and Ethics—The Thermodynamic Imperative[19]

The common view of science is that it has no place in the realm of values. While science is most useful and powerful in describing events in the physical universe, its results are supposedly neutral. Science can provide us with no sense of ought. It can warn us of the dangers of applying a scientific principle or of disregarding the laws of nature, but it appears to be silent on what we ought to do. Apparently there are no scientific imperatives like the Golden Rule or Kant's categorical imperative.

---

[19] Much of the material in this section is based on R. B. Lindsay's superb book *Science in Civilization*, Harper & Row, Publishers, pp. 290–298, 1963. Reading this book will be a rewarding experience for the reader interested in the relationships between science and the humanities, philosophy, history, communication, technology, government, and human behavior.

Some scientists have opposed the idea that science is ethically neutral. One prominent physicist, Robert B. Lindsay, has proposed that an ethical imperative can be derived from thermodynamics. He has termed this ethical rule *the thermodynamic imperative*.

We have seen that according to the second law there is a natural or spontaneous tendency toward increasing disorder in the system plus surroundings. However, practically every element in man's struggle to develop and maintain civilization can be interpreted as either an instinctive or a deliberate attempt to introduce order into a chaotic environment. Man creates order by building houses and roads, by cultivating land, building factories, and creating language, art, and music. Most of man's activities within the system consume entropy rather than produce it. We have seen previously that this is not in violation of the second law, because this entropy consumption is at the expense of an even greater production of entropy in the surroundings.

Lindsay feels that this suggests a new kind of imperative that can form the basis for an ethical code. The resulting *thermodynamic imperative* is as follows: *"All men should fight always as vigorously as possible to increase the degree of order in their environment, i.e., to consume as much entropy as possible."* In other words, *men should fight continually and vigorously against the Second Law of Thermodynamics.*

One obvious objection to this rule is that the second law always wins in the end through death of the organism. We are "dust" and to "dust" we ultimately return. Some would ask why fight the inevitable? Relax and try to enjoy the "thermodynamic ride." To many, life takes on meaning only when we struggle to preserve the gift of life and to be creative by bringing order and beauty out of chaos. The meaning of life and the embodiment of what it means to be a man lies in struggle and creativity—not resignation. This is the intent of the thermodynamic imperative. Many poets and writers have expressed the idea that man should oppose the destructive aspects of time and death.

Old age should burn and rave at close of day;
Rage, rage against the dying of the light.

Dylan Thomas

and

"In the darkness with a great bundle of grief the people march."

Carl Sandburg
The People Will Live On

Another difficulty arises in trying to define just what is meant by order. What is it we are asked to strive for and increase and how do we recognize it? This difficulty is, of course, common to all ethical rules. The command to love one another raises the question of what we mean by love. The reader will be able to set up everyday ethical situations where the command to love and the command to increase order are opposed to one another, and the individual must decide which one he should follow. For example, bringing another human being into the world is an act that creates order in terms of the person and it is normally an act of love. However, in terms of the present situation on spaceship Earth, each additional passenger threatens the life-support system for all passengers on an already overcrowded and over-polluted ship. What does the thermodynamic imperative tell us to do—be fruitful and multiply or be wise and not multiply? These problems of situational ethics are not to be dismissed lightly, but they are problems inherent in all ethical rules. Objection to the thermodynamic imperative on these grounds is an objection to all ethical rules.

It has also been objected that a literal adherence to the thermodynamic imperative can lead to absurdity. (Many would argue that the literal adherence to the Golden Rule or other imperatives can also lead to an absurd situation.) Using the thermodynamic imperative, man is to consume as much entropy as possible. As we saw above, one way to consume entropy and to produce order is to promote the production of living things. The conclusion one might then be tempted to reach is that the imperative requires us to populate this planet with as many people as possible. I hope that you are now convinced that following this dictum would be the height of ecological stupidity.

There are two problems with this literal interpretation of the thermodynamic imperative. First, it assumes that the most desirable way to create order is to create more human beings. There are many ways of creating order with existing life instead of creating new life. Man as a rational being should not blindly create order but instead increase order to improve the quality of life for everyone. Blindly crowding the planet with life results in disorder, not order.

Second, the literal interpretation assumes that order should increase indefinitely. This ignores the fact that life is a steady state between system and surroundings and that in the most important form of order for survival an organism is in a homeostatic equilibrium with its surroundings. Lindsay's thermodynamic imperative in this sense is far too simplistic. His suggestion for continually increasing order ignores the fact that the most interesting, challenging, and creative life is dynamic—a flowing state in which the order of one second is not the order of the next. The goal is a dynamic, always changing tension between order and disorder. Simple entropy fighting may be valid but dull—necessary, but not sufficient, for a full and useful life. The inner-stabilized individual who is changing and moving, nevertheless at ease in and contributing to a rapidly changing pluralistic society—this, rather than a robot-like entropy fighting machine, is the ideal.

The real imperative that we should learn from thermodynamics is simple and profound: *whenever you do anything, be sure to take into account its present and possible future impact on the surroundings, or environment.* This is an ecological imperative that must be learned and applied now if we are to survive or at least to prevent a drastic degradation in the quality of life for all passengers on our fragile craft.

Study Question 12-2 Comment on the validity or lack of validity of the thermodynamic imperative. Do you have additional objections? What are its limitations?

Show how it could be applied to three important life situations to tell you what you ought to do. Try to find a situation in which it apparently does not work.

12-5 How to Enjoy Life in Spite of the Second Law

The second law takes much of the joy out of life. While it does provide us with the joy and challenge of struggle, the problem is that there is no half-time period—the second law never takes a rest. It means that if we want to get anything done, we have to get and stay highly organized.

Most people adopt one of three life styles. First, there is the *swinging hedonist*, who decides to quit fighting the second law altogether and ride with the law in an orgy of disorder. Many think of trying this approach, but they lack either the guts, the money, or both. Many who have tried it find that it lacks meaning—there is no sense of struggle.

The second life style is the *automated man approach*—the opposite of the first style. In this case, the thermodynamic imperative is taken literally. One strives to become highly organized and never ceases to struggle against the second law. Life is a series of schedules and a constant frenzy to get more and more done in a shorter and shorter time. For most, this also lacks meaning, because there is no time to seek the joy and beauty of life—one does not take the time to smell the flowers.

The third style is the one followed by most people. This is the *guilty-automated-hedonist approach*. Obviously, this implies that one has attempted to seek some balance or steady state between order, or organization, and disorder—a balance between joy and struggle. This type of person has learned some of the techniques of order, but he has not perfected them very well. He tries hard but he doesn't really get a lot done; he is always behind. Throwing up his hands, he periodically decides to take off—to quit the struggle and find some joy in life. The problem is that all the time he is "taking off" he feels guilty, because he knows he is getting further and further behind. A black cloud hangs over his head, and he can never really relax and enjoy life.

The purpose of this last section is to suggest a fourth life style—one that allows the enjoyment of life in spite of the second law. In this as in the third style one seeks a balance or steady state between order and disorder—between the meaning found in both struggle and joy. The secret here is that one must spend part of this life perfecting the techniques of order and organization. One must learn how to organize and schedule his life to a very high degree. He must be able to "crank it out" with great efficiency and dispatch. However, one of the major purposes of learning these techniques of order is to obtain longer periods of joyful disorder without feeling guilty. The truly creative person often achieves this balance. In the act of creating, one wallows in the joy of letting his mind and thoughts go. But once the creative idea is conceived, the translation of the idea into an actual theory, book, or work of art, and so forth, requires the intensive and efficient use of discipline and order. Creative achievements are conceived in disorder but developed through the discipline of order.

We have to date chopped away only a few fragments from the
mountain of knowledge—fragments that have changed our way of life.
But looming ahead of us, practically intact, lies a huge mass of
fundamental facts, any one of which if uncovered could change our
civilization.

Charles Kettering

12-6 Review Questions

12-1. Do you believe there is any hope? Why or why not?

12-2. Describe the three levels of environmental awareness. What are some of the dangers of remaining only at the first level? Where are you?

12-3. What are the four major options we have for dealing with the ecological crisis? Which option should we be engaged in instituting now? Why? How?

12-4. What is a population implosion?

12-5. Criticize these statements made by one of the President's top advisers.
a. The U.S. does not have a population problem because its projected growth of 100 million by the year 2000 may turn out to be only about 75 million.
b. The only population problem in the U.S. is one of population distribution.

12-6. List the major causes of the environmental crisis. Show in a diagram how they are related.

12-7. Give several examples showing why it is essential that we deal simultaneously with all of these variables instead of focusing on one or two.

12-8. Why are approaches such as family planning, legalized abortion, and better methods for birth control very important but not enough?

12-9. Explain how we are using technology unwisely. Give several examples. What do you think we should do about this?

12-10. Explain why dismantling our technology and science would be disastrous.

12-11. Give reasons why our present political and economic systems are not suited for dealing with the environmental crisis. Suggest some changes.

12-12. What things do you think we should do to deal with our ecological crisis?

12-13. Distinguish between the first and second thermodynamic revolutions.

12-14. Study the specific proposals for what we must do that were listed in Section 12-2. Which ones do you disagree with? Why? Suggest some additional proposals.

12-15. Give the pros and cons of continuing to support major efforts in our space program.

12-16. List some specific things you *could* do personally to help deal with the problems of pollution and overpopulation. List the things you are actually going to do.

12-17. If you were in a position of sufficient power in the U.S., what specific things would you do about the following factors?
a. overpopulation (consider both the U.S. and the world),
b. energy sources,
c. pollution,
d. food production,
e. space program,
f. SST (supersonic transport).

Selected Bibliography[1]

A. Chemical Thermodynamics

1. S. W. Angrist and L. G. Hepler, *Order and Chaos*, Basic Books, Inc., New York, 1967.
2. L. E. Strong and W. J. Stratton, *Chemical Energy*, Reinhold Publishing Corp., New York, 1965 (paperback).
3. J. J. Thompson, *An Introduction to Chemical Energetics*, Houghton Mifflin Co., New York, 1967 (paperback).
4. G. C. Pimentel and R. D. Spratley, *Understanding Chemical Thermodynamics*, Holden-Day, Inc., San Francisco, 1969 (paperback).
5. J. A. Allen, *Energy Changes in Chemistry*, Allyn and Bacon, Inc., Boston, 1966 (paperback).
6. L. K. Nash, *Elements of Chemical Thermodynamics*, 2nd edition, Addison-Wesley Publishing Co., Reading, Mass., 1970 (paperback).
7. D. C. Firth, *Elementary Chemical Thermodynamics*, Oxford University Press, New York, 1969 (paperback).
8. B. H. Mahan, *Elementary Chemical Thermodynamics*, W. A. Benjamin, Inc., New York, 1963 (paperback).
9. W. F. Luder, *A Different Approach to Thermodynamics*, Reinhold Publishing Co., New York, 1967 (paperback).
10. J. Waser, *Basic Chemical Thermodynamics*, W. A. Benjamin, Inc., New York, 1966.
11. H. A. Bent, *The Second Law*, Oxford University Press, New York, 1965.
12. J. H. Linford, *An Introduction to Energetics*, Butterworths, London, 1966.
13. J. R. W. Warn, *Concise Chemical Thermodynamics*, Van Nostrand Reinhold Company Ltd., London, 1969.
14. J. R. Goates and J. B. Ott, *Chemical Thermodynamics*, Harcourt, Brace, Jovanovich, Inc., New York, 1971, (paperback).

B. Chemical Kinetics

1. E. L. King, *How Chemical Reactions Occur*, W. A. Benjamin, Inc., New York, 1963 (paperback).
2. J. L. Latham, *Elementary Reaction Kinetics*, Butterworths, London, 1962 (paperback).
3. J. A. Campbell, *Why Do Chemical Reactions Occur?* Prentice-Hall, Inc., Englewood Cliffs, N.J., 1965 (paperback: *Kinetics and Thermodynamics*).
4. B. Stevens, *Chemical Kinetics*, Franklin Publishing Co., Inc., New York, 1965.

[1] This bibliography is highly recommended for the reader who wishes to continue his study of chemical dynamics and life. Each book is approximately at the same or a slightly higher level than this book. I am deeply indebted to the authors of these books for their many insights and ideas.

C. Energy and Entropy in Biological Systems

1. A. L. Lehninger, *Bioenergetics*, 2nd Edition, W. A. Benjamin, Inc., New York, 1971 (paperback).
2. J. J. W. Baker and G. E. Allen, *Matter, Energy and Life*, 2nd Edition, Addison-Wesley Publishing Co., Inc., Reading, Mass., 1970 (paperback).
3. I. M. Klotz, *Energy Changes in Biochemical Reactions*, Academic Press, New York, 1967.
4. R. A. Goldsby, *Cells and Energy*, The Macmillan Co., New York, 1969 (paperback).
5. D. A. Coult, *Molecules and Cells*, Houghton-Mifflin Co., New York, 1966 (paperback).
6. I. Asimov, *Photosynthesis*, Basic Books, Inc., New York, 1968.
7. A. L. Lehninger, *Biochemistry*, Worth Publishers, Inc., New York, 1970.
8. H. J. Morowitz, *Entropy for Biologists*, Academic Press, Inc., New York, 1970.
9. H. J. Morowitz, *Energy Flow in Biology*, Academic Press, Inc., New York, 1968.
10. T. P. Bennett and E. Frieden, *Modern Topics in Biochemistry*, The Macmillan Co., New York, 1966 (paperback).
11. J. D. Watson, *Molecular Biology of the Gene*, 2nd Edition, W. A. Benjamin, Inc., New York, 1970 (paperback).
12. S. Rose, *The Chemistry of Life*, Penguin Books, London, 1966 (paperback).

D. Cybernetics

1. N. Wiener, *The Human Use of Human Beings*, Doubleday and Co., Inc., New York, 1954 (paperback).
2. J. R. Pierce, *Symbols, Signals and Noise*, Harper and Row, New York, 1961 (paperback).
3. J. Singh, *Great Ideas in Information Theory, Language and Cybernetics*, Dover Publications, Inc., New York, 1966 (paperback).
4. L. L. Langley, *Homeostasis*, Reinhold Publishing Corp., New York, 1965 (paperback).
5. A. Porter, *Cybernetics Simplified*, Barnes and Noble, Inc., New York, 1969 (paperback).

E. Applications of Entropy

1. R. B. Lindsay, *The Role of Science in Civilization*, Harper and Row, New York, 1963.
2. H. F. Blum, *Time's Arrow and Evolution*, 3rd edition, Princeton University Press, Princeton, N.J., 1968 (paperback).
3. J. T. Fraser, ed., *The Voices of Time*, George Braziller, Inc., New York, 1966.
4. M. Calvin, *Chemical Evolution; Molecular Evolution towards the Origin of Living Systems on Earth and Elsewhere*, Oxford University Press, London, 1969 (paperback).

5. D. H. Kenyon and G. Steinman, *Biochemical Predestination*, McGraw-Hill Book Co., Inc., New York, 1969 (paperback).

6. R. Arnheim, *Entropy and Art*, University of California Press, Berkeley, 1971.

F. Population, Pollution, and Environment

1. René Dubos, *So Human An Animal*, Charles Scribner's Sons, New York, 1968.

2. P. R. Ehrlich, *The Population Bomb*, Ballantine Books, New York, 1968 (paperback).

3. R. Rienow and L. T. Rienow, *Moment in the Sun*, Ballantine Books, New York, 1969 (paperback).

4. P. R. Ehrlich and H. R. Ehrlich, *Population, Resources and Environment*, W. H. Freeman and Co., San Francisco, 1970.

5. Rolf Edberg, *On the Shred of A Cloud*, University of Alabama Press, Birmingham, Ala., 1969.

6. D. W. Ehrenfeld, *Biological Conservation*, Holt, Rinehart, and Winston, Inc., New York, 1970 (paperback).

7. R. L. Smith, *Ecology and Field Biology*, Harper and Row, New York, 1966.

8. E. J. Kormondy, *Concepts of Ecology*, Prentice-Hall, Inc., Englewood Cliffs, N.J., 1969 (paperback).

9. G. de Bell, ed., *The Environmental Handbook*, Ballantine Books, Inc., New York, 1970 (paperback).

10. H. W. Helfrich, Jr., ed., *The Environmental Crisis*, Yale University Press, New Haven, Conn., 1970 (paperback).

11. H. W. Helfrich, Jr., ed., *Agenda For Survival: The Environmental Crisis-2*, Yale University Press, New Haven, Conn., 1971 (paperback).

12. H. D. Johnson, ed., *No Deposit—No Return*, Addison-Wesley Publishing Co., Reading, Mass., 1970 (paperback).

13. P. Appleman, *The Silent Explosion*, Beacon Press, Boston, 1966.

14. G. Borgstrom, *The Hungry Planet*, The Macmillan Co., New York, 1967.

15. The National Academy of Sciences, National Research Council, *Resources and Man*, W. H. Freeman and Co., San Francisco, 1969 (paperback).

16. R. M. Linton, *Terracide—America's Destruction of Her Living Environment*, Little, Brown and Co., New York, 1970.

17. E. P. Odum, *Ecology*, Holt, Rinehart and Winston, Inc., New York, 1966 (paperback).

18. K. Reid, *Nature's Network—The Story of Ecology*, Natural History Press, New York, 1970.

19. René Dubos, *Reason Awake—Science for Man*, Columbia University Press, New York, 1970 (paperback).

20. J. W. Gardner, *The Recovery of Confidence*, W. W. Norton and Co., Inc., New York, 1970.

21. Ian L. McHarg, *Design With Nature*, The Natural History Press, New York, 1969.

22. W. Jackson, *Man and Environment*, Wm. C. Brown Co., Dubuque, Iowa, 1971 (paperback).
23. R. Theobald, *The Economics of Abundance*, Pitman Publishing Corp., New York, 1970.
24. J. McHale, *The Ecological Context*, George Braziller, Inc., New York, 1970.
25. Report of the Study of Critical Environmental Problems, *Man's Impact on the Global Environment*, The MIT Press, Cambridge, Mass., 1970 (paperback).
26. P. Rienow and L. T. Rienow, *Man Against His Environment*, Ballantine Books, Inc., New York, 1970 (paperback).
27. P. R. Ehrlich and R. L. Harriman, *How to be a Survivor*, Ballantine Books Inc., New York, 1971 (paperback).
28. P. Swatek, *The User's Guide to the Protection of the Environment*, Ballantine Books, Inc., New York, 1970 (paperback).
29. P. Farb, *Ecology*, Time-Life, Inc., New York, 1970.
30. V. R. Potter, *Bioethics*, Prentice-Hall, Inc., Englewood Cliffs, N.J., 1971 (paperback).
31. A. S. Boughey, *Man and Environment*, The Macmillan Co., New York, 1971 (paperback).
32. R. Disch, ed., *The Ecological Conscience*, Prentice-Hall, Inc., Englewood Cliffs, N.J., 1970 (paperback).
33. G. Borgstrom, *Too Many*, The Macmillan Co., New York, 1969.
34. W. Petersen, *Population*, 2nd edition, The Macmillan Co., New York, 1970.
35. J. Ridgeway, *The Politics of Ecology*, Dutton, New York, 1970.
36. R. Saltenstall, *Your Environment and What You Can Do About It*, Walker and Company, New York, 1971.
37. G. Hardin, ed., *Population, Evolution, and Birth Control*, W. H. Freeman and Co., San Francisco, 1969 (paperback).
38. R. H. Wagner, *Environment and Man*, W. W. Norton and Co., Inc., New York, 1971.
39. Scientific American, *The Biosphere*, W. H. Freeman and Co., 1971 (paperback).
40. W. W. Murdoch, ed., *Environment*, Sinauer Associates, Inc., Stamford, Conn., 1971 (paperback).
41. J. Harte and R. H. Socolow, *Patient Earth*, Holt, Rinehart, and Winston, Inc., New York, 1971 (paperback).
42. R. A. Falk, *This Endangered Planet*, Random House, Inc., New York, 1971.
43. M. Terry, *Teaching for Survival*, Ballantine Books, Inc., New York, 1971 (paperback).
44. M. A. Strobbe, *Understanding Environmental Pollution*, C. V. Mosby Co., St. Louis, 1971 (paperback).
45. G. Calliet, P. Setzer and M. Love, *Everyman's Guide to Ecological Living*, The Macmillan Co., New York, 1971 (paperback).
46. J. P. Holdren and P. R. Erlich, eds., *Global Ecology*, Harcourt, Brace, Jovanovich, Inc., New York, 1971, (paperback).

Answers to Selected Study Questions

Chapter 1

1–1. None. A hole does not contain any dirt.

1–4. a. 1.7×10^3, or approximately 1700 people per hour.
      b. 4.1×10^4, or approximately 41,000 people per day.
         2.9×10^5, or approximately 290,000 people per week.
      c. 2.3 years, or approximately 2 years and 4 months.
      d. Approximately 2 weeks.

1–6. The reaction is thermodynamically feasible but the rate of reaction is so slow that it is *not* kinetically feasible. See page 15.

Chapter 2

2–1. a. No. A forbidden reaction cannot be made to occur.
      b. Yes. It can be kinetically nonspontaneous (too slow).
      c. Yes, if outside energy is used and it is kinetically feasible.

2–3.

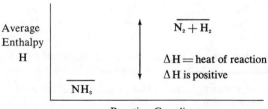

2–4.   a. Endothermic because the reaction requires heat.

b. +

c. −

d. Reverse reaction because heat is evolved.

e. Forward reaction.

f. No. It *will* occur only if it is also kinetically spontaneous.

2–6.   a. Deck of cards after shuffling.

b. Sugar cubes thrown on the floor.

c. Sugar cube dissolved.

d. Gas.

e. Liquid.

f. Gas.

2–7.   a. +   b. −   c. +   d. +   e. +

f. −   g. +   h. +   i. +   j. −

2–8.   a. No, because the ΔS for the surroundings must also be included to determine spontaneity.

b. No. Same reason as above.

2–9.   a. Not necessarily, unless the reaction proceeds at a satisfactory rate—that is, it is also kinetically spontaneous.

b. Yes. $\Delta S_{net} = -24 + 74 = +54$ entropy units. ΔS_{net} is +.

c. No. $\Delta S_{net} = -24 + 15 = -9$ entropy units. ΔS_{net} is −.

2–10. a. 298°K   b. 373°K   c. 100°C   d. 27°C   e. 400°K   f. −273°K

2–11. a. Spontaneous. ΔH is −.   b. Nonspontaneous. ΔS is −.

c. $\Delta G = \Delta H - T\Delta S$ T°K = 273 + 27 = 300°K

$\Delta G = -30{,}000 - (300)(-30) = -30{,}000 + 9000$

$\Delta G = -21{,}000$ calories

d. ΔH factor predominates because it is favorable and larger. Spontaneous.

e. ΔG is − for a spontaneous reaction and + for a nonspontaneous reaction.

2–12. $\Delta G = \Delta H - T\Delta S$ or $T\Delta S = \Delta H - \Delta G$

$$\Delta S = \frac{\Delta H - \Delta G}{T} = \frac{-5000 - (-4400)}{300} = \frac{-600}{300} = -2 \text{ cal/°K}$$

Decrease in disorder (ΔS is −).

2–13. No. It must also be kinetically feasible.

2–14. The reaction is in a state of chemical equilibrium (see Chapter 4).

2–15.

	ΔH factor	$T\Delta S$ factor	Spontaneous at Room Temperature	High Temperature
a.	Yes	Yes	Yes	Yes
b.	Yes	Yes	Yes	Yes
c.	No	No	No	No
d.	Yes	Yes	Yes	Yes
e.	No	No	No	No
f.	No	Yes	No	Yes
g.	Yes	No	Yes	No
h.	Yes	No	Yes	No

2–16. When the $T\Delta S$ factor is favorable (ΔS is +) and the temperature is high so that the $T\Delta S$ factor outweighs the unfavorable ΔH factor.

2–17. Because of the first law no energy can be lost or gained and the heat removed by the cooling system of the refrigerator will be exhausted out the back of the refrigerator into the environment. (This is why the back of window air-conditioners must stick outside of the room to be cooled.) Because of the second law, no process is 100%

efficient and the net change in entropy will be +. Thus, the increase in entropy of the surroundings will be greater than the decrease in entropy of the system (the front of the refrigerator).

Chapter 3

3–1. a. 1. 12×10^{23} b. 1. 60×10^{23} c. 1. 6×10^{23}
           2. 60×10^{23} 2. 60×10^{23} 2. 60×10^{23}
           3. 12×10^{23} 3. 18×10^{23} 3. 60×10^{23}
       d. 1. 2 moles   e. No. 4, 20 moles of water, H_2O
           2. $\frac{1}{2}$ mole
           3. 20 moles

3–2. No. 1: $3O_2(g) \rightleftharpoons 2O_3(g)$
      No. 2: $C(graphite) + O_2(g) \rightleftharpoons CO_2(g)$
      No. 3: $2C_8H_{18}(l) + 25O_2(g) \rightleftharpoons 16CO_2(g) + 18H_2O(g)$
      No. 4: $2NO(g) + O_2(g) \rightleftharpoons 2NO_2(g)$
      No. 5: $SO_2(g) \rightleftharpoons S(s) + O_2(g)$
      No. 6: $4NH_3(g) + 3O_2(g) \rightleftharpoons 2N_2(g) + 6H_2O(g)$
      No. 7: $2H_2(g) + O_2(g) \rightleftharpoons 2H_2O(l)$
      No. 8: $N_2(g) + 3H_2(g) \rightleftharpoons 2NH_3(g)$
      No. 9: $6CO_2(g) + 6H_2O(l) \rightleftharpoons C_6H_{12}O_6(s) + 6O_2(g)$
      No. 10: $C_6H_{12}O_6(s) + 6O_2(g) \rightleftharpoons 6CO_2(g) + 6H_2O(l)$

3–3. a. Spontaneous since ΔG is $-$.
      b.

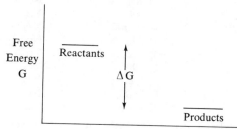

Reaction Coordinate

      c. Exergonic
      d. 10,000 calories less, since this is the maximum work.
      e. $\Delta S = \dfrac{\Delta H - \Delta G}{T} = \dfrac{-19,000 - (-10,000)}{300}$

          $\Delta S = \dfrac{-9000}{300} = -30$ calories $/°K$

      f. Reactants. ΔS is $-$ and there is a decrease in disorder.
      g. $\Delta G = \Delta H - T\Delta S = -19,000 - (800)(-30)$
          $\Delta G = -19,000 + 24,000 = +5000$ cal
          Nonspontaneous (ΔG is $+$)
      h. $\Delta G = \Delta H - T\Delta S$
          $0 = \Delta H - T\Delta S$

          $T = \dfrac{\Delta H}{\Delta S} = \dfrac{-19,000}{-30} = 633°K$

          $°C = 633 - 273 = 360°C$

3–4. a. $(5)(-94,050) = -470,250$ cal

b. −26,420 cal; 13,210 cal

c. 3×10^{23} molecules

d. $CO_2(g)$, since it has a more negative ΔH_f°.

3–5.

Reaction No.	$\Delta H_{298\,^\circ K}^\circ$ in cal	Favorable?
1	+68,000	No
2	−94,000	Yes
3	−2,428,000	Yes
5	+71,000	No
6	−304,000	Yes
7	−136,000	Yes
8	−22,000	Yes
9	+667,000	No
10	−667,000	Yes

3–6.

Reaction No.	$\Delta S_{298\,^\circ K}^\circ$ cal/$^\circ$K	Favorable?
1	−33	No
2	+2	Yes
3	+229	Yes
5	−2	No
6	+31	Yes
7	−77	No
8	−47	No
9	−63	No
10	+63	Yes

3–7. a.

Reaction No.	$\Delta G_{300\,^\circ K}^\circ$ cal	Spontaneous
1	+77,900	No
2	−94,600	Yes
3	−2,496,700	Yes
5	+71,600	No
6	−397,000	Yes
7	−112,900	Yes
8	−7,900	Yes
9	+675,900	No
10	−675,900	Yes

b. and c.

Reaction No.	More or less spontaneous at High T	Reason: TΔS factor is:	Approximate *T$_x$ $^\circ$K	$^\circ$C
1	Less	Unfavorable	None	
2	More	Favorable	None	
3	More	Favorable	None	
5	Less	Unfavorable	None	
6	More	Favorable	None	
7	Less	Unfavorable	1766	1493
8	Less	Unfavorable	1468	1195
9	Less	Unfavorable	None	
10	More	Favorable	None	

Assumption: ΔH° and ΔS° do not vary with temperature.

* T$_x$ only applies to yes-no or no-yes type reactions

e. (1) No, reaction is thermodynamically nonspontaneous at 27°C (ΔG° is −). Energy from the sun could be used to provide energy to drive the reaction.

(2) Gasoline (−2,496,700 cal vs − 94,600 cal)

Gasoline $\left(\dfrac{-2,496,700}{114} = -20,900 \text{ cal/g vs} \dfrac{-94,600}{12} = -7883 \text{ cal/g} \right)$

(3) $S(s) + O_2(g) = SO_2(g)$ $\Delta G^\circ_{300\,°K} = -71,600$ cal. Spontaneous, more spontaneous as T increases.

No. ΔG is plus and unfavorable. It would not be thermodynamically feasible at any temperature but it would be less nonspontaneous at a low temperature.

(4) Yes. $\Delta G^\circ_{300\,°K}$ is $-$. Higher T would help since the $T\Delta S$ factor is favorable.

(5) Yes. $\Delta G^\circ_{300\,°K}$ is $-$. No, the $T\Delta S$ factor is unfavorable.
The reaction is kinetically nonspontaneous. It would not be thermodynamically feasible at 27°C ($\Delta G°$ is $+$) but it would become feasible at high T because the $T\Delta S$ factor is favorable ($+$).

(6) Thermodynamically feasible only at low T. Reverse reaction would be thermodynamically feasible only at high T (favorable $T\Delta S$ factor).

(7) Respiration because $\Delta G°$ is $-$. It is driven by solar energy.

Chapter 4

4–1.   d and f

4–2.   c and e

4–3.   c, e, f, and i

4–4.

Reaction No.	Q_p	Reaction Quotient	Q_c
1	$P_{O_3}^2/P_{O_2}^3$	$[O_3]^2/[O_2]^3$	
2	P_{CO_2}/P_{O_2}	$[CO_2]/[O_2]$	
3	$P_{CO_2}^{16}P_{H_2O}^{18}/P_{O_2}^{25}$	$[CO_2]^{16}[H_2O]^{18}/[O_2]^{25}$	
4	$P_{NO_2}^2/P_{NO}^2P_{O_2}$	$[NO_2]^2/[NO]^2[O_2]$	
5	P_{O_2}/P_{SO_2}	$[O_2]/[SO_2]$	
6	$P_{N_2}^2P_{H_2O}^6/P_{NH_3}^4P_{O_2}^3$	$[N_2]^2[H_2O]^6/[NH_3]^4[O_2]^3$	
7	$1/P_{H_2}^2P_{O_2}$	$1/[H_2]^2[O_2]$	
8	$P_{NH_3}^2/P_{N_2}P_{H_2}^3$	$[NH_3]^2/[N_2][H_2]^3$	
9	$P_{O_2}^6/P_{CO_2}^6P_{H_2O}^6$	$[O_2]^6/[O_2]^6[H_2O]^6$	
10	$P_{CO_2}^6P_{H_2O}^6/P_{O_2}^6$	$[O_2]^6[H_2O]^6/[O_2]^6$	

4–5.   a. $Q_c = [NH_3]^2/[N_2][H_2]^3 = (10^{-2})^2/(1)(1)^3 = 10^{-4}$
$\Delta G_{300\,°K} = \Delta G^\circ_{300\,°K} + 4.6\, T \log Q_c$
$\Delta G_{300\,°K} = -7900 + (4.6)(300) \log 10^{-4}$
$= -7900 + (4.6)(300)(-4) = -7900 - 5520$
$= -13,420$ cal. More spontaneous.

b. $Q_c = (10^2)^2/(10^{-2})(10^{-2})^3 = 10^4/10^{-8} = 10^{12}$
$\Delta G_{300\,°K} = -7900 + (4.6)(300) \log 10^{12} = -7900 + (4.6)(300)(12)$
$= -7900 + 16,560$
$= 8660$ cal. Nonspontaneous.

4–6.   a. $\Delta G^\circ_{373\,°K} = -4.6\, T \log K_{eq} = -(4.6)(373) \log 10^{-36}$
$= -(4.6)(373)(-36) = +61,769$ cal. Nonspontaneous.

b. $\Delta H^\circ_{300\,°K} = -26,000 + 0 - (-94,000)$
$= -26,000 + 94,000 = +68,000$ cal
$\Delta S^\circ_{300\,°K} = 47 + 49/2 - 51 = 20.5$ cal/°K
$\Delta G^\circ_{300\,°K} = 68,000 - (300)(20.5) = +61,850$ cal

c. At 100°C, since $\Delta G°$ at 100°C is less positive.

	K_{eq}	Relative Equilibrium Position	$\Delta G^{\circ}_{300\,^{\circ}K}$ Cal/Mole
4-7.	10^{-100}		$+138,000$
	10^{-10}		$+13,800$
	10^{-1}		$+1,380$
	0		0
	10^{1}		$-1,380$
	10^{10}		$-13,800$
	10^{100}		$-138,000$

	Reaction No.	Relative Equilibrium Position	Relative Value of K_{eq}
4-8.	1		<1
	2		>1
	3		>1
	4		>1
	5		<1
	6		>1
	7		>1
	8		>1
	9		<1
	10		>1

	Reaction No.	$\Delta G^{\circ}_{300\,^{\circ}K}$ in Cal	K_{eq}	Relative Equilibrium Position
4-9.	1	$+77,900$	10^{-56}	
	2	$-94,600$	10^{68}	
	3	$-2,496,700$	10^{1809}	
	4	$-17,500$	10^{13}	
	5	$+71,600$	10^{-52}	
	6	$-397,000$	10^{288}	
	7	$-112,900$	10^{82}	
	8	-7900	10^{6}	
	9	$+675,900$	10^{-490}	
	10	$-675,900$	10^{+490}	

Chapter 5

	Set No.	Q_c	$\Delta G_{300\,^{\circ}K}$ (Cal)	Spontaneous to the right	System at Equilibrium
5-1.	4	10^{21}	$+11,480$	No	No
	5	10^{-10}	$-21,200$	Yes	Yes

5-2. a. Left b. Right c. Right
      d. Left e. Left

5-3. b. $T_x = 1468^{\circ}K = 1195^{\circ}C$

5-6. a. Reaction 1: $\Delta G^{\circ}_{300\,^{\circ}K} = -71,600$ cal. Spontaneous.
      Reaction 2: $\Delta G^{\circ}_{300\,^{\circ}K} = -16,250$ cal. Spontaneous.
      More $SO_2(g)$, because it has a more negative ΔG° value.
      b. $S(s) + 3/2O_2(g) \rightleftharpoons SO_3(g)$
      c. $\Delta G^{\circ}_{300\,^{\circ}K} = -71,600 - 16,250 = -87,850$ cal
      Yes.

5–7.   Reaction 1: $(-2x)$ $2C_8H_{18}(l) \rightleftharpoons 16C(s) + 18H_2(g)$

$\Delta H_1^\circ = (2)(60{,}000)$

Reaction 2: $(9x)$ $18H_2(g) + 9O_2(g) \rightleftharpoons 18H_2O(g)$

$\Delta H_2^\circ = (9)(-116{,}000)$

Reaction 3: $(-16x)$ $16\ C(s) + 16O_2(g) \rightleftharpoons 16CO_2(g)$

$\Delta H_3^\circ = (16)(-94{,}000)$

Net: $2C_8H_{18}(l) + 25O_2(g) \rightleftharpoons 16CO_2(g) + 18H_2O\ (g)$

$\Delta H_{net}^\circ = \Delta H_1^\circ + \Delta H_2^\circ + \Delta H_3^\circ = -2{,}428{,}000$ cal

Chapter 6

6–1.   The energy is used to increase the positional entropy by breaking the bonds between molecules rather than to increase the motional entropy.

6–2.   a. Decrease   b. Increase   c. Increase
        d. Increase   e. Increase

6–3.   No. The second law only applies to the net entropy change of system plus surroundings.

6–4.

Reaction No.	Predicted ΔS	Actual ΔS (Cal/°K)
2	~0	+2
3	+	+229
4	−	−35
5	~0	−2
6	+	+31
7	−	−77
8	−	−47
9	−	−63
10	+	+63

6–5.   a. and b. No. It would probably be nonspontaneous at low T because of the unfavorable ΔH factor and spontaneous at high T because of the favorable $T\Delta S$ factor.

6–6.   a. False   b. True   c. False   d. False   e. False   f. False
        g. False   h. True   i. True   j. False   k. True

6–7.   e., or Reaction 5, because both factors are unfavorable.

6–8.   a. Three microstates

$$W_I = \frac{4!}{1!\ 3!\ 0!\ 0!} = \frac{4 \times 3 \times 2 \times 1}{(1)(3 \times 2 \times 1)(1)(1)} = 4$$

$$W_{II} = \frac{4!}{3!\ 0!\ 0!\ 0!} = \frac{4 \times 3 \times 2 \times 1}{(3 \times 2 \times 1)(1)(1)(1)} = 4$$

$$W_{III} = \frac{4!}{2!\ 1!\ 1!\ 0!} = \frac{4 \times 3 \times 2 \times 1}{(2 \times 1)(1)(1)(1)} = 12$$

Microstate III is the most probable.
        b. Four microstates

$$W_I = \frac{4!}{0!\ 4!\ 0!\ 0!} = \frac{4 \times 3 \times 2 \times 1}{(1)(4 \times 3 \times 2 \times 1)(1)(1)} = 1$$

$$W_{II} = \frac{4!}{1!\ 2!\ 0!\ 0!} = \frac{4 \times 3 \times 2 \times 1}{(1)(2 \times 1)(1)(1)} = 12$$

$$W_{III} = \frac{4!}{2!\,1!\,0!\,1!} = \frac{4 \times 3 \times 2 \times 1}{(2 \times 1)(1)(1)(1)} = 12$$

$$W_{IV} = \frac{4!}{2!\,0!\,2!\,0!} = \frac{4 \times 3 \times 2 \times 1}{(2 \times 1)(1)(2 \times 1)(1)} = \frac{24}{4} = 6$$

Microstates II and III are the most probable.

6–9.

Microstate	$S = 4.6 \log W$
I	$(4.5)(10) = 46$
II	$(4.6)(23) = 105.8$ or 106
III	$(4.6)(100) = 460$

6–10. $\Delta S = 4.6 \log (10^{10}/10^{16}) = 4.6 \log 10^{-6} = (4.6)(-6) = -27.6$ cal/°K.
No.

6–11.
a. HI
b. PH_3
c. HF
d. F_2
e. CS_2
f. CCl_4
g. $H_2(g)$ at 1000°K
h. 2N(g)
i. $I_2(g)$

Chapter 7

7–1. b.

7–2. a.

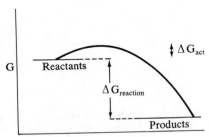

Reaction Coordinate

b.

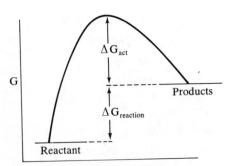

Reaction Coordinate

c.

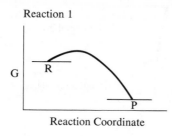

Reaction 1

G

R

P

Reaction Coordinate

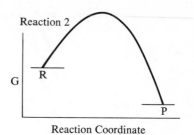

Reaction 2

G

R

P

Reaction Coordinate

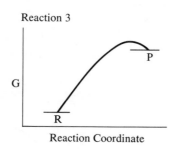

Reaction 3

G

P

R

Reaction Coordinate

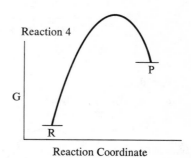

Reaction 4

G

P

R

Reaction Coordinate

d. Reaction 1 at low T
Reactions 1 and 2 at high T. Reactions 3 and 4 would occur if the TΔS factor is favorable.

7–3. Factor of four. Rate = K[A][B] = K(2)(2) = 4

7–4. Because the TΔS is unfavorable and it predominates at high T.

7–5.

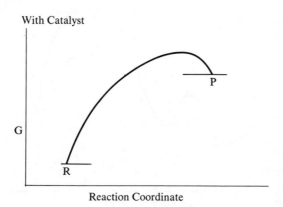

With Catalyst

G

R

P

Reaction Coordinate

7–6.

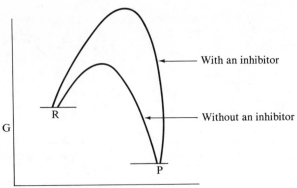

7–7.　e

7–8.　b, c, d, e, f

7–9.　a.　$A + 2B \rightarrow E + 2F$

　　　b.

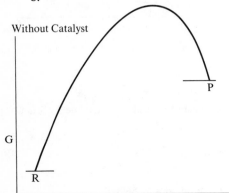

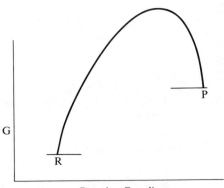

　　　c. and d.

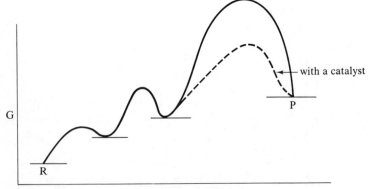

　　　e.　Rate = K[C][E]

7–10. a. Step 4
    b. Put more workers on more efficient machinery at Step 4 and cut the number of workers at Step 5 in half.

Chapter 8

8–1. Same as Figure 8–2 except information feedback is positive instead of negative.

8–2.

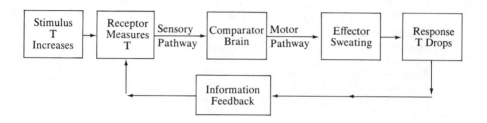

8–4.

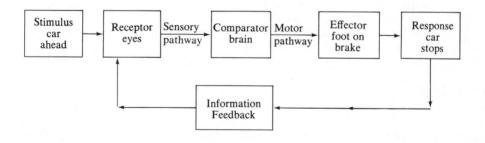

8–5. a.

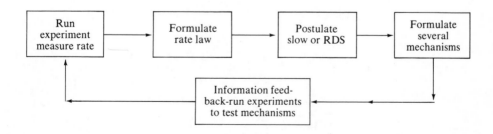

8–8. a. AU   UG   GC   CA

           AG   UC   GU   CG

           AC   UA   GA   CU

           AA   UU   GG   CC

Chapter 9

9–1. See Figure 11–9 on page 308

9–2. f. None of these

9–3. a. Steady state   b. Steady state   c. Steady state

       d. Approaching equilibrium   e. Dynamic pyuilibrium   f. Steady state

9–4. a. Reaction 1: $\Delta G^{\circ}_{300°K} = -191,700$ cal feasible

           Reaction 2: $\Delta G^{\circ}_{300°K} = -313,300$ cal feasible

      b. Reaction 1: $\Delta G^{\circ}_{273°K} = -191,727$ cal feasible

           Reaction 2: $\Delta G^{\circ}_{273°K} = -312,463$ cal feasible

      c. Reaction 1: $\Delta G^{\circ}_{363°K} = -191,637$ cal feasible

           Reaction 2: $\Delta G^{\circ}_{363°K} = -315,253$ cal feasible

      d. $CH_4(g) + NH_3(g) \rightleftharpoons HCN(g) + 3H_2(g)$

           $\Delta G^{\circ}_{300°K} = +46,800$ cal not feasible

           $\Delta G^{\circ}_{273°K} = +47,988$ cal not feasible

           $\Delta G^{\circ}_{373°K} = +43,588$ cal not feasible

9–5. a. Reaction 1: $\Delta G^{\circ}_{273°K} = +47,988$ cal not feasible

                       $\Delta G^{\circ}_{373°K} = +43,588$ cal not feasible

                       $\Delta G^{\circ}_{1500°K} = -6,000$ cal feasible

         Reaction 2: $\Delta G^{\circ}_{273°K} = +10,000$ cal not feasible

                       $\Delta G^{\circ}_{373°K} = +10,000$ cal not feasible

                       $\Delta G^{\circ}_{1500°K} = +10,000$ cal not feasible

      b. $\Delta G^{\circ}_{273°K} = -3163$ cal barely feasible

Chapter 10

10–1. $$\Delta S^{\circ} = \frac{\Delta H^{\circ} - \Delta G^{\circ}}{T} = \frac{-5000 - (-7300)}{300} = \frac{2300}{300} = 7.7 \text{ cal/}°K$$

Increase in disorder. More spontaneous at higher T because of the favorable TΔS factor. Above 37°C the chemicals or enzymes for the reaction might be destroyed.

10–2. See Figure 7–9 on page 179.

Index

Exercise=(e), Figure=(f), Footnote=(n)

Index